BEDFORD'S
BASIC PRINCIPLES OF
VENTILATION AND HEATING

BEDFORD'S BASIC PRINCIPLES OF VENTILATION AND HEATING

THIRD EDITION

Edited by F. A. CHRENKO, Ph.D., B.Sc.(Hons), M.I.Biol., M.I.H.V.E.

With 127 illustrations

LONDON
H. K. LEWIS & Co. Ltd.
1974

First published 1948
Second Edition 1964
Third Edition 1974

I.S.B. No. 0 7186 0370 2

Printed in Great Britain by The Whitefriars Press Ltd.,
London and Tonbridge

PREFACE TO THIRD EDITION

It is for me a privilege and a pleasure to edit and revise the book written by my former colleague, the late Dr Bedford. He was the first scientist in this country to make a comprehensive study of the effects of temperature and other physical factors of the internal built environment upon personal feelings of warmth. Subsequently, much work has been done on this subject. In recent years, the work of Professor Fanger in this field has been outstanding and special attention in this edition has been given to the researches which he has described in his latest monograph.

Many years ago, Bedford drew attention to and distinguished between two subjective factors in thermal comfort, namely, feelings of warmth and impressions of freshness. He noted the confusion which had occurred between considerations of warmth and freshness. It still exists today. Thanks to the generosity of the Electricity Council, I have been able to conduct some research into the freshness factor, and some of the results, hitherto unpublished, are given in this edition. Comparatively little work has been done on this subject which is increasing in importance as sealed windows and air-conditioning become more widespread in Great Britain.

The purpose of this book remains unchanged. It is to deal with ventilating and heating in terms of basic human needs. It is addressed not only to heating and ventilating engineers or environmental engineers as some of them prefer to be called but to others who are also concerned with the creation and maintenance of a healthy environment in which people live and work including architects, industrial medical officers, occupational hygienists, ergonomists, medical officers of health, public health inspectors, factory inspectors and those involved in industrial management.

And so, as in previous editions, there are no details of engineering practice; these may be found in numerous textbooks now available to engineers.

Although the Heating and Ventilating and Construction Industries have now adopted the 'Système International d'Unites', I did not consider it necessary or desirable to convert British units to metric throughout the book. The question of units of measurement is not of primary importance in a book dealing with principles rather than practice. But I felt that it might be useful to include some information on S.I. units and this has been given in an Appendix.

I have tried to make any other amendments to thc book in the same spirit as Bedford himself would have done.

Bedford's book has become and is likely to remain one of the standard textbooks on ventilation and the book as it now stands is still his.

1973 F.A.C.

PREFACE TO FIRST EDITION

ELEVEN years ago I had the privilege of giving some public lectures under the Heath Clark Bequest to the National Institute of Industrial Psychology, and the subject matter was published later under the title *Modern Principles of Ventilation and Heating.* In preparing those lectures it seemed preferable that they should not be heavily laden with technical matter, but, even apart from the desirability of simple treatment, it would manifestly have been impossible to deal with the subject in much detail in the course of only three lectures.

Since the account of those lectures was published it has often appeared to me that there was need for a book which would treat the fundamentals of ventilation and heating in greater detail. Consequently, when during the later stages of the war the publishers asked whether I would consider revising the earlier book for a second edition, I proposed that instead I should prepare a fuller treatise on the subject, and my suggestion was welcomed.

In the preface to my earlier book I remarked that there were available a number of textbooks in which the technical details of heating and ventilating practice were admirably described. Further excellent books on heating, ventilating and air-conditioning have appeared, but the basic principles of the subject seem to receive inadequate attention. My purpose in preparing this volume has been to deal with ventilation and heating in terms of basic human needs. I have made no attempt to give a detailed treatment of engineering practice, but I have included some discussion of the application of principles to practice. It is my hope that the book will be found useful as a complement to the valuable works on engineering which are now available, and that it will be of service not only to engineers and architects, but also to public health officers, industrial medical officers and others concerned with industrial management, and, indeed, any who are interested in the control of the environment in which men live and work.

It is a pleasant duty to acknowledge my indebtedness to friends for assistance given by them in the preparation of this book—to Mr. A. C. Pallot, M.B.E., for valuable adviçe concerning Chapter XV, to Mr. C. N. Davies, M.Sc., for reading the proofs, and to Mr. F. A. Chrenko for much help in the preparation of diagrams. My thanks for permission to use various illustrations are due to the Controller of H.M. Stationery Office, the American Society of Heating and Ventilating Engineers, the Institution of Heating and Ventilating Engineers, the *Journal of Clinical Investigation,* the *Journal of the Royal Institute of British Architects,* the *Gas Journal,* and *Country Life.* I am also indebted to the Controller of H.M. Stationery Office, Messrs. Davidson and Co. Ltd., Messrs. Mitchell Russell and Co., The Eagle Range and Grate Company, Ideal Boilers and Radiators, Ltd., The Wolverhampton Corrugated Iron Co. Ltd., Messrs. Keith Blackman, Ltd., and the Editors of *Country Life,* the *Gas Journal,* and the *Journal of the Institution of Heating and Ventilating Engineers,* for the loan of blocks.

T.B.

February, 1948

CONTENTS

CHAPTER XIX

CHAPTER XX

CHAPTER XXI

CHAPTER XXII

CHAPTER I

INTRODUCTION

MAN is an adaptable creature: he can endure the climate of the equator or can live within the Arctic Circle. Yet, from the earliest times, he has sought protection from the rigours of winter and the heat of summer.

Proper supplies of food, air and water are essential to life, but the maintenance of bodily warmth is also of vital importance. Indeed, at different periods in the history of mankind much ingenuity has been expended in improving the means of warming dwellings, while little regard was paid to ventilation before the nineteenth century. The provision of warmth was the paramount consideration, and ventilation was merely a by-product.

The vital processes of the animal body are accompanied by considerable energy changes, all the energy being derived from the oxidation of foodstuffs. A man doing moderate work has total energy requirements amounting to roughly 3,000 kg calories, or about 12,000 Btu/day. A large proportion of this energy is evolved as heat, and serves to maintain the body temperature. Even when the body is completely at rest and in warm surroundings, the heat production does not fall below a certain minimum level–the basal metabolism–and when physical work is being performed the heat production rises considerably. The gross mechanical efficiency achieved varies with the type of work, and is for many tasks of the order of 20 per cent, the remaining 80 per cent of the energy expended appearing as heat. Thus it is that in a cold environment we can keep warm by taking exercise.

One of the physiological requirements for health is the maintenance of a practically constant internal body temperature. In man, the average mouth temperature is about 98.4° F, and the temperature of the deep tissues is about 99° F, and in health these temperatures vary but slightly. In order that the body temperature may remain constant it is necessary that the rate of heat production within the body and the rate of heat loss from it shall be equal, and elaborate provision is made to secure this balance. Physiological mechanisms regulate the heat production within the body and the rate of heat loss from its surface, and it is through the operation of these mechanisms that the temperature remains nearly constant when the body is exposed to varying external temperatures. But the range of ambient temperature in which the nude body can maintain its temperature at a constant level is not very great, and at lower temperatures the rate of heat loss must be limited by the wearing of clothing.

The Effects of Climate on Man

That the warmth of man's surroundings has a profound influence on his well-being is no new idea. Hippocrates, in his book on Airs, Waters and Places, says that climate has great effects on the physique, intelligence and temper of peoples. Those who live in temperate climates, he says, have the best physique, while a variable climate brings out the higher mental qualities of men. Inhabitants of a region which is mountainous, rugged, high and watered, where the changes of season exhibit sharp contrasts, are likely to be of big physique, well-adapted for endurance and courage, and to possess not a little wildness and ferocity. On the other hand, the inhabitants of a region that is stifling, with more hot winds than cold, will be neither tall nor well-made: bravery and endurance are not by nature part of their character. Galen, also, remarks that the inhabitants of temperate regions surpass those of the torrid and frigid zones in mental qualities.

In the eighteenth century, Falconer discussed the influence of climate and other factors on mankind. He remarked that the inhabitants of temperate climates are more inclined to activity and exertion of the powers of both mind and body than are those people who live in climates where either heat or cold predominates in great degree. Politeness and elegance of behaviour, Falconer says, have always attained to the greatest perfection in temperate climates.

Civilization is produced by the efforts of man, and it is therefore reasonable to assume that where men enjoy conditions that suit them best there are present the essentials of civilization. The effects of climate on civilization have been studied by climatologists in recent years. Ellsworth Huntington (1915) concludes that shifts in the centres of civilization during the course of history can be accounted for by a theory of climatic pulsation.

Markham (1942) agrees that great climatic changes have undoubtedly taken place in the course of the world's history, but, he urges, these have been so slow that to hold them responsible for great national upsurges in modern times is to lay too much to Nature's account. He is convinced that one of the basic reasons for the rise of a nation in modern times is its control over climatic conditions. Civilization, he holds, depends to a great extent upon climate control, *i.e.* upon the control of the indoor climate in a good natural climate. Taking as indices of national energy, death-rates, infantile mortality rates, national incomes, and world trade, Markham finds that these all show a close relationship with the climate ranks of the various countries. The leaders are those countries with an almost ideal summer, and an easily controlled spring, autumn and winter climate.

It has been said that human culture began with the discovery of fire. To save the labour that primitive methods of making fire entailed, men kept fires burning day and night in public buildings. In earlier days, Egyptians had a fire in every temple: Greeks, Romans and Persians had public fires in all towns and villages whilst in America the Maya and Peruvians had their national fires burning on large pyramids.

Civil and political interests grouped themselves round the Prytaneion which, in ancient Greece, was a temple, a town hall, a tribunal and a resort for gossip where all public and much private business was conducted by the light and warmth of the communal fire.

The first guardians of the tribal fire were the earliest public servants and the Hellenes, like the Aztecs, received foreign ambassadors in the temples of fire and feasted guests at the national hearth.

Primitive methods of heating rooms by means of open fires remained and even as late as the eighteenth century, open braziers burning charcoal or coke were to be found in the House of Commons.

The flue had been invented some time before but it was only in the nineteenth century that the effects of man's control of his private climate or internal environment on his external environment began to be measured.

J. F. Campbell, who invented an instrument known at the time as Campbell's registering sundial, conducted an experiment from Christmas 1857 to 22nd June 1858 by exposing three dials, two in London and one at a distance of about 30 miles from it. The result was that the dial furthest from the centre of London was most marked, while those in London were least marked on that side where the sun was most obscured by smoke; showing that the smoke of London exercises a considerable influence on its climate.

The disastrous smog episode of December 1952 which was responsible for some 4,000 extra deaths of human beings and a number of show cattle at Olympia led to the first Clean Air Act in 1956. The result of smoke control has been an improved atmosphere and in Central London winter sunshine has increased by 50 per cent and is now almost the same as at Kew.

One notes that nearly 100 years were to elapse from the time of Campbell's experiment and the coming into law of the first Clean Air Act.

But, as Stamp (1964) pointed out, far too many writers on the relationship between man and his climate appear to ignore or overlook the fact that a large proportion of life is lived indoors in a man-made climate.

Stamp recalled that he discussed a proposal to build a completely air-conditioned university in West Pakistan where the average outside temperature in July is 32°C. Within, students would remain for the entire academic year and he asked the following questions: What should be the ideal climate? A seasonal rhythm? A daily rhythm? Attuned to the outside world or quite independent? He said that he was far from satisfied that we yet have the full answers and the present writer is inclined to agree.

CHAPTER II

HEAT TRANSMISSION

INTRODUCTION

HEAT is a form of energy, and like other forms of energy it is a measurable quantity. Temperature, which measures the hotness of bodies, also measures thermal potential, for heat is transferred from hotter bodies to cooler ones. The transfer of heat occurs in three ways: by conduction, by convection, and by radiation.

Conduction is the flow of heat through an unequally heated substance from places of higher to places of lower temperature. It is the main process by which heat is transmitted through solids, but it also occurs in liquids and gases.

Generally, when a fluid is heated its density is reduced. If such a fluid is brought into contact with a hot surface that part of it which touches the surface is warmed, its density is lowered, and it is displaced by the cooler and denser portion of the fluid. Thus a current is set up within the mass of the fluid. The transfer of heat from the hot surface to the fluid immediately in contact with it is by conduction, but the term *convection* is applied to the process by which the diffusion of heat is facilitated by the motion of the warmed fluid.

By *radiation* energy is transferred from one body to another without sensibly warming the intervening space. Whereas continuity of matter is necessary for conduction and convection to take place, since in those processes the heat is transmitted from molecule to molecule, radiant heat can pass through a vacuum–the sun's rays warm us although they have passed millions of miles through space.

Another form of energy exchange to which reference may conveniently be made in this chapter is that associated with *evaporation* and *condensation.* When water is evaporated from a surface some energy is required to convert the liquid water to water vapour, and this latent heat of vaporisation is taken from the surface, which is thereby cooled. Conversely, when water vapour condenses on a surface the latent heat of condensation is yielded up to the surface.

CONDUCTION OF HEAT

Whenever different parts of a body differ in temperature, heat flows from the hotter parts to the cooler ones. Thus, in the case of a hot-water radiator, the hot water in the radiator heats the inner surface of the radiator casing, heat is conducted through the thickness of the casing to the outer surface, and is thence partly carried off by convection currents set up in the air of the room and partly radiated to the other surfaces in the room. The rate at which heat is transferred through the metal casing depends on the conducting power, or thermal conductivity, of the metal, on the difference between the temperatures of the two surfaces, on the surface area of the radiator, and on the thickness of the casing. With a given temperature difference the rate of conduction diminishes as the thickness of the casing is increased, and with a given thickness the conduction increases in proportion to the temperature difference.

In c.g.s units the thermal conductivity of a substance, usually denoted by the letter k, is expressed as the number of gram-calories conducted per sq. cm/sec across a slab of the substance 1 cm thick, per 1°C difference between the two surfaces. In heating engineering calculations k is usually given in terms of the number of British thermal units per sq. ft/hour, per inch thickness, per 1°F temperature difference. Values of k given in these British units can be converted into c.g.s. units by multiplying by 0.0003445. A few typical thermal conductivities are shown in the Table opposite. The reciprocal of k measures the thermal resistivity of the substance.

Metals are good conductors of heat, and they are also good electrical conductors. For most metals and alloys the ratio of the thermal conductivity to the product of the electrical conductivity and the absolute temperature is approximately

Thermal Conductivities of Certain Substances

Substance	*Thermal conductivity, k, in Btu/sq. ft/hour/°F diff./in thickness*	*Authority**
Silver at 18°C	2,828	1
Copper at 18°C	2,665	1
Aluminium at 18°C	1,463	1
Wrought iron at 18°C	418	1
Steel at 18°C	334	1
Brickwork	8.0	2
Glass	7.3	2
Wood	1.0	2
Cork, granulated	0.33	2
Slag wool	0.3	2
Fibre board	0.38	2
Woollen cloth	0.29	3
Cotton cloth	0.49	3
Air at 22°C	0.167	1
Water at 20°C	4.15	1

* Taken from (1) Kaye and Laby's *Tables of Physical and Chemical Constants,* (2) *The Computation of Heat Requirements for Buildings,* Institution of Heating and Ventilating Engineers, (3) W. H. Rees, *Shirley Institute Memoirs,* Vol. 18, 1941.

constant. Generally, the thermal conductivity of pure metals decreases slightly with rise of temperature, while the reverse holds true for alloys.

Non-conducting substances, which are poor electrical conductors, are also poor thermal conductors. Air is a poorer conductor than any solid, so that it is to be expected that porous non-metallic substances which contain much air within their structure will have lower thermal conductivities than non-porous substances such as glass. The table shows that substances such as cork, slag wool, fibre board and textile fabrics have conductivities of only about one-twentieth of that of glass: their thermal resistances are high and they are therefore useful as heat insulators.

CONVECTION

When a liquid or gas is in contact with a surface at a temperature different from its own, heat transfer by convection takes place. Even in an otherwise still fluid currents are caused by the gravitational effects due to differences in density. These natural convection currents, as they are called, cause the heat transferred from the surface to be much greater than it would be merely by conduction in a perfectly still fluid.

Completely still air is rarely encountered. Even in a closed compartment variations in the temperature of the walls and other surfaces set up air currents so that there is some movement; while when fans are employed, or when there are openings to the external air, the air movement may be considerable. Such air currents substantially increase the heat transferred by convection from hot surfaces. In other systems, such as heating or cooling circuits, the fluid medium is in motion, and this motion must be taken into account in computing the convection transfer. In this chapter the consideration of convection will be confined to convection in air. For a more complete discussion of convection in other gases and in liquids reference should be made to the works of Fishenden and Saunders (1932), and McAdams (1933).

Natural Convection

On the whole, the heat loss from an object by natural convection is proportional to the 5/4th power of the difference between the temperature of the object and that of the surrounding air. If the convection heat loss per unit of surface area is H_c and the temperature difference θ,

$$H_c = c\theta^{5/4}$$

where c is a constant depending on the shape, size and orientation of the object.

Griffiths and Davis (1931) summarized the convection data for the important cases of horizontal cylinders and vertical planes in air. They showed the great effect of size on the convection loss. As the diameter of the cylinder or the height of the vertical plane diminishes below about 2 ft the convection loss from unit surface area increases rapidly. The variation of the convection factor, c, with size is shown in Fig. 1.

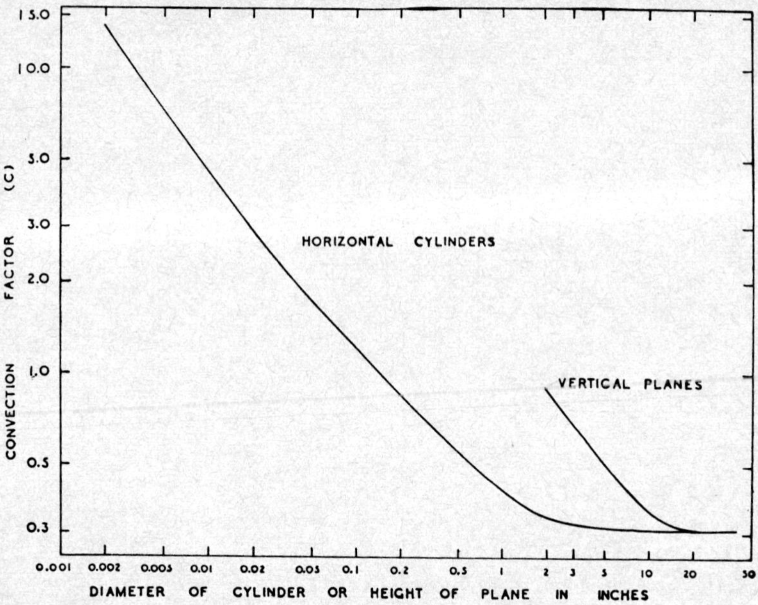

Fig. 1. Variation of rate of heat loss by natural convection with diameter of cylinder or height of vertical plane. (Griffiths and Davis.)

For thermal engineering calculations a general equation for free convection, which takes into account size and shape, is

$$H_c = C\left(\frac{1}{D}\right)^{0.2}\left(\frac{1}{T}\right)^{0.181}\theta^{1.27}$$

where H_c is the heat transmission by convection, in Btu/ft²/hour;

C is a constant depending upon the shape and orientation of the surface;

D is the diameter of the pipe or circular duct, or height of the plane vertical surface, in inches;

T is the mean of the temperatures of the surface and of the air, in °F abs. (*i.e.* t°F + 460); and θ is the temperature difference, in °F.

For vertical surfaces C is 1.394; for a horizontal plate which is warmer than the air the value of C is 1.79 when the plate is facing upwards and 0.89 when it faces downwards; whilst for horizontal cylinders the value of C is 1.016 (A.S.H.V.E. Guide, 1944).

The value of $(1/D)^{0.2}$ is halved as the diameter of a pipe, or the height of a vertical surface, is increased from 1 to 24 in.

Forced Convection

As the speed of a stream of air passing over a heated surface is increased the rate of heat loss from the surface also increases. As with natural convection, the shape, size and orientation of the surface exert their influence on the rate of heat loss. In forced convection the rate of heat loss is directly proportional to the temperature difference multiplied by some power of the velocity which varies with the size of the cooling body. In the case of a cylinder with the direction of air flow normal to the axis the power of the velocity (V) is about 0.5 for fine wires, and increases with increasing diameter to about 0.7 for cylinders of 10 in or more in diameter. In the case of a plane surface when the stream of air flows parallel to it, the exponent of V is about 0.8 at high velocities, while at low velocities the heat loss is proportional to V. The influence of size on the rate of heat loss is illustrated by the graph in Fig. 2, which is drawn from data tabulated by Fishenden and Saunders.

RADIATION

All kinds of matter give off electro-magnetic waves which are excited by temperature. These are known as thermal radiation. The intensity of the radiation and its wavelength distribution depend on the nature and temperature of the matter.

A perfectly opaque material with a completely non-reflecting surface, *i.e.* a 'black body', emits radiation at the maximum possible rate for any given temperature. Stefan showed that for such a body the rate at which radiant energy is emitted depends only on the fourth power of the absolute temperature. Stefan's law is expressed:

$$H_R = \sigma T^4$$

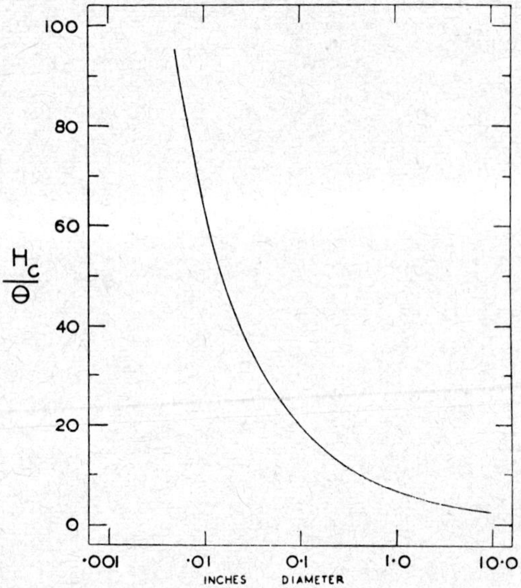

Fig. 2. Coefficients of forced convection for air flow normal to axis of cylinder. Velocity 600 ft/min (H_c = *rate of heat loss, Btu/sq. ft/hour;* θ = *temperature difference in °F).*

where H_R is the total energy radiated in unit time by unit area of the surface;
T is the absolute temperature of the surface; and
σ is Stefan's constant.

As the temperature of the radiating body rises the wavelength of maximum radiation intensity becomes shorter, and the distribution changes so that a greater proportion of the energy is radiated in the shorter waves. When λ_m is the length of the waves with maximum emissive power (in cm), and T is the absolute temperature on the Fahrenheit scale, Wien's displacement law states that $\lambda_m T = 0.52$. Wien also found that the intensity of the emitted radiation at the wavelength of maximum emission is proportional to the fifth power of the absolute temperature. These relations are shown graphically in Fig. 3.

For a perfect radiator the distribution of the energy radiated

Although at 1,500° F there is a distinctly red colour, it will be noticed from Fig. 4 that the peak of the energy emission is not in the red region of the spectrum. The proportion of energy radiated at wavelengths within the visible part of the spectrum is very small, but it is enough to stimulate the retina and evoke a sensation of red light.

Radiation Exchanges

So far we have considered only the full radiation from a black body, but not all substances emit full radiation. It has long been known that bright metallic surfaces emit only a small fraction of the energy thrown off by black bodies at the same temperatures. Thus, 140 years ago, Leslie found that the emission from a bright tin surface was only 12 per cent of that from a lampblack surface, but if the polished tin surface was covered with a very thin coating of isinglass the emissivity was greatly increased. Good emitters are also good absorbers of radiation, while poor emitters are good reflectors and poor absorbers. This follows from Prevost's theory of exchanges as amplified by the results of later researches, notably those of Kirchhoff and Balfour Stewart.

Prevost's theory is that all bodies, whatever their temperature, are constantly radiating heat. When a system of bodies is in thermal equilibrium there is equality of heat exchange, and when that equilibrium is upset it is restored by unequal exchanges. Thus, the radiation absorbed by any body placed within an enclosure which is impervious to radiation and which is uniformly at the same temperature as the body, is exactly compensated by the radiation emitted by it, so that no heat transfer takes place. When an opaque body is in equilibrium with the enclosure in which it is placed the proportion of the incident radiation which it absorbs is exactly compensated in quality as well as quantity by the proportion emitted. A good reflector, such as a bright metallic surface, absorbs only a small fraction of the incident radiation and reflects the rest, and this reflected portion together with the radiation emitted is equivalent to the total incident radiation. When the body is to some extent transparent to the radiation it receives, the sum of the radiation transmitted, that absorbed, and that reflected is equal to the incident radiation.

While bright metallic surfaces are good reflectors and poor absorbers, non-conductors of electricity, including most building materials, are generally good absorbers and poor reflectors of radiant heat. The reflecting power of a substance varies with the wavelength of the radiation. Platinum black has an emissivity of 93 per cent of that of a black body at 100° F and one of 96 to 97 per cent at temperatures of 500° F and over. Highly polished silver and gold have emissivities of only 2 or 3 per cent up to temperature of about 2,000° F, so that for most wavelengths 97 or 98 per cent of the incident radiation is reflected. Silver has an emissivity of 11 per cent for solar radiation.

At low temperatures, or for radiations of long wavelength, pigments generally have emissivities of 90 to 100 per cent, but their absorbing powers for solar radiation vary with their colour. Griffiths and Davis (1931) showed that for temperatures below 212° F various paints ranging from dead black to white enamel were indistinguishable as regards heat radiation; but for solar radiation emissivities vary from 97 per cent for black and dark blue paints to 12 to 14 per cent for various white paints. At temperatures up to 212° F, aluminium and other metallic paints have an emissivity only about one-half that of a black surface. (Excellent tables of the emissivities of various substances are given by Fishenden and Saunders.)

For most of the conditions that need to be considered in connection with the heating of buildings the exchange of radiant energy between a body and its surroundings is expressed by the equation:

$$H_R = \sigma E(T_1^4 - T_2^4)$$

where E is the emissivity of the body expressed as a fraction of unity;

T_1 and T_2 are the absolute temperatures of the body and of the surroundings respectively, expressed in °F abs;

σ is Stefan's constant, 17.3×10^{-10}; and
H_R is the heat exchange by radiation, in Btu/sq. ft/hr.

When the body is large in relation to the size of the enclosure this formula needs to be modified.

Clothing and the human skin radiate virtually as black surfaces. For radiations of the wavelengths encountered in buildings, they absorb almost as black objects. Indoors, white clothing has no advantage over black; but out-of-doors, in the sun, white clothing reflects much of the sun's radiation while still radiating heat outwards freely, whereas black clothing absorbs the sun's rays.

Some substances, such as glass, rock salt, liquids and gases, are more or less transparent to radiation, or at any rate to radiation of certain wavelengths. Glass is transparent to wavelengths within the visible range, but it absorbs radiation in the infra-red region. Rock salt transmits much infra-red radiation.

EVAPORATION

Atmospheric air contains a variable quantity of water vapour. The maximum amount of vapour that can be present in a given volume of air varies with the temperature: it increases as the temperature is raised. The water vapour contributes its part towards the total pressure exerted by the atmosphere, so that the maximum partial pressure of water vapour varies with temperature, as is shown in Fig. 5.

When a wet surface is exposed to the air water will be evaporated from it if the partial pressure of water vapour in the air is below that of air saturated at the temperature of the surface, and the rate of evaporation will depend on the difference between the vapour pressure of the surface and that of the air immediately adjacent to it. The film of air adjoining the wet surface tends to be more nearly saturated than the general atmosphere, and evaporation is thus impeded, but as the air velocity is increased this film effect is lessened and the rate of evaporation is increased. The energy required to change the water from the liquid to the vapour state—the latent heat of

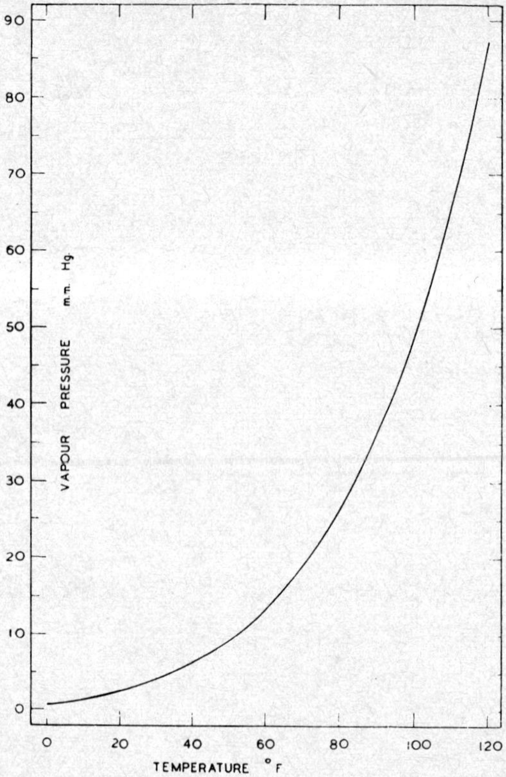

Fig. 5. Partial pressure of water vapour in saturated air.

vaporization—is taken from the wet surface, so that an unheated wet surface from which evaporation is taking place has its temperature lowered below that of the surrounding air. When such a surface reaches a steady temperature, the rate of heat loss by evaporation is balanced by the heat gained by convection and radiation. The latent heat of vaporization varies slightly with temperature. At 60° F it is 1060 Btu/lb of water vaporized.

Evaporation from plane and cylindrical surfaces has been studied by Powell and Griffiths (1935). From general considerations they conclude that the rate of evaporation should depend on: (i) the geometrical form of the surface; (ii) the nature of the air flow over the surface; (iii) the velocity of air relative to the surface; (iv) the difference between the saturation pressure corresponding to the temperature of the surface and the actual vapour pressure in the air stream; and, in the case of an unsaturated surface, (v) on the resistance to the flow of water from the interior of the solid to the surface. For a vertical cylinder of 8.6 cm diameter and 19 cm length, they found that the rate of evaporation, E, in g/sec/cm length of the cylinder, was represented by the equation

$$E = 1.83 \times 10^{-6} V^{0.61} (p_w - p_a)$$

where V is the wind speed in cm/sec and $(p_w - p_a)$ is the vapour pressure difference in cm of mercury.

At the slower wind speeds especially, the setting up of natural convection currents owing to variations in air density resulted in a variation in the rate of evaporation at different heights in the cylinder wall. Thus as the length of cylinder exposed was decreased from 18 cm successively to 11.7, 7.9, and 3,7 cm, the rate of evaporation per sec/cm of length increased from 62 to 66, 76, and 82 units respectively.

When cylinders of different diameters are considered the rate of evaporation per unit area increases rapidly as the diameter decreases, being about 2.3 times as great for a cylinder 1 cm in diameter as for one with a diameter of 8.6 cm. As the diameter is further increased beyond 8.6 cm the rate of evaporation remains fairly constant, so that for a cylinder of 37 cm diameter it is only about 12 per cent less than for one with a diameter of 8.6 cm.

The evaporation rate from a plane horizontal surface is substantially less than that for a cylindrical surface.

Further studies on the evaporation from several surfaces of various dimensions are reported by Powell (1940). The surfaces studied were unheated, and included spheres, cylinders set transverse and parallel to the air-stream, and planes set at various angles to the direction of the stream. For each type of surface it was possible to correlate the results obtained for a range of surface lengths (or diameters), l, and air speeds, u, by plotting $El/(p_w - p_a)$ against ul. The value of E is increased as l is decreased, and for surfaces of the same area it is greatest when the surface is in the form of a sphere.

CONDENSATION

When air already saturated with water vapour is cooled, some of the vapour is condensed. Thus, when on a humid day the temperature out-of-doors falls considerably, vapour condenses on the nuclei of suspended matter in the atmosphere, and a mist is formed. Indoors, if moisture-laden air is brought into contact with a cool surface, say a window or a cold wall, so that it is cooled below the temperature at which it becomes saturated (*i.e.* cooled below its dew point), moisture is deposited on the cold surface. When such condensation occurs the latent heat of the condensed vapour is yielded up to the surface on which deposition takes place.

CHAPTER III

THE THERMAL ENVIRONMENT AND ITS MEASUREMENT

THE four individual factors of the environment which affect the rate or mode of heat-loss from the body and stimulate its heat-regulatory mechanism, or influence sensations of warmth, are the radiation from the surroundings, and the temperature, velocity, and humidity of the air. If the thermal environment is to be completely specified all these factors must be taken into account.

AIR TEMPERATURE

In studies of ventilation and heating the air temperature is most commonly measured by means of ordinary mercury-in-glass thermometers. If the thermometer is simply suspended in the air it should be protected from radiant heat, for if this precaution is not taken the reading of the thermometer will be influenced by the radiant heat from the surroundings as well as by the temperature of the ambient air. The bulb of the thermometer may be covered with metallic foil, but it is better to use a cylindrical shield of polished metal to protect the bulb. If such a shield is used it should be long enough to afford complete protection, and its diameter should be large enough to permit free circulation of air around the bulb.

The thermometer reaches equilibrium sooner, and readings are taken more speedily, if a rapid current of air is passed over the bulb. With the aspirating thermometer this is contrived by enclosing the bulb of the thermometer in a tube through which air is drawn, while the sling, or whirling, thermometer is rotated quickly, like a rattle, so that the bulb passes through the air at a rapid rate. With either of these types of ventilated thermometer the rapid air current increases convection effects, and thus decreases errors due to radiation. Further, the tube enclosing the bulb of the aspirating thermometer is usually of polished metal, so that the protection of this instrument against radiation is fairly complete.

In many circumstances it is convenient to use thermocouples for the measurement of air temperature. Since the thermal capacity of the thermocouple is very small in comparison with that of a thermometer, especially when the couple is made from fine-gauge wire, it is generally possible to take readings quickly, even without arranging for rapid air movement over the junction. If the test junction is polished and kept clean radiation effects are not likely to upset the readings unless the radiation is intense. The wire from which thermojunctions are made should not be thicker than 30-gauge, and should preferably be of 34- or 36-gauge.

AIR VELOCITY MEASUREMENTS

For the measurement of air velocity various types of instrument are available, and the choice of instrument is governed mainly by the situation in which the measurement is to be made. In studies of the thermal environment it is the velocity in the free space of the apartment that is required, but in the examination of ventilation systems it is necessary to measure the speed of air flow in ducts and at air inlets and outlets.

Measurements in Ducts

Measurements in ducts are commonly made by means of Pitot tubes. The pressure due to velocity is measured by connecting the tubes with a U-tube manometer, preferably a tilting micro-manometer, and the velocity is then calculated from the expression:

$$V = 1096\sqrt{\frac{h}{d}}$$

where V is the velocity in ft/min;
h is the velocity pressure, in in. of water; and
d is the density of air, in lb/sq. ft.

Since the velocities at various points in the cross-section of a duct vary considerably it is necessary to make many measurements at different points to determine the mean velocity in the duct.

Measurements at Ventilation Openings

At ventilation openings the instrument commonly used for velocity measurements is the rotating vane anemometer. On the face of the instrument are a number of dials, similar to those on a gas meter, and these record the number of linear feet of air which pass through the vane aperture in the measured period of observation. By dividing the number of feet recorded by the number of minutes during which the observation is made the velocity in ft/min is ascertained. Most of these instruments are not sensitive enough to give accurate results at velocities lower than 200 ft/min, but the Owers type will measure distinctly lower velocities.

Measurements in Free Air

The air movement with which we are concerned in the measurement of the thermal environment is not the movement within ducts or at ventilation openings, but that in the free atmosphere of the room or other occupied space. Whereas the movement in ducts is substantially unidirectional, the motion of the air in rooms consists very largely of eddies, and it is this eddying movement that we want to measure. In occupied rooms the air velocity is most usually not more than 100 or 150 ft/min, and it is commonly of the order of 20 to 40 ft/min, while velocities as low as 10 ft/min are frequently encountered in closed rooms. The most suitable instruments for measuring these low velocities are (i) the kata thermometer, (ii) the electrically-heated thermometer, and (iii) the hot-wire anemometer.

THE KATA THERMOMETER

In his classical work, *An Experimental Inquiry into the Nature and Propagation of Heat,* Leslie (1804) described experiments on the cooling effects of wind. He wrote: 'From the same principle we derive the construction of a new and very simple kind of *anemometer.* It is in reality nothing more than a thermometer, only with its bulb larger than usual. . . . The bulb of the thermometer ought to be more than half an inch in diameter, and may, for the sake of portability, be filled with alcohol, tinged, as usual, with archil.' The thermometer was to be warmed in the hand, and the time of cooling noted.

The kata thermometer was introduced by Sir Leonard Hill (1914) primarily as a measure of warmth, but its principal use nowadays is as an anemometer. It is an alcohol thermometer with a bulb 4 cm long and 1.8 cm in diameter. On the stem there are two graduations 5° F apart. At the top of the stem the bore is enlarged to form a small reservoir, so that the instrument may be heated several degrees above the higher graduation without risk of breakage. On the back of the stem is marked the factor of the thermometer (*e.g.* F.475). This figure represents the amount of heat, in milli-calories/sq. cm of the bulb surface, lost as the thermometer cools through the range of 5° F marked on the stem.

In use, the thermometer is heated, generally by immersing the bulb in hot water, until the alcohol reaches the reservoir at the top of the stem. Then the bulb is carefully dried, and the thermometer is suspended in the air, preferably in a clamp. As the thermometer cools, the time taken by the alcohol meniscus to fall from the upper graduation to the lower is measured with a stop watch. Two or three such observations are made and the average time taken. The factor of the particular kata thermometer used is then divided by the average cooling time, expressed in seconds, and the quotient, known as the 'cooling power', is the rate of heat-loss from the bulb of the instrument expressed in milli-calories/sq. cm/sec. At the same time the temperature of the air is measured, and from the air temperature and the cooling power the air velocity can be ascertained, either by direct calculation or by use of a nomogram.

The cooling range of the standard kata thermometer is 100° to 95° F. At temperatures much higher than 80° F the standard kata thermometer cools very slowly and readings become tedious. Therefore, for use in hot atmospheres, what is known as the high temperature kata thermometer, with a cooling range of 130° to 125° F, was introduced.

In calculating the air velocity from kata thermometer readings, the air temperature and cooling power only are used. Although the polished glass bulb of the kata thermometer has an emissivity of 0.9, and is therefore much influenced by radiant heat, the air velocity formulae contain no term in radiation. Allowance for radiation is made by the constants in the equations, on the assumption that the mean temperature of solid surroundings is the same as the air temperature. If the surroundings are at a temperature appreciably different from that of the air the calculated air velocity is subject to error. With the surroundings warmer than the air the velocity is underestimated, the percentage error being especially great at low velocities.

For the measurement of air velocities even in the presence of considerable radiant heat, kata thermometers with silvered bulbs are now available (Bedford and Warner, 1933). With these instruments the radiation effects are reduced to negligible proportions. Since the silvered surface reduces the radiation loss from the bulb the silvered kata thermometer takes longer to cool than the standard kata. For this reason it is not generally necessary to use silvered katas with a cooling range of 100° to 95° F except in cool environments. At ordinary room temperatures satisfactory results can be obtained with the silvered high temperature instrument with a cooling range of 130° to 125° F. In especially hot situations instruments with a still higher range are useful, and it is now possible to obtain silvered kata thermometers with a cooling range of 150° to 145° F (Bedford, 1964*a*). When silvered kata thermometers are used care should be taken that the bulbs are free from tarnish and well polished.

For the various types of kata thermometer different equations are necessary for the calculation of air velocities. Although equations can be derived which will give reasonably accurate results at high and at low velocities, calculations are made much simpler by using for each type of instrument two equations, one for velocities below about 200 ft/min, and the other for higher velocities. These equations are of the form:

$$V = \left[\frac{\dfrac{H}{(T-t)} - a}{b} \right]^2$$

where V is the air velocity, in ft/min;
t is the air temperature, in °F;
H is the kata thermometer cooling power, *i.e.* the kata factor divided by the average cooling time (in sec); and
T a and b are constants depending on the type of instrument used.

These last constants are shown in the Table below:

Constants for Kata Thermometer Equations

Cooling range °F	*Bulb surface*	*T*	*Low velocities*		*High velocities*	
			a	*b*	*a*	*b*
100- 95	Glass	97.7	0.111	0.01584	0.0586	0.02012
130-125	Glass	127.5	0.118	0.0139	0.064	0.0184
100- 95	Silver	97.7	0.056	0.0158	–	–
130-125	Silver	127.5	0.061	0.0139	0.011	0.0170
150-145	Silver	147.5	0.074	0.0184	0.018	0.0223

The low velocity equations enable velocities to be ascertained with considerable accuracy down to about 10 ft/min. The high velocity equations cover wind speeds up to 1,000 ft/min, and can be used for somewhat higher velocities with reasonable safety.

By the use of a chart such as that in Fig. 6, air velocities can be evaluated without laborious calculations. By drawing a line, or better by stretching a thread, from the point on the left-hand vertical scale which corresponds to the factor of the kata thermometer used, through the point on the left-hand inclined scale which represents the observed cooling time, and extending

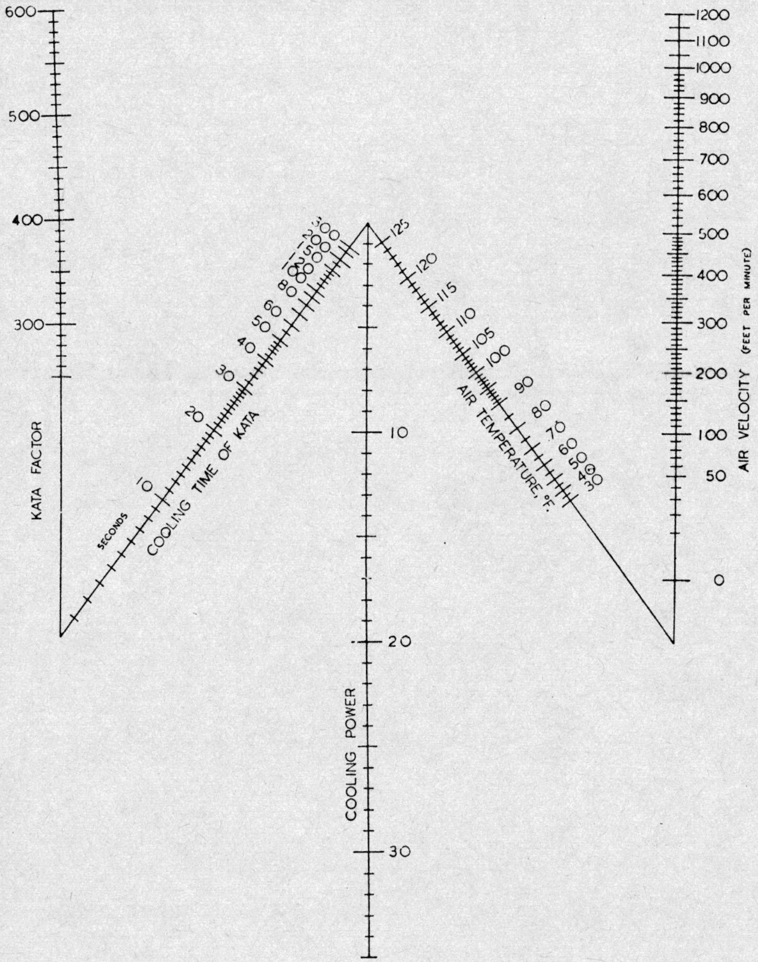

Fig. 6. Chart for the estimation of air velocities from readings of the silvered high temperature kata thermometer (cooling range 130° to 125° F). (M.R.C. War Memo. No. 17, H.M.S.O.)

the line until it cuts the middle vertical scale, the cooling power is found. Then by taking a fresh line from the point just found on the cooling power scale, through the appropriate point on the temperature scale, and producing it to cut the air velocity scale, the air velocity can be read off. The Chart in Fig. 6 applies to the silvered high temperature kata thermometer with a cooling range of 130° to 125° F. Other charts are necessary for the other types of kata thermometers.*

THE HEATED-THERMOMETER ANEMOMETER

The electrically-heated thermometer anemometer of Yaglou (1938) is capable of measuring velocities of 10 to 6,000 ft/min It is in effect a form of hot-wire anemometer, which possesses much of the simplicity of the kata thermometer. The temperature of the wire is only 10° to 40° F above air temperature, and it is indicated by a small-bulbed thermometer around which a coil of manganin wire is wound.

The heat input may be supplied by small dry cells or other suitable source of electricity. Other apparatus required for controlling the electrical input consists of an accurate voltmeter for measuring the voltage applied to the heating element, and two wire-wound rheostats in parallel, about 200 ohms each, for regulating the voltage. The accessories can be assembled compactly in a small box, and any number of heated thermometers can be connected to a single control box for simultaneous observations of velocities in different positions.

The principle of the instrument is that the temperature of the heated bulb is reduced when it is exposed to air currents, and the extent of the cooling, as indicated by the difference between the temperature of the heated thermometer and that of a similar unheated instrument, varies with the air velocity. The temperature of the heated thermometer depends on: (*a*) the electrical input to the heating coil; (*b*) the ambient air temperature; and (*c*) the velocity of the air current to which it is exposed. The relationship between the net heat input to the bulb, the temperature difference, and the air velocity is of the same form as that for the kata thermometer, *viz.*:

$$\frac{\text{net heat input}}{\text{temperature difference}} = k_0 + k_1\sqrt{v}$$

* Charts for use with silvered kata thermometers with cooling ranges of 130° to 125° F and 150° to 145° F are included in the Supplement to the Medical Research Council's War Memorandum No. 17 (H.M. Stationery Office).

where k_0 and k_1 are instrumental constants determined by calibration. In the laboratory a sensitivity within 2 per cent is attainable at all velocities by using precision voltmeters and thermometers, and in field instruments using a 3-in hand-calibrated voltmeter and standard thermometers the sensitivity may be between 2 and 4 per cent.

The instrument automatically compensates for ordinary variations of temperature, and is negligibly affected by humidity, radiant heat, and currents due to its own convection. It is especially useful for measuring the air movement in rooms, or the wind at ventilation openings.

THE HOT-WIRE ANEMOMETER

Various patterns of this instrument are used, but the principle of all of them is that the exposed fine resistance wire is heated by the passage of an electric current, and the effect of air movement on its temperature is measured.

A completely new type of hot-wire anemometer for the measurement of very low air velocities has been put on the market by the well-known Danish firm DISA. It has a scale calibrated in cm/sec for velocities from 0 to 30 cm/sec when used in one mode of operation. In a second mode of operation, the instrument is used as a conventional constant-current anemometer using a Wheatstone bridge circuit which is brought to balance at zero velocity. In this way the range of air velocity is extended to about 2 m/sec.

It is more expensive than the kata thermometer but more accurate than that instrument and it may be connected to a pen recorder for continuous measurements of air velocity.

ATMOSPHERIC HUMIDITY

There are various ways of expressing the state of humidity of the atmosphere. We have remarked that the maximum quantity of water that can be contained by a given quantity of air varies with the temperature, and that the partial pressure of water vapour at saturation therefore varies with the temperature. For thermodynamic calculations the water content of the air is conveniently expressed in terms of the weight of water vapour associated with unit weight of dry air. This quantity can be calculated, if the atmospheric pressure and the partial pressure of water vapour are known, from the formula:

$$g = \frac{4354x}{p - x}$$

where g is the number of grains of water vapour per lb of dry air;

x is the partial pressure exerted by the water vapour present in the atmosphere; and

p is the barometric pressure.

The two pressures must be given in the same units, *e.g.* mm of mercury.

Since 1 lb contains 7,000 grains, the expression that the water content is 50 grains/lb means that 7,050 grains of the moist air comprises 7,000 grains of dry air and 50 grains of water vapour.

As the water content at saturation diminishes with reduction of temperature, air that is only partly saturated with water vapour at one temperature will become saturated as it is cooled to a certain lower temperature which depends on the water content. The temperature at which it becomes saturated is known as the *dew point.* The dew-point temperature is thus another way of expressing the quantity of water vapour present in the air. It may be defined as the temperature of saturated air which has the same vapour pressure as the air under consideration.

For many purposes it is desirable to express the moistness of the atmosphere as a fraction, or as a percentage, of the saturation value. The actual water content given as a percentage of the water content of air saturated at the same temperature is known as the *percentage saturation.* Saturated air at 60° F consists of a mixture of 78.4 grains of water vapour with 1 lb of dry air; and when air at 60° F has a water content of only 39.2 grains/lb it is 50 per cent saturated.

Another, and very similar, expression is the *relative humidity*. The relative humidity is the ratio of the actual vapour pressure to the saturation vapour pressure for the same air temperature, expressed either as a fraction or as a percentage. The relative humidity and the percentage saturation are not identical, but at ordinary temperatures they are very nearly so. At 40° F saturated air has a water content of 36.9 grains/lb, while the moisture content of air at the same temperature with a relative humidity of 50 per cent is 18.3 grains/lb—slightly less than half the saturation value. The discrepancy is greater at higher temperatures. Thus at 100° F the water content of air with a relative humidity of 100 per cent is 305 grains/lb, while with 50 per cent relative humidity it is 147 grains/lb, so that in this instance a relative humidity of 50 per cent represents about 48 per cent saturation.

Wet- and Dry-Bulb Temperatures

The most convenient and most widely used instrument for the measurement of atmospheric humidity is the wet- and dry-bulb hygrometer. This consists of two thermometers, one of which—the wet bulb—has its bulb covered with wet muslin. The dry-bulb thermometer measures the air temperature. If the air is completely saturated with water vapour the temperature recorded by the wet-bulb thermometer is the same as the dry-bulb temperature, but when, as is commonly the case, the air is unsaturated, the wet-bulb temperature is below the dry-bulb temperature. The drier the air the more rapidly is water evaporated from the wet bulb, and the more is that bulb cooled. Hence, at a given air temperature, the difference between the dry- and wet-bulb temperatures—sometimes referred to as the *wet-bulb depression*—increases as the humidity of the air diminishes.

The older form of wet- and dry-bulb hygrometer, sometimes called Mason's hygrometer, had its two thermometers mounted on a frame, with a water reservoir and wick to keep the wet bulb moistened. The hygrometer was hung in a suitable position and readings taken as required. This form of instrument, where the bulbs may be exposed to stagnant air or to air moving at a high velocity, is a notoriously unreliable one. In calm air the wet bulb becomes invested by a film of air which is much more highly charged with water vapour than is the general atmosphere, and the wet-bulb temperature then recorded is erroneously high.

If the construction of the instrument is so modified that air at a definite velocity is drawn past the thermometer bulbs it can be converted into an instrument of satisfactory precision (Griffiths, 1921). Satisfactorily ventilated hygrometers are of two main types. One is the Assmann hygrometer, in which the two thermometers are fixed alongside each other in two tubes, and a fan driven either by clockwork or an electric motor, draws air past the bulbs. The other form is the whirling, or sling, hygrometer (or psychrometer). In this instrument the frame in which the thermometers are mounted is provided with a handle. By rotating the instrument about the handle, the thermometers can be whirled rapidly so that their bulbs pass through the air at a high velocity. Rapid whirling immediately before the thermometers are read is essential, and the velocity of the air past the wet bulb should be about 800 ft/min.

By the use of suitable tables or charts, the vapour pressure, the water content of the air, the relative humidity, the percentage saturation, and the dew-point temperature can be ascertained if the wet- and dry-bulb temperatures are known. Amongst the tables which are available are *Tables of Hygrometric Data for Air,* issued by the Institution of Heating and Ventilating Engineers, and C. F. Marvin's *Psychrometric Tables,* published by the U.S. Department of Commerce. A short table of relative humidities is given on p. 33.

It is sometimes easier to read humidity values from a chart than to make the interpolations which are often necessary when tables are used. Various charts, more or less similar in construction, are available. Figure 7 illustrates a chart which is contained in the Supplement to the Medical Research Council's War Memorandum No. 17.

The horizontal scale and the vertical lines represent dry-bulb temperature; the wet-bulb temperature lines run diagonally

Table of Relative Humidities

Dry-bulb temperature, °F	*Relative humidity when depression of wet bulb (°F) is:* 1	2	4	6	8	10	12	14	16	20	24
40	92	84	68	52	37						
44	93	85	71	56	43	30					
48	93	86	73	60	48	36	24	12			
52	94	88	75	63	52	41	30	20	10		
56	94	88	76	66	55	45	35	25	16		
58	94	88	77	67	57	47	37	28	19		
60	94	89	78	68	58	49	40	30	22		
62	94	89	79	69	59	50	41	33	24		
64	95	90	80	70	60	52	43	35	27	12	
66	95	90	80	71	62	53	45	37	29	15	
68	95	90	81	72	63	54	46	39	31	17	
70	95	90	81	72	64	56	48	40	33	20	
72	95	91	82	73	65	57	49	42	35	22	
74	95	91	82	74	66	58	51	44	37	24	12
76	96	91	82	74	67	59	52	45	38	26	14
80	96	92	83	76	68	61	54	47	41	29	24
84	96	92	84	77	70	63	56	50	44	33	22
88	96	92	85	78	71	64	58	52	46	35	25
92	96	93	85	78	72	65	60	54	48	38	28
96	96	93	86	79	73	67	61	55	50	40	31
100	96	93	86	80	74	68	62	57	52	42	33
104	97	93	87	81	75	69	63	58	53	44	35
108	97	93	87	81	75	70	65	60	55	46	37
112	97	94	87	82	76	71	66	61	56	47	39
116	97	94	88	82	77	71	66	62	57	48	41
120	97	94	88	83	77	72	67	63	58	50	42

downwards from left to right; relative humidity is shown by the curved lines rising from left to right; the vertical scale on the right-hand side of the chart and the horizontal lines represent the water content of the air in grains of water vapour per lb of dry air; and on the left-hand side is given a scale of vapour pressure.

If the dry-bulb and wet-bulb temperatures are known, the relative humidity can be ascertained by following the dry-bulb line until it cuts the appropriate wet-bulb temperature line, and then estimating the relative humidity from the distance of the

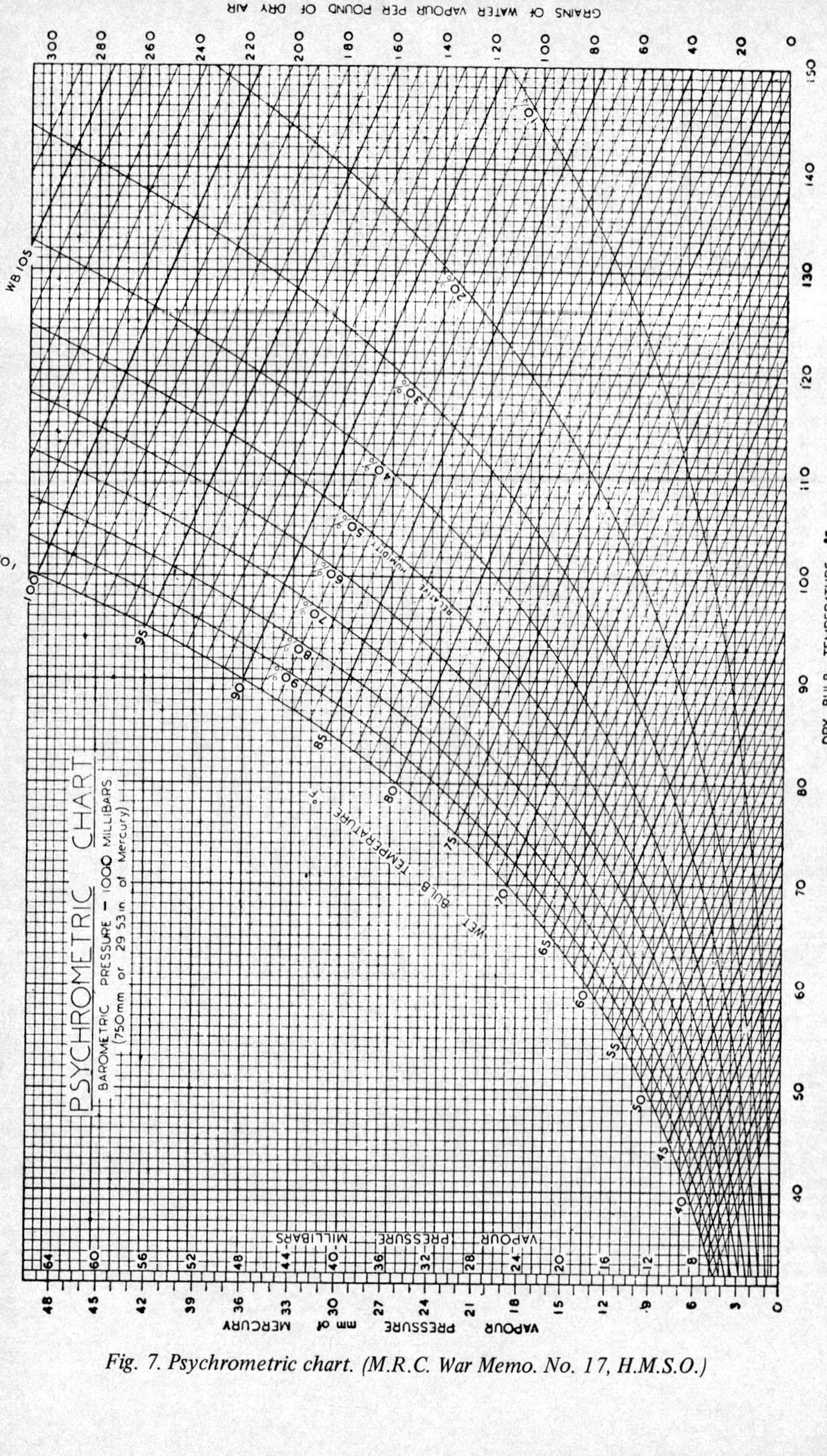

Fig. 7. Psychrometric chart. (M.R.C. War Memo. No. 17, H.M.S.O.)

point of intersection from one or other of the relative humidity lines. If, from the point of intersection of the dry-bulb and wet-bulb temperature lines, a horizontal line is drawn coincident with, or parallel to, the horizontal ruling on the chart, the water content can be read off from the right-hand scale, or the vapour pressure from the left-hand scale.

The weight of water vapour present in the air has been calculated for a barometric pressure of 1,000 millibars. For other barometric pressures the values given in the chart can be corrected to a close approximation by multiplying them by the ratio 1,000/*p* where *p* is the barometric pressure in millibars.*

Dew-Point Apparatus

The dew-point temperature may be determined directly by observing the temperature at which a mist appears on a highly polished metal surface as it is cooled, and by noting the temperature at which the mist later disappears as the surface is allowed to warm. This method has been used as the standard of reference in calibration work on hygrometry (Griffiths, 1921).

The most convenient form of dew-point apparatus is Regnault's hygrometer, or some modification of it (Fig. 8). A test tube has its lower part removed and replaced by a thin, polished silver thimble. The tube is closed with a cork through which two holes are bored. A narrow glass tube passing to the bottom of the silvered tube is fitted in one of the holes, and a thermometer is fitted in the other so that its bulb almost touches the bottom of the tube. A side tube connects the silvered tube to the aspirator. The silvered tube is half-filled with ether. When the aspirator is opened air bubbles through the ether and causes it to be evaporated. This cools the remaining ether, and when its temperature is reduced to the dew point a mist forms on the silvered thimble. The aspirator is then shut off and the temperature of the ether rises. By careful manipulation of the aspirator it can be contrived that the temperatures at which the mist is observed first to appear and

* A barometric pressure of 1,000 millibars equals 29.5306 in of mercury.

then to disappear are almost identical. The mean of these two temperatures is taken as the dew point.

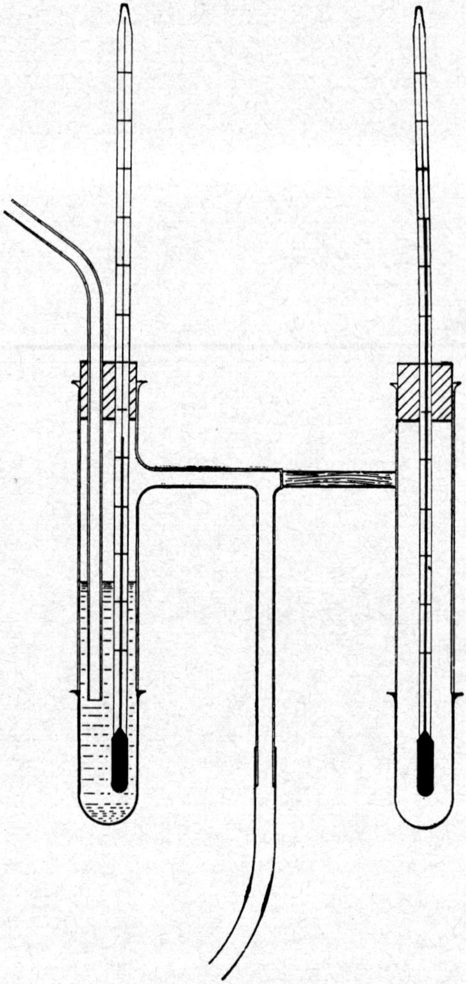

Fig. 8. Regnault's dew-point apparatus.

Chemical Absorption

The moisture content of the air can be measured by the chemical drying of a known quantity of air and subsequent weighing of the absorbed moisture. The method requires special apparatus and much technical skill, and is too tedious for routine use.

MEAN RADIANT TEMPERATURE

The expression 'mean radiant temperature' is used to indicate a measurement of the intensity of radiant heat in the environment, but there has been confusion as to its exact significance. The term has commonly been used to indicate the mean energy flux at a point, or the intensity of radiation as it affects a small sphere, in which case radiation of a given intensity over a given solid angle carries equal weight whatever its direction; but it has also been used as indicating a measure of radiation as it affects an object of the shape of the human body. In this latter case the direction of radiation may exert a strong effect.

Mean Radiant Temperature at a Point

The mean radiant temperature as it affects a small sphere can be measured directly by means of a sensitive radiation thermopile. Bedford and Warner (1934*b*) used a Moll thermopile for this purpose. The thermopile was clamped on to a special stand provided with two graduated circular scales, one in a vertical and the other in the horizontal plane. The thermopile was pointed in turn to every part of the sphere surrounding the observation point, equal weight being thus given to each section of the surroundings in calculating the mean intensity of radiation at the point of observation. Forty-six readings were taken for each such calculation.

The mean intensity of radiation thus measured is expressed in terms of the mean radiant temperature, which is that uniform temperature at which a black surface would radiate with an intensity equal to the mean observed.

Although some form of radiometer is necessary for direct measurements, the mean radiant temperature can be ascertained more simply and easily by an indirect method in which the globe thermometer is used.

The Globe Thermometer

The globe thermometer (Fig. 9) consists of a hollow copper sphere coated with black paint, and containing an ordinary thermometer with its bulb at the centre of the sphere. The temperature of the instrument depends on the environment in which it is placed. If the mean radiant temperature is higher than the air temperature, the temperature recorded by the globe thermometer will be above air temperature; and, conversely, with the surroundings cooler than the air the globe thermometer temperature will be below air temperature.

The globe thermometer is also influenced by the velocity of air movement, so that with constant air and mean radiant temperatures the temperature of the globe will vary with the air

Fig. 9. The globe thermometer.

velocity; increase of the velocity will cause the globe thermometer temperature to approach nearer to the air temperature. The globe thermometer reaches approximate equilibrium with its environment after about 20 min exposure. It should be remarked that the temperature of the globe can be taken by means of a thermojunction instead of a mercury-in-glass thermometer.

Vernon (1930; 1932) introduced the globe thermometer as a means of assessing the influence of radiation on comfort, and Bedford and Warner (1934*b*) used it for the measurement of mean radiant temperature. These observers used a globe 6 in. in diameter. More recently the instrument has been recommended for field use by the League of Nations Housing Commission (1937) and by the American Society of Heating and Ventilating Engineers (1942). The latter body recognizes the globe thermometer as the standard instrument for field use, but recommends that an 8-in. globe be used for that purpose. In this book, references to the globe thermometer apply to the 6-in. diameter instrument unless it is expressly stated otherwise.

When the globe thermometer is in equilibrium with its environment the effects of radiation and convection balance each other. If the temperature of the globe exceeds that of the air, the radiation gain, H_R, is given by

$$H_R = 0.95 \times 1.73 \times 10^{-9}(T_S^4 - T_G^4)$$

and the convection loss H_C from the 6-in. globe has been found by Bedford and Warner to be

$$H_C = 0.169\sqrt{v}(t_g - t_a)$$

where H_R and H_C are in Btu/sq. ft/hour;

T_S and T_G are the mean radiant temperature and the temperature of the globe thermometer, respectively, both expressed in °F abs;

t_g is the temperature of the globe thermometer in °F;

$$(T_G = t_g + 460)$$

t_a is the temperature of the air in °F; and

v is the air velocity in ft/min.

By writing these equations as a heat balance and transposing we get

$$T_S^4 \times 10^{-9} = T_G^4 \times 10^{-9} + 0.1028\sqrt{v}(t_g - t_a).$$

Thus, if the air temperature, the globe thermometer temperature, and the air velocity are known, the mean radiant temperature can be calculated. The air velocity is conveniently measured by means of the silvered kata thermometer. Calculations involving the fourth powers of absolute temperatures are troublesome, and much labour can be saved by the use of a suitable alignment chart. The two charts given in Figs. 10 and 11 include a range of temperature which will be sufficient for most purposes.*

The method of using the charts is as follows. Through the appropriate points on scales *A* and *B* a line is drawn, or a thread stretched, to cut scale *C*. The intersection on scale *C* gives the value of $(T_S^4 - T_G^4) \times 10^{-9}$. From this intersection another line is drawn through the appropriate point on scale *D*. The intersection of this line with scale *E* gives the mean radiant temperature, or mean temperature of the surroundings.

As an example, suppose a set of observations gives an air temperature of 60° F, a globe thermometer reading of 65° F and an air velocity of 50 ft/min. The value of $(t_g - t_a)$ is +5° F. Drawing the lines as described it is found that the mean radiant temperature is 71° F.

Despite this clear explanation given by Bedford, the present writer has encountered a report written by engineers as recently as 1971, in which the authors imagine that the globe thermometer temperature is the same as the mean radiant temperature. As if this error were not sufficient, the authors stated that the mean radiant temperature at a given point calculated from surface temperatures and solid angles subtended at the point by heated surfaces differs from the globe thermometer reading by nearly 4° F in some cases!

It must be emphasized that the reading of the globe

* Large scale charts of convenient size for calculations are included in the Supplement of the Medical Research Council's War Memorandum No. 17.

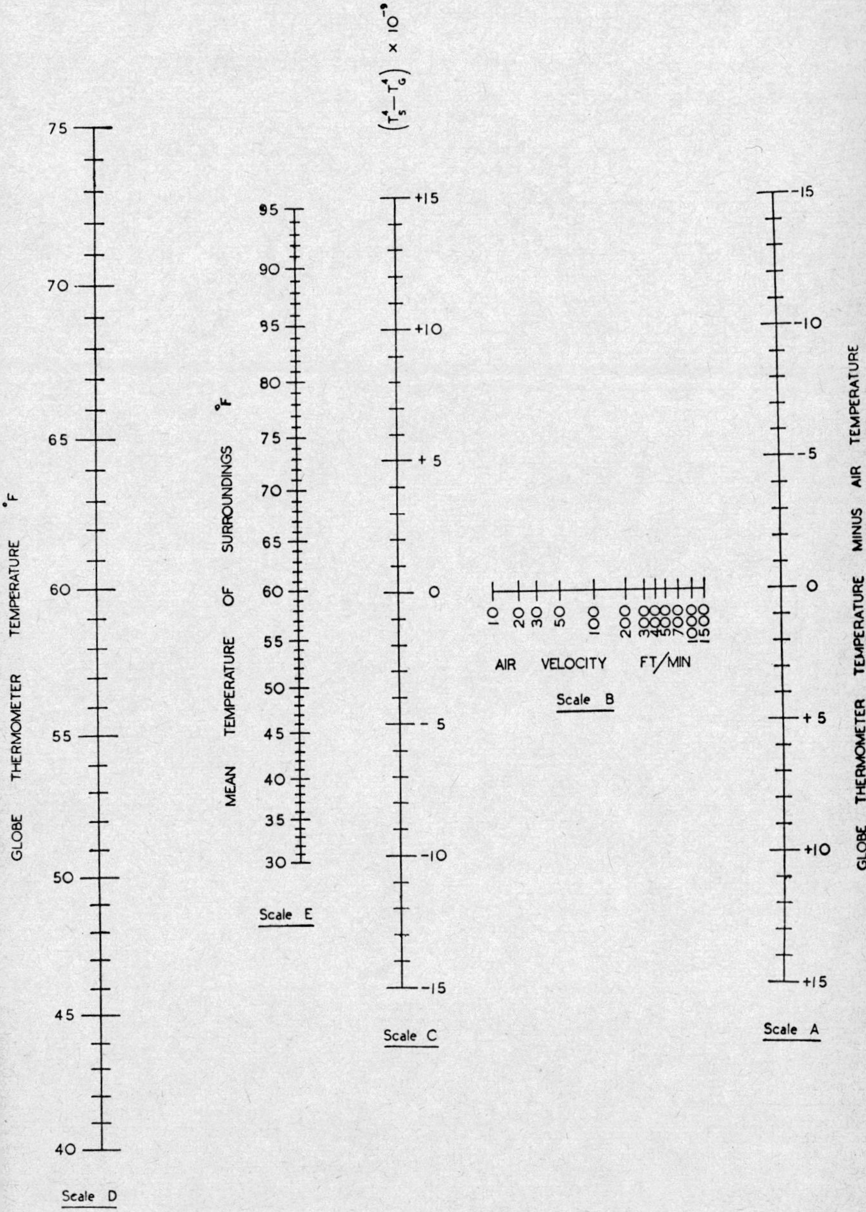

Fig. 10. Chart for the estimation of radiation from globe thermometer readings. (M.R.C. War Memo. No. 17, H.M.S.O.)

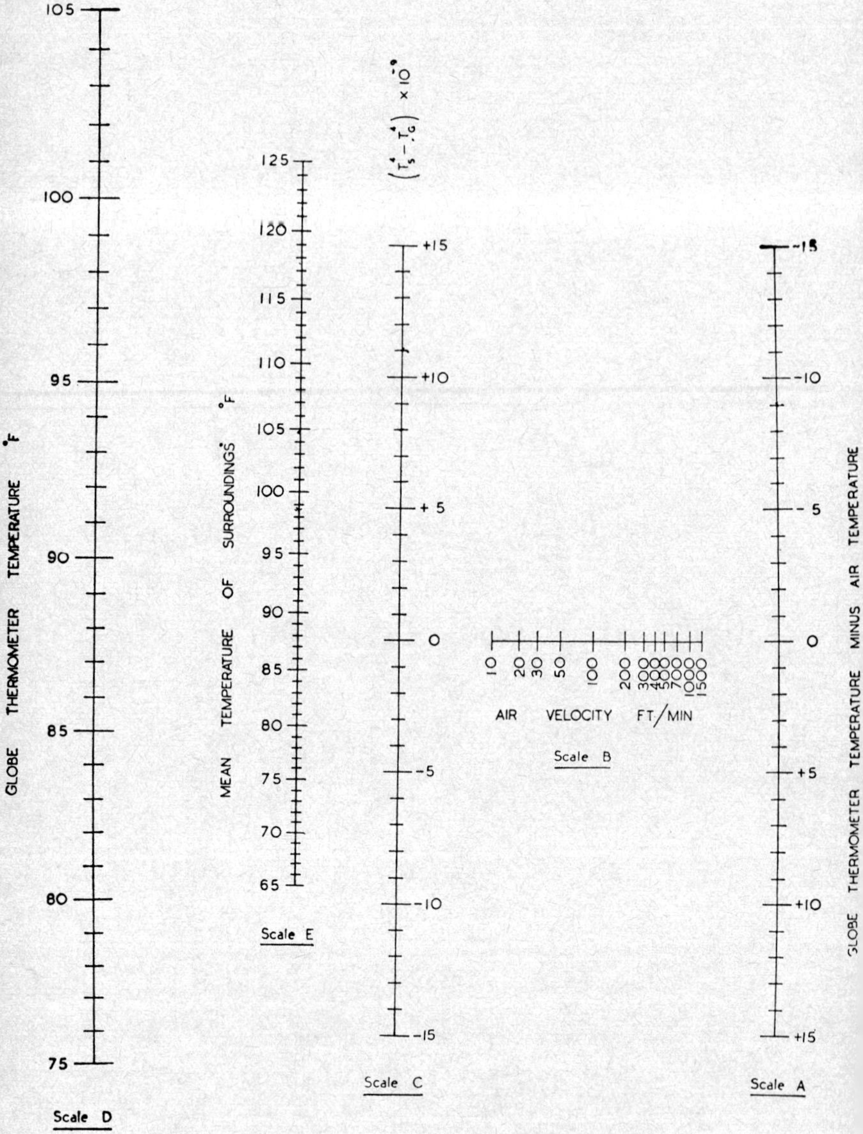

Fig. 11. Chart for the estimation of radiation from globe thermometer readings.

thermometer taken by itself is only an *indicator* of the radiant temperature. The radiant temperature may be measured directly with a calibrated radiometer.

Mean Radiant Temperature Referred to a Man

Raber and Hutchinson (1947) have provided graphs from which one can derive shape factors for use in computing the mean radiant temperature as it affects a man. Dufton and Marley (1942) described the construction of a diagram which facilitates the portrayal of the various surfaces of a room on a plane chart so that each area of the diagram affords a measure of the effect of the radiation from the corresponding surface upon a body shaped like the eupatheoscope. The equation used by these authors can be transformed to enable one to construct a similar chart to refer to a man, and thus from a knowledge of the temperatures of the boundary surfaces of the room and of the position of the man in the room one can compute the mean radiant temperature.

Estimation of the mean radiant temperature as it affects a sphere by means of the globe thermometer is a simple procedure, and it was desirable that we should learn how well the mean radiant temperature as it affects a man can be ascertained by the use of that instrument. In the radiant wind tunnel at the Tropical Research Unit, Singapore, experiments were made in which estimates made from globe thermometers were compared with those arrived at by calculation and also from observations made on some 'metal men' (Newling, 1954). These last were rather crude effigies of men and were used in pairs, one having a surface of polished metal and the other a black surface. From the heat input necessary to bring the surface temperature of the polished model up to that of the black one the radiation exchange can be calculated, and the mean radiant temperature as it affects this roughly human shape determined.

The two sides and the ceiling and floor of the experimental tunnel were heated so that in conditions in which the radiation was by no means uniformly distributed the mean radiant temperature was sometimes as much as 20° F above the air temperature. Globe thermometers were placed at heights of 2 ft, 3 ft 9 in, and 5 ft 6 in above the floor. The mean radiant temperatures estimated from the individual globe thermometer readings differed on occasions by as much as 3°F from the mean radiant temperature as ascertained directly from the metal men, but when the readings of the three globe thermometers at the different levels were averaged the mean radiant temperature agreed with that derived from the metal men to within about 1° F.

In most indoor situations it is most unlikely that the excess of the mean radiant temperature over the air temperature will be as high as in Newling's experiments, and the results of those experiments suggest that for most conditions likely to be met the mean radiant temperature estimated from globe thermometer observations will be accurate enough for practical purposes. Nevertheless the point is one of some importance in connection with panel heating, and a comparison of the results obtained by estimating mean radiant temperatures in various ways was undertaken by Tredre (1960). The measurements were made in a small room warmed by a heating panel placed horizontally overhead at heights of 8 and 10 ft or on one wall. In some experiments the temperature of the heating panel exceeded 120° F.

The mean radiant temperature with reference to a point 45 in above the floor was measured directly by scanning with a thermopile. Measurements were also made by means of globe thermometers placed at heights of 24 in, 45 in and 64 in, and the mean radiant temperatures at points at those heights were also calculated from measurements of the boundary surface temperatures and calculations of the solid angles subtended at the points by the various elements of surface.

The mean radiant temperature as it would affect a standing man was assumed to be that derived from measurements made on the metal models of men used by Newling. From the measurements of the boundary surface temperatures and the geometrical projections of those surfaces calculations were also made of the mean radiant temperatures as it would affect a standing cylinder 68 in. in height and 12 in. in diameter.

Under conditions in which the intensity of radiation from the heated panel was sometimes distinctly greater than was comfortable the differences between the mean radiant temperatures derived from the metal men and those obtained (*a*) by direct measurement with the thermopile, (*b*) from globe thermometer measurements, and (*c*) from surface temperature measurements and calculation of solid angles, were all small. There were occasions when the average of the mean radiant temperatures referred to points at the three heights from the floor agreed more closely with the results from the metal men than did the value referred to a point at a height of 45 in, but again the differences were small. The average difference between the values derived from the metal men and the average of three globe thermometer estimates was less than 0.3° F, while that between the values from the metal men and those derived from single globe thermometer measurements at a height of 45 in from the floor was less than 0.5° F.

Hence, Tredre concludes that in indoor environments where the more usual forms of heating are used the simple measurement of the mean radiant temperature with reference to a point at a height of about 45 in from the floor will generally give a satisfactory indication of the radiant heat as it affects a man. On the other hand, if there is intense radiation from sharply localized sources, and more especially out of doors where there is direct solar radiation, the measurement of the mean radiant temperature at a single point may be inadequate.

Surface Temperatures

A globe thermometer temperature which is higher than the air temperature indicates that the mean temperature of the solid surroundings is higher than the air temperature. This may be because the boundary surfaces of the compartment are uniformly warmer than the air, or, as often happens, it may be that while some of the surfaces are approximately at air temperature, one or two surfaces are much warmer than the air. Although the globe thermometer enables the mean radiant temperature to be calculated, it does not enable one to determine which are the sources of heat. When, in a warm environment, the globe thermometer records a temperature appreciably higher than the air temperature, further exploration should be made of the temperatures of the various boundary surfaces. It will often be found, in factories and in ships, for example, that the warmth of the environment can be substantially reduced by the lagging or screening of hot surfaces.

For the measurement of surface temperatures thermocouples are desirable, but fairly accurate results can be obtained by mercury-in-glass thermometers suitably used. One useful type of thermometer for this purpose has a flattened bulb, and is so made that on one side the flat bulb surface is in the same plane as the stem of the instrument, thus making it easy for the thermometer to be applied to the surface for a few minutes before the temperature is read. The thermometer can be fixed in position by means of adhesive tape or Plasticine.

If the temperature of a surface is to be measured by means of a thermometer it is essential that the bulb shall make good contact with the surface. The flattened bulb just described makes better contact than the usual cylindrical bulb. When a thermometer with a cylindrical bulb must be used, greater accuracy can be achieved by using a special holder. Such a holder can be made from a block of balsa wood or cork, by cutting in one face a groove which has a depth very slightly less than the diameter of the thermometer bulb, so that when the block is applied to the surface the bulb makes close contact with the surface. Heat conduction between the surface and the thermometer can be increased if before the bulb of the thermometer is inserted the groove is filled with putty or with grease which has been heavily impregnated with aluminium powder.

It should be emphasized, however, that for accurate work fine-wire thermocouples are necessary.

CHAPTER IV

THERMAL ENVIRONMENTAL INDICES

IN THE past it was common practice to specify the thermal environment in terms of air temperature only, or of the air temperature and atmospheric humidity. It would be useful if the effects of the four physical factors namely, thermal radiation from the surroundings, together with the temperature, velocity and humidity of the ambient air were combined into a single index. This would facilitate the comparison of different thermal environments. The thermal environment makes certain thermal demands on the human body; different thermal environments could make different thermal demands yet certain combinations of the four physical variables might make the same thermal demand. Hence one such index could be the thermal demand or the thermal load of the given environment. However, the calculation of such an index, especially if solar radiation is taken into account, is a very complicated matter indeed. Burton and Edholm (1955) devote a chapter of their book to a discussion of this particular problem and they conclude that whilst the physical factors mentioned, including thermal radiation from the sun and terrain, may be included in a single index, the **factor dominating man's comfort is his own rate of heat production** which *cannot* be included in such a climatic index.

Some ingenious attempts have been made to overcome the difficulties presented by these problems and since these attempts are closely linked with studies of thermal comfort, a short account of some of the earlier work is of considerable interest. Incidentally, it would not be out of place to remind the student of today that many of the earlier workers did not have access to computers: they had not been invented.

HISTORICAL

That we become chilled by exposure to wind is a matter of common experience. Two hundred years ago, Arbuthnot (1733) explained that the wind causes this chilling by dispersing the layer of heated, moist air which invests the body. Leslie (1804) concluded from his cooling experiments that 'even a moderate wind will quadruple the waste of heat, and that a vehement hurricane is capable of increasing the rate of dissipation perhaps fifteen or twenty times. Hence also the keen impression of frost winds on our feelings, and the prodigious effects on chilling the surface of the ground. We thus perceive, in a strong light, the vast utility of shelter, conspiring with the genial influence of the sun.'

Some years later, Heberden (1826) remarked that the thermometer was a very imperfect measure of *sensible cold,* for the sensations of the body depended on the rapidity with which its own heat was removed. In order to estimate the *sensible cold,* Heberden heated a thermometer to well above 100° F and read off the temperature at 10- or 20-sec intervals after the temperature had fallen to 100° F. Leslie had recommended his large-bulbed thermometer as a means of measuring air velocity, whereas Heberden used the rate of cooling of his thermometer as an indication of the combined effects of temperature and air velocity on the human body. Reid (1844) also stressed the unreliability of the thermometer alone as an index of comfort and insisted that the air movement must be taken into account.

All are familiar with the fact that the radiation from the sun, or from an open fire, feels warm. The chilling effect of the radiation loss to cold surfaces is perhaps less obvious, but this also has been known for a long time. The effect of radiant heat on comfort has long been recognized by writers on warming and ventilation. Tredgold (1824) mentioned as an advantage of an apartment heated chiefly by radiation that the objects receiving heat from the fire always feel warmer than the air. The General Board of Health Commissioners (1857) advocated, as one of the requirements for comfortable conditions, that the walls of a room be at least as high in temperature as the general temperature of the room, and they made many observations of the temperatures of wall surfaces. Galton (1880) wrote that, with radiant heating, the temperature of the walls is higher than that of the air, while with any type of convection heating the reverse is true.

Discussing meteorological observations, Aitken (1887) alluded to the need for an instrument which would show the effects of radiation as tempered by the wind. He used a blackened, hollow, metal sphere fitted with a thermometer with its bulb at the centre of the sphere. He referred to readings with globes 15 to 40 cm in diameter, and remarked that the size of the ball, if not too small, did not seem to affect the readings much. His paper described a method by which, from the temperature of this black globe and that of a vacuum thermometer, one could obtain a figure which would give an indication of the combined effects of solar radiation and wind. Aitken does not appear to have used his instrument for the study of indoor conditions.

KATA THERMOMETER COOLING POWER

The kata thermometer has been referred to as an instrument for use in measuring air velocities, but it was first introduced by Hill for measuring the physiological effects of the environment. When the kata thermometer is used as an anemometer its bulb is carefully dried after the thermometer has been heated, and the rate of heat loss thus ascertained is known as the dry-kata cooling power. For many years this cooling power was used as an index of warmth. Hill also advocated, that, especially in hot environments, observations should be made with the bulb of the kata thermometer covered by a wet silk finger stall; this wet-bulb reading yielded what was known as the wet-kata cooling power. The dry-kata cooling power combines the heat losses from the instrument due to radiation and convection, and the wet-kata cooling power adds to these the loss by evaporation. When the cooling power is to be used as an index of warmth the instrument which must be used is the standard 100° to 95° F kata with its polished glass bulb.

The term 'cooling power', which has become widely used in connection with the kata thermometer, is an unfortunate one. It lends itself to misconstruction, for it is easy to fall into the error of assuming that the cooling power represents the power of the environment to cool the human body. That is not so. The temperature of the instrument in the middle of its cooling range is roughly that of the human body, but the cooling power only measures the rate at which heat is lost from the bulb of the thermometer. Because of the much smaller size of the kata thermometer, air currents exert a greater cooling effect on it than on the human body. The sensitivity of the kata thermometer to wind, which makes the instrument an excellent anemometer, renders it unreliable as a measure of warmth.

EFFECTIVE TEMPERATURE

A thermal index which is widely used, especially in the United States and in Canada, is the effective temperature scale (Houghton and Yaglou, 1923, 1924; Yaglou and Miller, 1925). The scale takes into account the temperature and humidity of the air and its rate of movement, and alignment charts are available from which the effective temperature corresponding to any combination of these three factors can be ascertained. The effective temperature of an environment connotes that temperature of still air, saturated with water vapour, in which an equivalent sensation of warmth was experienced by the subjects of the long series of experiments carried out in the laboratory of the American Society of Heating and Ventilating Engineers at Pittsburgh. Two scales of effective temperature have been derived—one for persons stripped to the waist and the other for people normally clad. The first of these is known as the basic scale and the second as the normal scale. The two effective temperature charts are shown in Figs 12 and 13.

In either of these charts there are two vertical scales, for dry-bulb and wet-bulb temperature. Curves drawn downwards from right to left show the effective temperatures corresponding to various air velocities, the effective temperature for any velocity and a particular combination of dry- and wet-bulb

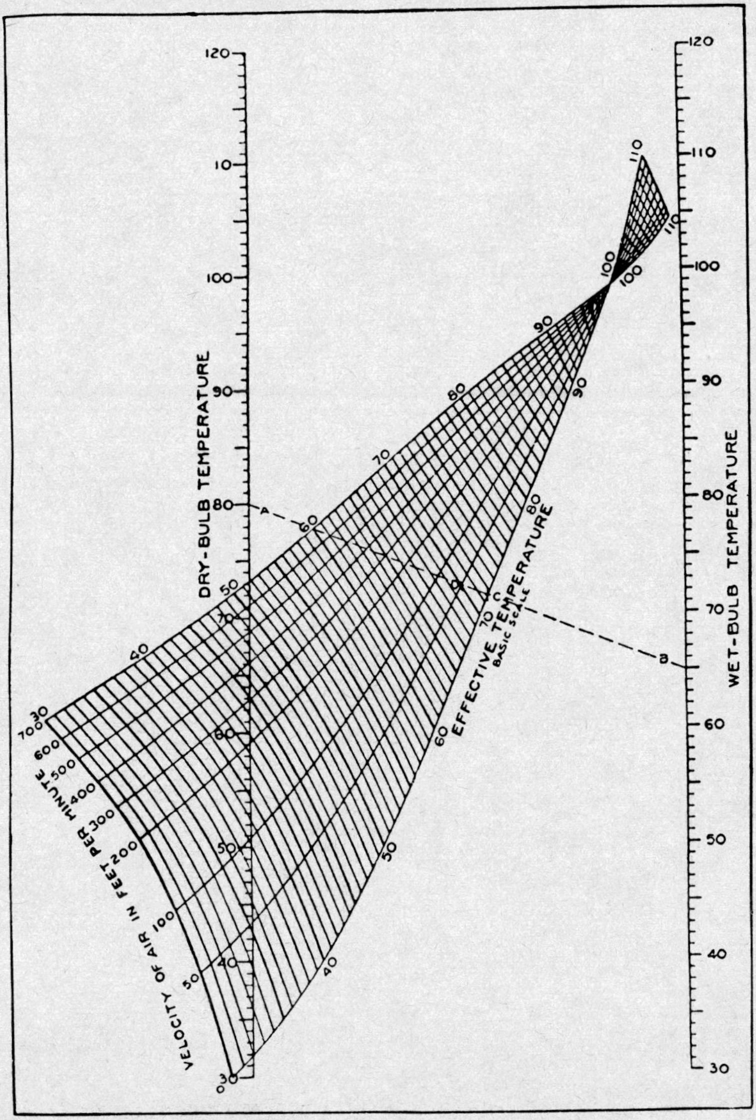

Fig. 12. Effective temperature chart (basic scale). (Courtesy A.S.H.V.E.)

temperatures being shown at the point where the line joining the appropriate points on the two vertical temperature scales cuts the air velocity line. Thus, on the basic scale (Fig. 12) the line drawn from 80°F dry-bulb to 65°F wet-bulb cuts the

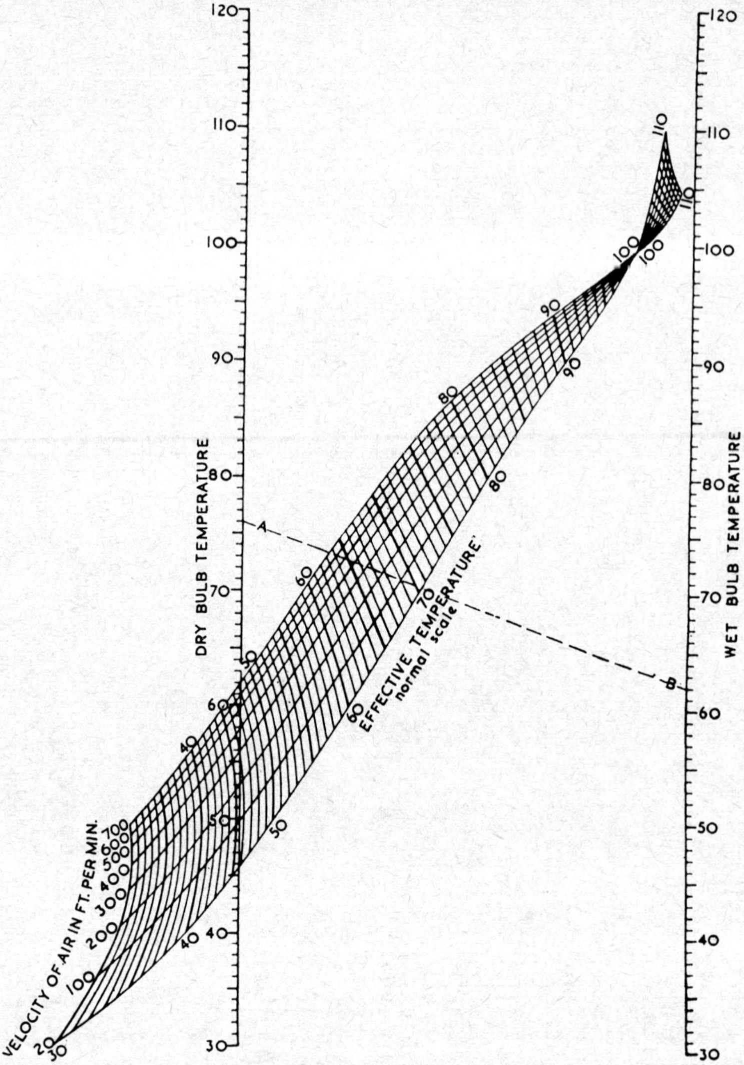

Fig. 13. Effective temperature chart (normal scale). (Courtesy A.S.H.V.E.)

50 ft/min velocity line at a point corresponding to an effective temperature of 70° F. Points on the different velocity lines indicating the same effective temperatures are joined together to form the grid shown on the charts, so that values can be interpolated for intermediate velocities.

Although the effective temperature scales were based on subjective impressions of warmth, data have been adduced to show that effective temperature correlates very well with various physiological reactions such as pulse rate and body temperature. The original observations were made on resting subjects, but in further experiments in which men worked at the rate of 90,000 ft/lb/hour there was close agreement between their physiological reactions and effective temperature, regardless of the dry-bulb temperature, humidity, or air velocity (Yaglou, 1927). There is now some evidence which suggests that at very high temperatures the scale needs adjustment, but at the temperatures which must be considered for most problems of heating and ventilation the scale is probably very reliable.

A criticism of the effective temperature scale is that it makes no allowance for radiation. If the temperature of the solid surroundings differs appreciably from that of the air, some correction for radiation is necessary. Another criticism is that the scale is based on the assessments of *three* observers (*vide* Provins and Bell, 1962).

The basic and normal scales of effective temperature should not be confused. If of two environments one is at an effective temperature of, say, 70° F on the basic scale and the other is at 70° F on the normal scale that implies that in the first a man stripped to the waist would feel as warm as if he were similarly stripped in still and saturated air at a temperature of 70° F, while in the second a man wearing normal clothing would feel as warm as if he were similarly clad in still and saturated air at 70° F. It does not mean that a man would be as warm stripped to the waist in the one environment as he would be fully clothed in the other. Such confusion of interpretation has occurred.

CORRECTED EFFECTIVE TEMPERATURE

Vernon and Warner (1932) applied a radiation correction to effective temperature by using the reading of the globe thermometer in place of the dry-bulb temperature, and they adjusted the observed wet-bulb temperature to allow for the

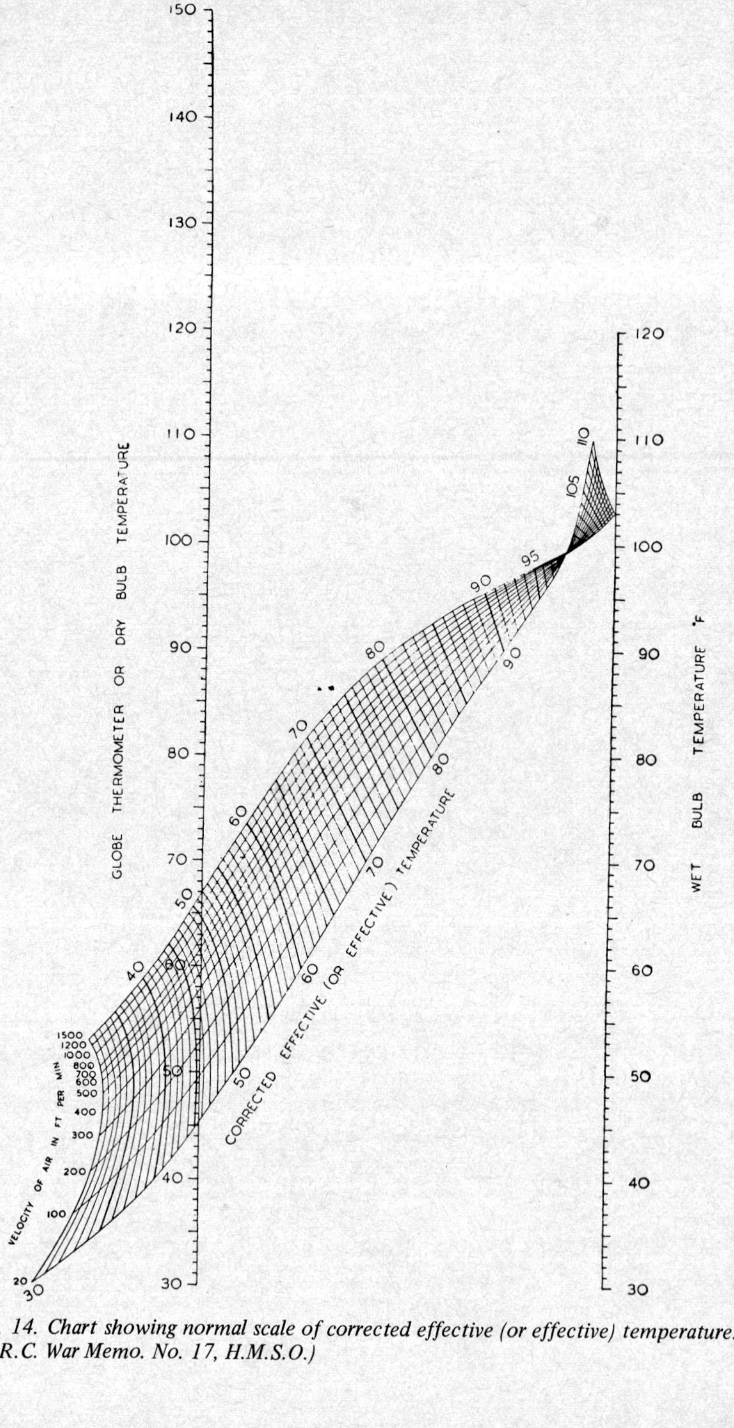

Fig. 14. Chart showing normal scale of corrected effective (or effective) temperature. (M.R.C. War Memo. No. 17, H.M.S.O.)

difference between the dry-bulb and globe temperatures. It later seemed likely that such adjustment of the wet-bulb temperature was of only secondary importance, and accordingly in the scale of Corrected Effective Temperature which has been provisionally adopted as the index of warmth to be used in H.M. ships the globe thermometer reading replaces the dry-bulb temperature and no adjustment is made to the wet-bulb temperature (Bedford, 1946*a*). It will be shown later that the allowance so made for radiant heat is adequate.

EQUIVALENT TEMPERATURE

The need for an instrument which, while taking account of radiant heat, would correspond approximately with the human body in its sensitivity to air currents, led Dufton to construct a blackened copper cylinder, the eupatheostat, for use as a radiation thermostat (Dufton, 1929). In the form of the *eupatheoscope,* or comfort indicator, this instrument has been developed as a method of assessing the thermal environment (Dufton, 1932).

The eupatheoscope is an electrically heated cylinder, 22 in high and 7½ in in diameter. In the earlier form of the instrument the heating current was controlled by a bimetallic strip regulator situated inside the cylinder. When the temperature of the thermostat reached 78° F the heating current was cut off, and as the thermostat cooled below that temperature the current was again switched on. When the environment was at a temperature of 78° F the temperature of the surface of the instrument was also 78° F, and in cooler environments the surface temperature was somewhat lower. In a still air enclosure at 65° F the surface temperature of the eupatheoscope was 75° F, which is about the average temperature of the clothed human body in a similar environment. The rate of heat loss from the surface of the eupatheoscope, and therefore its heat input, varies with the temperature and velocity of the surrounding air, and with the radiation from the surroundings. The reading of the instrument depends on the heat input, and therefore gives an indication of the combined effects of air

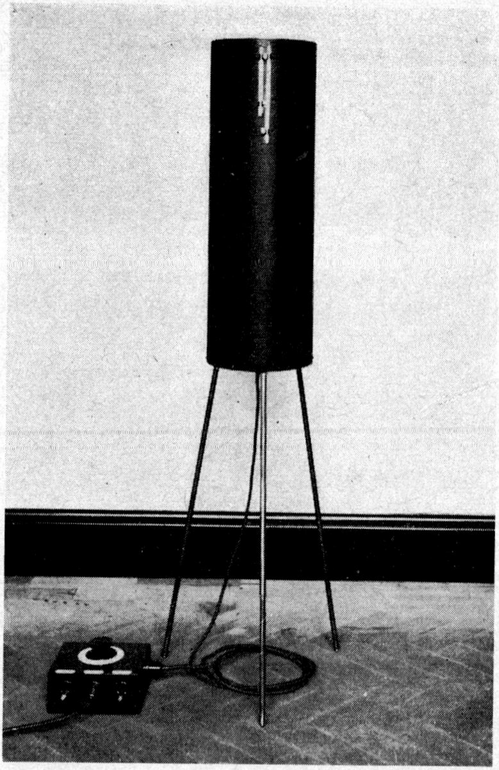

Fig. 15. The eupatheoscope. (Bldg. Res. Bd. (1938), H.M.S.O.)

temperature, air movement, and radiation. The readings are given on a scale of *equivalent temperature.*

Although in its original form the eupatheoscope had a surface temperature corresponding with that of the human body at comfortable temperatures, its temperature was less variable than that of the human body. In its present form (Mark II), however, the instrument has a surface temperature which is almost identical with that of the clothed human body over a considerable range of environmental temperatures (Dufton, 1936). The equivalent temperature, as indicated by the revised instrument, can be defined as: That temperature of a uniform enclosure in which, in still air, a black body of sufficient size would lose heat at the same rate as in the environment, the surface temperature of the body being one-third of the way between the temperature of the enclosure and 100° F.

It is not always possible or convenient to use the eupatheoscope for the measurement of equivalent temperature, and some other method of assessment is then desirable. Dufton (1933*a*) devised a method of using two specially made kata thermometers, one black and the other silvered. More recently (Build. Res. Bd., 1938) he has advocated the use of an 'Aitken' thermometer (*i.e.* a metal cylinder 1·2 in long and 4 in in diameter, holding a thermometer with its bulb enclosed within the cylinder) together with a kata thermometer.

Although for the precise measurement of equivalent temperature the eupatheoscope is required, measurements of the individual thermal factors which influence the equivalent temperature are also needed if one is to decide how to rectify unsuitable thermal conditions. If these thermal factors have been measured one can estimate the equivalent temperature by the use of the chart shown in Fig. 16 (Bedford, 1936). This chart is based on a study of thermal comfort (*vide* Chapter VI). Instruments with which these measurements can conveniently be made are the dry-bulb thermometer, the silvered kata thermometer, and the globe thermometer. If these instruments have been used the equivalent temperature can be estimated by means of Fig. 17.

A straight line is drawn from the appropriate point on the air temperature scale to that on the mean radiant temperature scale (or, in Fig. 17, the globe thermometer scale) and where this line cuts the air velocity line concerned the equivalent temperature is read off. Thus, with the air temperature 65° F and the mean radiant temperature 70° F, the equivalent temperature is 64.8° F when the air velocity is 25 ft/min, and 62.2° F when the velocity is 100 ft/min. Values can be interpolated for intermediate velocities.

Figure 16 is drawn from the equation:

Equivalent temperature

$$= 0.522t_a + 0.478t_w - 0.01474\sqrt{v}(100 - t_a)$$

where t_a and t_w are the air temperature and the mean

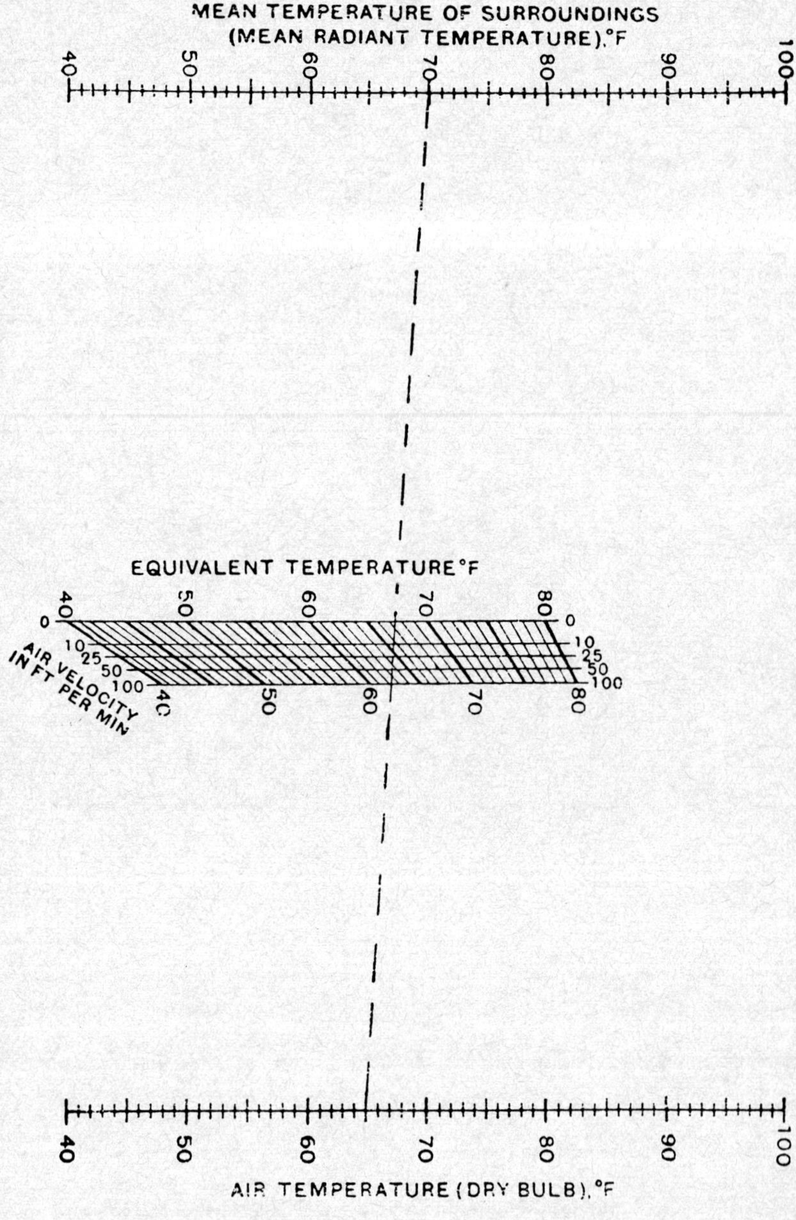

Fig. 16. Chart for the estimation of equivalent temperature for persons normally clad and doing sedentary or very light work. (I.H.R.B. Rept. No. 76, H.M.S.O.)

temperature of the surroundings respectively, both in °F, and v is the air velocity in ft/min.

The equation for Fig. 17 is:

Equivalent temperature

$$= 0.522t_a + 0.478t_g + \sqrt{v}(0.0808t_g - 0.0661t_a - 1.474)$$

where t_g is the globe thermometer temperature in °F.

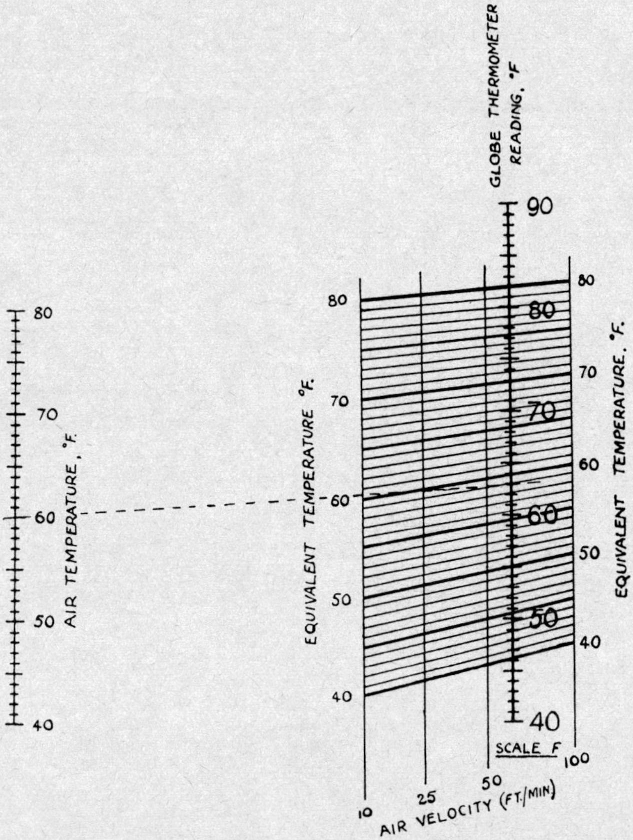

Fig. 17. Chart for the estimation of equivalent temperature from globe thermometer readings.

GLOBE THERMOMETER TEMPERATURE

Vernon (1930) introduced his globe thermometer as a means of indicating the combined effects of radiation and convection as they influence the human body. The temperature of the globe thermometer was termed by him (1932*b*) the 'radiation-convection temperature'.

By itself the globe thermometer is inadequate as an index of the thermal environment. For example, when the air and the surrounding surfaces are at the same temperature, the globe thermometer will record air temperature whatever the air velocity. Again, if the surroundings are cooler than the air, the globe thermometer reading will be below air temperature, but with air temperature and radiation constant the temperature of the globe will approach more nearly to that of the air as the velocity increases. Thus, of two environments, that which would cause the greater cooling of the human body would give the higher temperature reading on the globe thermometer.

This difficulty with the globe thermometer is only serious when the air velocity varies over a considerable range. For the calm conditions which usually prevail in occupied rooms the globe thermometer is a valuable index of comfort, for it makes adequate allowance for both the air temperature and the radiation from the surroundings.

For conditions in which the air is calm, and where humidity can be ignored, Missenard (1935) proposed the use of what he called the 'resultant dry temperature'. This is measured by means of a thermometer with its bulb enclosed in a blackened, hollow metal sphere of about 3½ in diameter, or in a hollow metal cylinder rather less than 3 in in diameter and 8 in in height. Missenard called this instrument the 'resultant dry thermometer'. The temperature recorded by it is very similar to that given by the globe thermometer.

RESULTANT TEMPERATURE

Missenard (1933) derived an equation for effective temperature, and then, from theoretical considerations, he made an adjustment to the equation so as to include a radiation term. This new measure he called *resultant temperature*. By using a function termed the 'fictitious air velocity' instead of the true velocity one could employ the effective temperature chart for arriving at the resultant temperature. This fictitious velocity is a function which takes account of the actual air velocity, the temperature of the air and of the surroundings, and also the mean surface temperature of the body (clothed or unclothed). Equations are given to enable the mean body surface temperature to be calculated for different conditions of clothing and activity. The calculation of this fictitious air velocity is troublesome, and the estimation of resultant temperature by its use is a laborious procedure. Missenard (1944) later constructed revised charts from which the resultant temperature can be determined with ease if the four thermal factors are known.

EQUIVALENT WARMTH

We have seen that 'effective temperature' takes into account the temperature, humidity and rate of movement of the air, but neglects radiation, while 'equivalent temperature' ignores atmospheric humidity, but takes count of radiation, air temperature and air movement. It was recommended by the Inter-Departmental Committee on Terminology, etc. in Heating and Ventilation Research (1932) that if the term 'warmth' were invested with a scientific meaning it should connote the proper assessment of all thermal factors, including humidity, which influence comfort. The term 'equivalent warmth', it was recommended, should be attached to measurements of warmth scaled by reference to standard conditions.

A provisional scale of equivalent warmth covering a limited range of environmental conditions has been constructed (Bedford, 1936). The standard conditions to which the scale is referred are those of a uniform enclosure in which the air and surroundings are at the same temperature, and in which the air is saturated with water vapour and quite still. Saturation was chosen as the standard of humidity with the object of simplifying comparison with effective temperature measurements, for, it will be remembered, complete saturation is the standard of humidity used in connection with that scale. Although the lead of American investigators was followed in this respect, it was recognized that there were objections to the use of this standard condition. In mining and in some other industrial situations almost saturated air may be encountered, but saturated atmospheres are rarely met with in heating and ventilating practice. In occupied rooms the relative humidity is commonly between 40 and 70 per cent, and it would have been reasonable to take some particular degree of humidity within that range as the standard condition for the scale of equivalent warmth. Since the scale was put forward, the Inter-Departmental Committee on Heating and Ventilation Research (Douglas, 1938) has recommended that for scales of equivalent warmth the standard of humidity should be a relative humidity of 50 per cent.

The equivalent warmth chart is shown in Fig. 18. The mean temperature of the surroundings is shown in scale *A,* scale *C* shows the absolute humidity, either as vapour pressure or as the weight of water vapour/lb of dry air, and scale *F* gives the dry-bulb temperature of the air. The equivalent warmth is given on scale *D* which really comprises five separate scales for various air velocities, the corresponding graduations on these scales being joined to facilitate interpolation. As an example, consider an environment in which the air temperature is 65° F, the vapour pressure 10 mm of mercury, the velocity 25 ft/min, and the mean temperature of the surroundings 70° F. A line is drawn from the graduation 70° F on scale *A* to that corresponding with a vapour pressure of 10 mm on scale *C,* and the point at which this line cuts scale *B* is noted. From this point of intersection on scale *B* a line is then drawn to the graduation representing 65° F on scale *F,* and the point where this line cuts the 25 ft/min line in scale *D* gives the equivalent warmth as 63.5° F. Thus, the feeling of warmth in an enclosure such as that specified would be the same as in a uniform enclosure of still and saturated air at 63.5° F.

To make easier the evaluation of readings taken with a sling hygrometer, a scale of wet-bulb temperature (scale *E*) is included in the chart. When it is desired to use this scale it is necessary to draw a straight line through the appropriate points on the dry- and wet-bulb temperature scales and extend it to cut scale *C.* The vapour pressure is then read off, and the operations described above are carried out.

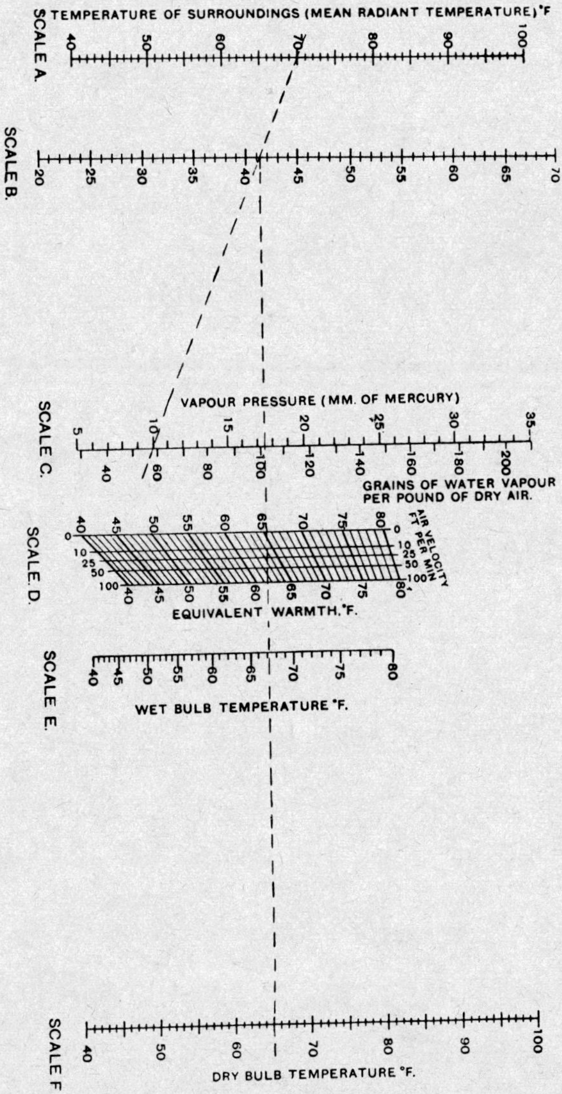

Fig. 18. Chart for the estimation of equivalent warmth conditions for persons normally clad and doing sedentary or very light work. (I.H.R.B. Rept. No. 76, H.M.S.O.)

Corrected Effective Temperature

The scales of effective temperature when corrected by the substitution of the globe thermometer reading for the dry-bulb temperature also take into account all four of the thermal variables.

WET-BULB GLOBE THERMOMETER INDEX (WBGT)

For the control of heat casualties at military training centres in the United States of America, Yaglou and Minard (1957) devised the wet-bulb globe thermometer index, a weighted mean of the wet-bulb and globe thermometer temperatures:

WBGT = 0.7 wet-bulb temperature + 0.3 globe thermometer temperature

The index takes into account the temperature and humidity of the air and the radiation from the sun and the terrain, and it makes some allowance for wind speed.

In a later form (Depts. of the Army, Navy and Air Force, 1957) the index has been defined as

$$\mathrm{WBGT} = 0.7\ \mathrm{WB} + 0.2\ \mathrm{GT} + 0.1\ \mathrm{DB}$$

where WB is the wet-bulb temperature, GT is the globe thermometer reading and DB is the dry-bulb temperature.

THE EQUATORIAL COMFORT INDEX

Webb (1952, 1959) studied the thermal comfort of persons living in Singapore, and from his results he derived an equation which expressed the observed dry-bulb temperature, vapour pressure and air speed in terms of the temperature of still and saturated air which would produce the same overall thermal sensation. The figure so obtained was referred to as the 'Singapore Index'. Its application was confined to the limited range of climatic conditions encountered in Singapore. To provide a scale applicable over a wider scale Webb (1960) also made use of the results obtained by various other workers, and from these composite results he derived a scale of warmth which he designated the Equatorial Comfort Index. The value of this index is read from a nomogram which includes scales for dry-bulb and wet-bulb temperatures and air speed.

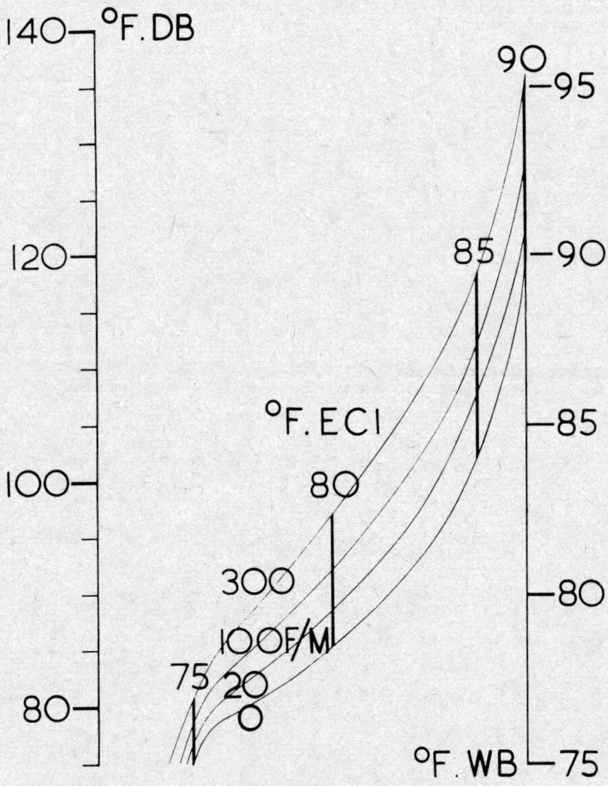

Fig. 19. The Equatorial Comfort Index. (Webb)

To use the nomogram join the points on the outer scales which indicate the dry-bulb and wet-bulb temperatures and mark the point of intersection with the curve which corresponds with the air speed. Estimate the value of the Equatorial Comfort Index (ECI) by interpolation between the vertical lines. The value is the temperature of still air saturated with water vapour that is equivalent physiologically to the climate under consideration.

Although no specific allowance is made for radiant heat it is thought that the scale will be applicable to indoor conditions over a wide area where the climate is warm, humid and oppressive, and where persons are acclimatized to wet-bulb temperatures of 75° F or above.

CHAPTER V

BODY HEAT PRODUCTION AND HEAT LOSS

THE MEASUREMENT OF HEAT PRODUCTION AND HEAT LOSS

DIRECT measurements of the total heat loss from the body are made by placing the subject in a form of calorimeter and measuring the amount of heat given off by him. Heat production is measured indirectly by measuring the oxygen consumption and the carbon dioxide production and calculating the heat production from these values. As early as 1777, Crawford had measured the heat produced by a guinea-pig in a calorimeter, and had concluded that the heat production was due to the transformation of pure air into 'fixed air' (carbon dioxide) and water. Three years later Lavoisier and Laplace repeated the animal experiments and reached a similar conclusion. It was Lavoisier who made the first respiration experiments on man.

The modern calorimeter enables heat production and heat loss to be measured simultaneously. The oxygen consumption and carbon dioxide output are measured, so that the heat production can be computed while the heat given off to the calorimeter is also measured. Although the direct measurement of the total heat loss by calorimetry is simple in theory it is more difficult in practice. Thousands of measurements and adjustments are needed during the course of a 3-hour experiment. But in spite of these difficulties, valuable researches have been carried on over a period of many years with the aid of the Sage Calorimeter at the New York Hospital (DuBois, 1937). For most purposes, however, body heat production is usually measured by the indirect method.

If the temperature of the body remained quite constant the heat production would always balance the heat loss, even in experiments of short duration. Actually, such equilibrium is probably rarely achieved, for changes in the rate of heat production or of heat loss are not instantly compensated, and slight variations in the temperature of the body result. Nevertheless, the temperature-regulating mechanism confines these changes within narrow limits, with the result that in prolonged experiments in the calorimeter it is found that measurements of heat loss and heat production agree very closely.

The heat production corresponding to a given oxygen consumption varies slightly with the amount of carbon dioxide produced. The ratio of the volume of carbon dioxide given off to the volume of oxygen consumed is known as the *respiratory quotient*. This varies between 1.00 and 0.71, the actual value depending on the nature of the foodstuffs that are being oxidized within the body. When the respiratory quotient is 1.00 the heat produced for each litre of oxygen consumed amounts to 5.05 kcal (20.05 Btu), while when the quotient falls to 0.71 the heat equivalent of 1 litre of oxygen is 4.7 kcal (18.66 Btu). For most purposes it is sufficient to take the round figure of 5 kcal/litre.

REGULATION OF BODY HEAT PRODUCTION

Lavoisier noticed that the heat production was greater in a cold environment than in a warm one. He found that a resting man in surroundings at 79° F absorbed 24 litres of oxygen per hour, whilst 27 litres were absorbed at a temperature of 54° F (Lusk, 1928). Thus at the lower temperature the heat production was increased by $12\frac{1}{2}$ per cent. Since Lavoisier's time many investigators have confirmed this increased metabolism in a cold environment. In really cold conditions the muscular activity associated with shivering accounts for the increase, and in less cold environments, where shivering is not noticeable, it is possible that the increased energy production is due to an imperceptible tensing of the muscles (DuBois, 1937). Tensing of muscles, even when no external work is done and no movement is perceptible, results in increased heat production: energy is required to effect such muscular tension and to maintain it. At high temperatures, also, the heat production tends to increase. This increase is due to the greater chemical activity of the body cells consequent on a slight rise in body temperature.

Rubner and Lewaschew (1897) found that in resting subjects clad in light summer clothing the metabolism was at a minimum at a temperature of 74° F. Later observations (Gagge, Winslow and Herrington, 1938) show that the metabolism is substantially the same for nude or clothed resting subjects at temperatures ranging from 68° to 102°F.

A person's heat production depends on his size. It is generally held that the rate of heat production of a person completely at rest and in warm surroundings—the basal metabolism—is a function of the surface area of the body, although some physiologists dispute this and claim that it is more nearly proportional to body weight. It is, at any rate, the general practice to express basal metabolic rates in terms of the number of heat units, calories or Btu, produced per hour/unit of body surface area.

With constant humidity and air velocity, the rate of body heat production varies with the temperature of the environment, but when the humidity and rate of air movement also vary their fluctuations affect the heat production. Then, body heat production is more closely related to a scale of warmth which takes not of the pertinent environmental variables than to air temperature alone. Thus, when temperature and humidity both vary over a wide range, body heat production is closely correlated with effective temperature. This is illustrated in Fig. 20 which is taken from the data of Houghten, Teague, Miller and Yant (1929) for clothed resting subjects.

Two curves are shown, one for a relative humidity of 95 per cent and the other for 20 per cent humidity. The curves are almost congruent, although, especially in the warmer environments, identical effective temperatures represent widely differing dry-bulb temperatures. These curves show minimal heat production at effective temperatures of about 67° to 84° F. In calm air at 50 per cent relative humidity these effective temperatures represent air temperatures of 68° to

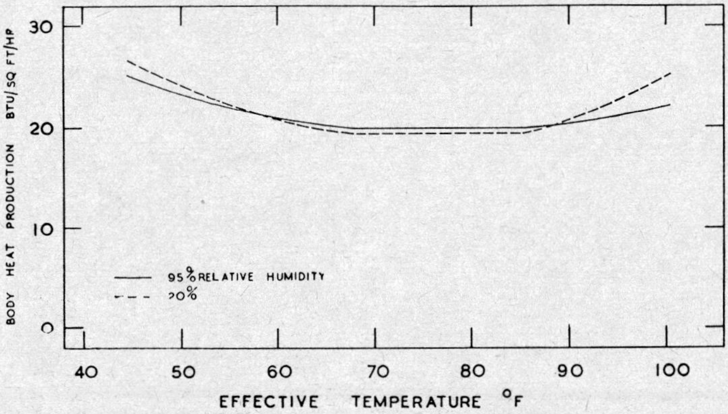

Fig. 20. Body heat production in relation to effective temperature. (Houghten, Teague, Miller and Yant, 1929.)

99° F, so that these data are in close agreement with the later findings of Gagge and his colleagues.

A person of average size, sitting at rest, has a total heat production of about 350 Btu/hour, but any form of activity, even talking, raises the energy exchange. It is the practice to take the heat production of sedentary persons at 400 Btu/hour, and that is a reasonably good figure. Standing makes further demands on energy, and these demands are increased by the performance of work. Persons doing very light work, standing, produce about 550 Btu/hour, and if the work entails moving about, even though it would still be classed as light, the heat production rises to 600 or 650 Btu/hour. More active work causes still greater heat production. Thus, walking at 3 m.p.h. results in an energy expenditure of 1,050 Btu/hour, while for walking at speeds of 4 and 5 m.p.h. the energy costs are about 1,390 and 2,330 Btu/hour.

BODY HEAT LOSS

Body Surface Area

Body heat production is commonly expressed in relation to the surface area of the body, and in any physical discussion of the heat loss from the body its surface area must be taken into account. A close approximation to the true surface area of the human body is given by the height-weight formula of DuBois and DuBois (1916):

$$A = W^{0.425} \times H^{0.725} \times 71.84$$

where A is the surface area in sq. cm;
W is the body weight in kg; and
H is the height in cm.

If A, W and H are given respectively in sq. ft, pounds and inches the formula becomes

$$A = W^{0.425} \times H^{0.725} \times 0.1087$$

Although this geometrical surface is that used for expressing the results of metabolism determinations, in studies of the effects of the thermal environment on body heat loss some allowance should be made for the influence of posture on the surface area which is effective for heat dissipation. When the body is in a crouching posture its effective radiating surface is less than when it is in a normal standing attitude, while when the legs and arms are outstretched the radiating surface is increased. Similarly the area exposed to the action of the surroundings air is reduced when the body is in a huddled position, so that the convection loss aiso is lessened.

Fishenden and Willgress (1925), discussing the DuBois equation, mention that the surface effective for radiation and convection is much less than the total surface. Barker (1931) assumes that a person whose total surface is 21 sq. ft has an effective radiating surface of 15 sq. ft, while he takes the whole 21 sq. ft as effective for convection. It may well be that the effective convection surface is greater than the surface which is effective for radiation, but in normal postures it will almost invariably happen that some areas of the body surface are in contact, and thus protecting each other from convection loss. Büttner (1932) takes the effective radiating surface to be from 10 to 15 per cent less than the geometric surface.

Bohnenkamp and his associates have emphasized the importance of ascertaining the effective radiating surface. They point out that the total loss by radiation should be calculated by integrating the radiation losses over the whole body surface, and describe a method by which the projection surfaces of the body can be estimated from photographs taken from various angles (Bohnenkamp and Ernst, 1931). This method involves much labour, and a simpler procedure is advanced in another paper (Bohnenkamp and Pasquay, 1931*a*). The electrical capacity of the subject is measured, and from this measurement the effective radiating surface is computed. Examples are given showing that for the same person very different values are obtained when the posture is varied. Deighton (1933) discusses this method critically, and draws attention to certain sources of error; while Bedford (1935) also shows that substantial errors may arise by its use.

Nevertheless, Bohnenkamp and Ernst (1931) concluded that the surface area of a man involved in heat transfer by radiation was 70 per cent of the area involved in convective heat transfer. If one assumes that the DuBois surface area is all available for heat transfer by convection, the result obtained by these workers only differs by about 2 per cent from results reported by Fanger (1970).

If the effective radiative surface area is divided by the DuBois surface area one obtains the 'effective radiation area factor', a term used by Fanger, whose results for ten male and ten female subjects are given in the Table:

Effective radiation area factor

Subjects	*Posture*	*Factor*
Female	sitting	0.692
Male	sitting	0.700
Female	standing	0.725
Male	standing	0.725

For the estimation of the body surface area of animals Deighton advocated an adaptation of Simpson's rule, with many measurements of perimeter, combined with photographs of side and plan views. Bedford suggested that for human subjects the effective radiating surface could be readily estimated by the simple application of Simpson's rule. From measurements made in that way, it appears that for a person standing in the 'attention' position the effective radiating surface when normally clothed is almost the same as the geometrical surface when nude. Gagge (1936) ascertains the radiating surface from direct measurements of heat loss.

Regulation of Body Heat Loss

The human body loses heat by radiation, convection and evaporation. As with an inanimate object, the rate of heat loss by these paths depends on the temperature, humidity, and velocity of the air, and on the radiation from the surroundings, but it is also largely governed by the physiological mechanisms which serve the body as thermostatic controls. These controls make possible the maintenance of a steady temperature in the vital organs in spite of considerable variation in the warmth of the surroundings.

One of these mechanisms is that which varies the blood supply to the skin. The reaction of the cutaneous circulation to thermal stimuli is complex, but in general the blood vessels of the skin constrict with cold and dilate with heat. Thus in cold surroundings the blood supply to the skin is diminished, and a lowered temperature of the skin surface results. Warmth provokes flushing of the skin with blood, and the surface temperature rises. These variations of the temperature of the skin surface, together with the environmental conditions, govern the rate of heat loss by radiation and convection.

Skin Conductance

It will be evident that the heat produced within the body can only pass to the surface, to be there eliminated, if the temperature of the skin surface is below that of the deeper tissues: there must be a temperature gradient to make possible the transfer of heat. The thermal conductivity of the skin and the adjacent tissues has been studied by a number of investigators. Temperature gradients were measured directly by Bazett and McGlone (1927), who inserted thermo-junctions into the tissues to various depths. Hardy and DuBois (1937) found that in nude subjects resting in the human calorimeter the conductance of the skin declined as the environmental temperature fell from 95° to 83° F, while at lower temperatures down to 73° F it remained steady. Changes in skin conductance were discussed more fully by Gagge, Winslow and Herrington (1938). They found that the conductance, expressed as the quantity of heat transmitted per unit of body surface area, reaches a minimum when the average skin temperature is 8° F below the temperature of the deep tissues as indicated by the rectal temperature. As the skin is further cooled the conductance remains the same, hence it appears that the constriction of the blood vessels in the superficial tissues is at a maximum when the temperature gradient is 8° F or more. As the temperature gradient is reduced below 8° F, by exposure to a warmer environment, the conductance increases rapidly. The increased conductance found with the lower temperature gradients is entirely due to an increase in the blood flow through the superficial tissues. The minimal conductance found by these workers, and by Hardy and DuBois, was approximately 10 kcal/m^2/hour/° C, or slightly over 2 Btu/ft^2/hour/° F.

Evaporation

Another way in which the body loses heat is by the evaporation of water from the skin and lungs. The expired air is saturated with water vapour, and the latent heat of vaporization, together with the heat required to warm the respired air to about 93° F, is taken from the respiratory surfaces. Water is lost from the skin by insensible perspiration and by sweating, and as the water is evaporated the latent heat is supplied by the skin surface.

The term 'insensible perspiration' is used to denote the loss of water from the skin when no sweating occurs, together with that from the lungs. It also includes the small loss of weight due to the slight excess of the weight of the carbon dioxide expired over that of the oxygen used. The insensible loss from the skin occurs even in a cool environment when the skin is apparently dry, but it is increased in warmer surroundings. It has been shown to be an osmotic process (Whitehouse, Hancock and Haldane, 1932).

Sweating is due to the activity of the sweat glands in the skin and may be regarded as the emergency mechanism which operates when the heat loss by other paths is insufficient to dissipate the heat produced by the body.

SKIN TEMPERATURE

Measurements of skin temperature are of importance in the physiological study of heating and ventilation problems, and it is appropriate at this point to describe the methods by which these measurements are made, and to indicate the variations which occur at different sites and under different environmental conditions.

Measurement of Skin Temperature

Of late years many studies of skin temperature have been made and various methods of measurement have been employed. Amongst those methods are the application to the skin of (i) some type of mercury-in-glass thermometer; (ii) thermo-junctions of various designs; and (iii) electrical resistance thermometers. Radiation thermopiles, which do not need to touch the skin, have also been used.

In the measurement of a surface temperature two principal difficulties arise. These are (1) that where an object of any considerable size is applied to the surface of a warm object the normal loss of heat from the covered area is immediately disturbed; and (2) that exposure of the thermometer bulb or other measuring apparatus to the air influences the temperature reading obtained, since the temperature attained by the instrument is the resultant of the effects of contact with the surface and with the air. In the presence of a strong air current the latter difficulty is liable to assume considerable importance. From the physical and engineering standpoints these difficulties have been discussed, and the literature reviewed, by Othmer and Coats (1928) and by Colburn and Hougen (1930). It has been shown that when a bare thermometer is applied to a rigid surface the temperature recorded by it differs from the air temperature by only 72 per cent of the true difference between the temperatures of the air and of the surface (Bailey, 1932). In measuring the temperature of the skin it is also essential that the pressure with which the measuring instrument is applied shall be kept as constant as possible: good contact with the skin is necessary, but excessive pressure will impede the cutaneous circulation.

Although the human skin reflects a proportion of the short wavelength radiation it receives from the sun, its emissivity for infra-red radiation is practically 100 per cent. Thus the skin radiates as a black body. Its radiation can be measured with a suitable radiometer and its temperature thus ascertained with accuracy. As the radiation thermopile does not actually touch the skin it does not interfere with the skin circulation, nor does it hinder the heat-loss from the surface. Observations are simply and rapidly made. With a sensitive instrument such as the Moll thermopile and a suitable galvanometer a reading can be taken in 4 to 6 sec. The results are not influenced by wind if the exposed junctions of the thermopile are screened, and as the temperatures are measured as differences from the known temperature of a black surface, transitory changes in the air temperature do not upset results. From a critical study of methods of measuring skin temperature, Bedford and Warner (1934*a*) concluded that the most accurate instrument is such a thermopile suitably screened (as by a rock-salt window).

Over 100 years ago, Davy (1814) used a mercury thermometer for measuring skin temperatures, and numerous workers have since used this method. It has been objected that pressure of the thermometer on the skin upsets the skin temperature, and that no mercury thermometer is rapid enough to record the true temperature of the skin. Careful tests show that by rolling a sensitive thermometer over a small area of skin the temperature can be measure with but very slight error. The method has the disability, however, that each observation takes from 1 to $1\frac{1}{2}$ min, so that accurate measurements cannot be obtained in special cases when the skin temperature is subject to rapid variation. In steady conditions accurate estimates can also be made by means of a clinical thermometer protected by a cork wedge, the thermometer bulb being inserted into a groove bored into the narrow edge of the wedge, but each observation takes from 4 to 5 min.

The most widely used method is the application of a thermo-junction to the skin. Forms of apparatus in which the junction makes only a light contact with the skin, and in which both the junction and its leads are exposed to the air, are liable to give very erroneous estimates of skin temperature. At ordinary room temperature, conduction along the leads may cool the junction, which is also exposed to direct cooling by contact with the air, even in a calm atmosphere, while when there is much air movement the cooling is increased. Observations show that the skin temperature may be underestimated by 5° or 6° F when such a junction is used. Many German physiologists have used thermo-junctions of this type, and the skin temperatures which they have reported are distinctly lower than would be expected from measurements made by other and more reliable methods.

When these sources of error are avoided, the thermo-junction method gives reliable results. A simple apparatus which compares favourably with more elaborate devices is the junction used by Sir Thomas Lewis (1930). The two wires (copper and constantan), of fine gauge, are twisted into a spiral for about $\frac{1}{4}$ in from their ends, the tip of the spiral is lightly soldered to form the junction, and the remaining length of wire carefully insulated. The junction is attached to the skin and protected by a strip of surgical tape.

Variation in Skin Temperature

It is a matter of experience that the temperature of the skin varies with the warmth of the environment, that in an overheated room the skin is flushed and warm, while outside on a cold day it is cool. Another fact with which all are familiar is that different parts of the body surface may be at different temperatures. The skin of the forehead, for instance, varies only over a narrow range of temperature, while the extremities may be very warm or quite cold.

The pallor and constriction of the skin caused by exposure to cold were referred to by medical writers of ancient Greece; and, later, Francis Bacon, in his *Novum Organum*, mentioned that: 'All animals are externally cold in winter and cold weather, but are thought to be internally warmer.' Bacon also noted that some parts of the body are warmer than others, and recommended that 'further inquiry be made as to the comparative heat in different parts and limbs of the same animal'.

Actual measurements of the temperature of the skin do not appear to have been made until long after Bacon's time. In a stimulating essay published in 1733, Arbuthnot described some observations of skin temperature which he made with Stephen Hales's thermometer. The first attempt to relate skin temperatures to the warmth of the environment appears to be due to Vincent (1890), who gave an equation expressing these temperatures in terms of the readings of meteorological instruments.

During the present century many investigators have made observations of skin temperature, usually that of the forehead, in relation to air temperature. The published results show considerable differences in the forehead temperatures observed at a given air temperature. For example, with an air temperature of 68° F, different observers record mean temperatures of the forehead, or of the forehead and cheeks, ranging from 89° to 95° F in clothed persons. Factors such as the amount of clothing worn, and individual differences between subjects, are likely to have contributed to the variation noted, but much of that variation is undoubtedly due to differences in the methods of measurement employed. The higher values are probably the more accurate.

The early observations relate to individuals or to small groups of subjects, and the results are not of general application. Observations on a larger scale were made a few years ago on industrial workers engaged on sedentary or very light work and wearing normal indoor clothing, and these were related to the warmth of the environment (Bedford, 1936). The temperatures of the forehead and of the hands were measured by means of a Moll thermopile, and for the temperature of the feet a thermo-junction mounted on a special holder that could be inserted inside the shoe was used. The hand temperature was measured on the open palm, and the foot temperature on the inner aspect of the foot about 1 in below the ankle.

The results of about 2,500 sets of observations, made on between 1,000 and 2,000 persons, are shown in Fig. 21. On the

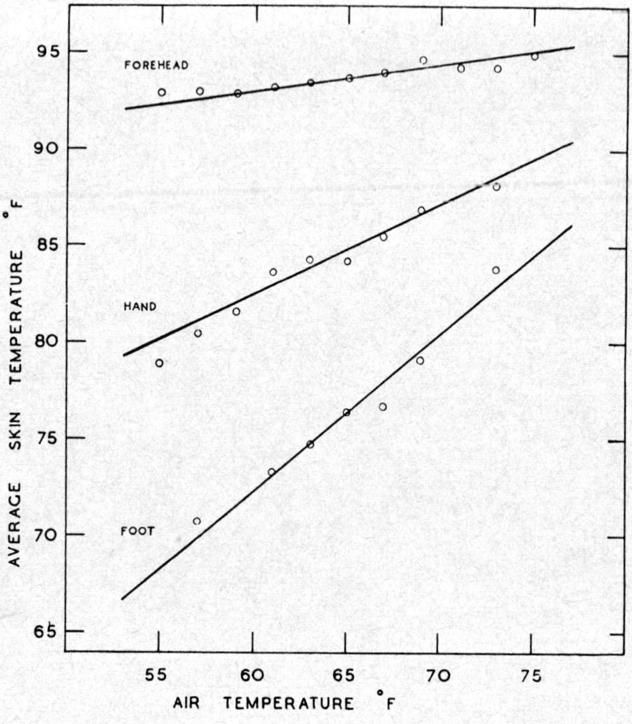

Fig. 21. Average skin temperature in relation to air temperature.

whole, a change of 1° F in the air temperature was accompanied by a change, in the same direction, of 0.14° F in the temperature of the forehead, about 0.46° F in that of the hand, and of 0.8° F in that of the foot. At an air temperature of 65° F the skin temperatures averaged 93.7° F on the forehead, 84.8° F on the hand, and 76.4° F on the foot. There was considerable variation in the skin temperatures recorded for different people at any particular air temperature.

Since the temperatures of the skin at different sites are each related to air temperature, it is to be expected that there will be some relation between the temperatures in different parts. Statistical examination of the data shows that the association between the skin temperatures of any two of the sites mentioned is not as close as that between the several skin temperatures and air temperature. Thus, the temperature of a person's foot, for example, can be estimated with slightly greater accuracy from a knowledge of the air temperature than from a knowledge of his hand temperature or his forehead temperature.

The preceding paragraphs relate to skin temperatures observed under steady, or virtually steady, environmental conditions. When the body is exposed to sudden chilling, whether by the impact of wind or by a change in air temperature, the skin temperature falls rapidly. This quick response of the circulation serves as a defence in helping to prevent excessive loss of body heat. An example of such adjustment of the skin temperature is shown in Fig. 22.

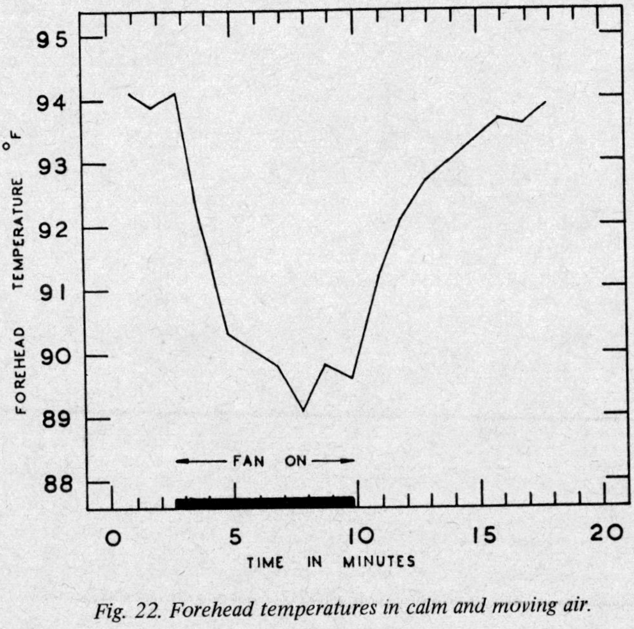

Fig. 22. Forehead temperatures in calm and moving air.

The observations were made at room temperature of 72° F on a subject wearing normal indoor clothing. For the first few minutes the air was practically still, an electric fan was then switched on and a wind of about 350 ft/min passed over the forehead. Then, after a further 7 min, the fan was stopped. The impact of the stream of air on the skin caused its temperature to fall rapidly to the extent of 4° to 5° F. When the fan was shut off the skin temperature rose, but its rate of recovery was distinctly slower than that of the initial fall (Bedford and Warner, 1934*a*).

Pickering (1932) concluded from his experiments that the immediate response to chilling of the body surface is a reflex constriction of the vessels in the skin. After a few minutes' exposure to cold the reflex vasoconstriction begins to weaken and the skin temperature rises somewhat. Heat applied to the skin produces dilation of the vessels only after a delay of a minute or more. This response to heat, and the continuation of vasoconstriction during prolonged exposure to cold, appear to be due to the action of a central mechanism which is determined by the temperature of the blood. This mechanism seems to be the body's chief defence against the prolonged effects of cold, and the only defence against excessive heat. It responds when the temperature of the tissues is changed by an amount varying from 0.01° to 0.04° C in 2 to 4 min. The immediate protection against cold is the central reflex which brings about the initial contraction of the skin vessels.

TRANSMISSION OF RADIATION THROUGH THE SKIN

Human skin emits radiation almost like a black body. For radiation of long wavelength, such as that which it emits, the skin likewise absorbs virtually as a black body. Radiation in the visible and shorter infra-red regions of the spectrum is to some extent reflected by the skin. Figure 23 shows the reflection of white and negro skin as determined by Hardy and Muschenheim (1934). At wavelengths longer than about 2 microns only a small percentage of the incident radiation is reflected.

It is well known that radiation of some wavelengths can be

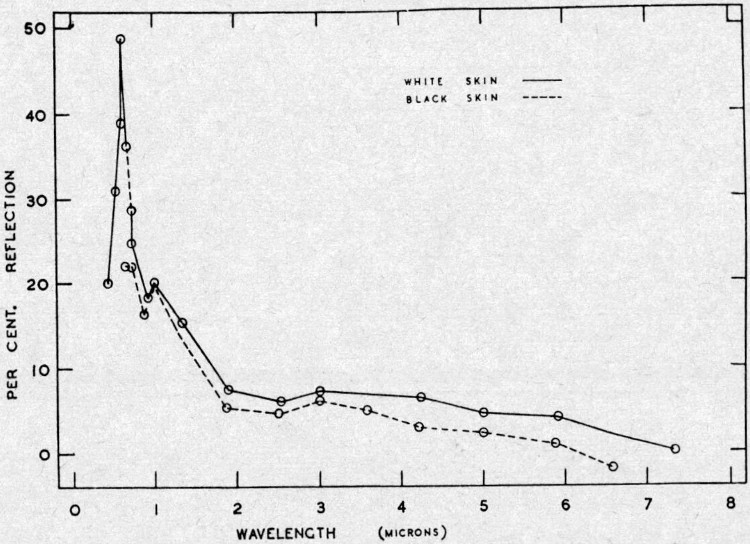

Fig. 23. Reflection of white and negro skin. (Hardy and Muschenheim.)

transmitted through human tissues, X-rays, for example, can pass through considerable thicknesses. Radiant heat, likewise, can penetrate the skin, but it is absorbed within a very short distance of the surface. In Fig. 24, Hardy and Muschenheim's (1936) curve for the transmission through a layer of human epidermis, 0.03 mm thick is shown, while Fig. 25 gives their

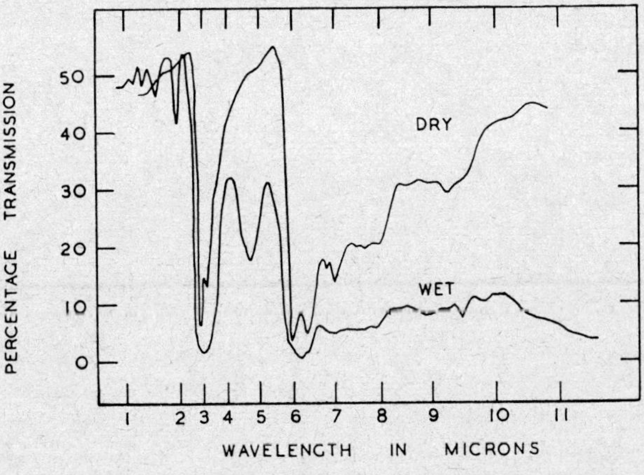

Fig. 24. Transmission spectrum of human epidermis. (Hardy and Muschenheim.)

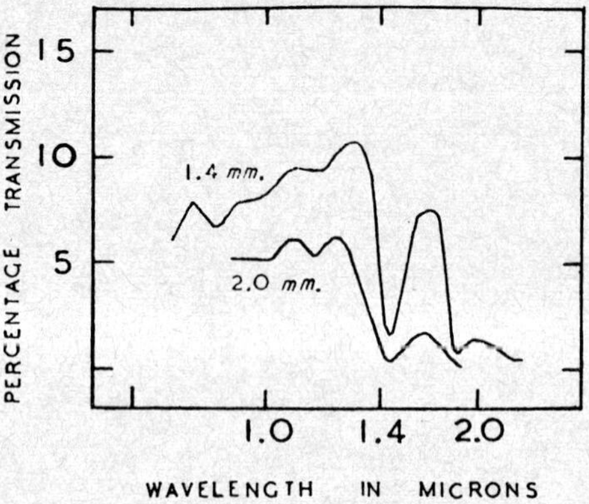

Fig. 25. Transmission curves for thick specimens of human skin. (Hardy and Muschenheim.)

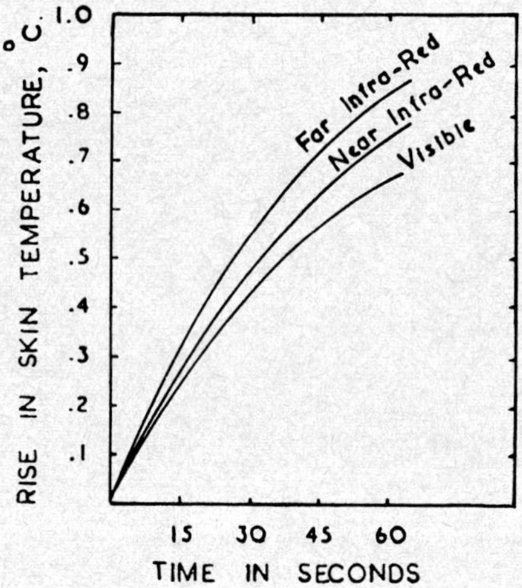

Fig. 26. Rise of skin surface temperature produced by the absorption of the same rate of radiation. (Oppel and Hardy.)

curves for the transmission through layers of skin 1.4 and 2.0 mm thick.

Substantial amounts of radiation penetrate the thin layer of the epidermis, but no radiation of wavelength longer that 2 microns penetrates as far as 2 mm, and virtually all the radiation is absorbed within 3 mm of the surface. One of the most penetrating of the infra-red radiations is that of wavelength 1.2 micron. Of the incident energy of this wavelength, 21 per cent is reflected by the surface, 56 per cent penetrates the outer horny layer, 50 per cent passes through the Malpighian layer to the subcutaneous tissue, but 99 per cent is absorbed within 3 mm of the surface.

Radiation which penetrates the skin exerts less heating effect on the skin surface than non-penetrating radiation. This is clearly shown by the observations of Oppel and Hardy (1937*a*). Figure 26 shows the rise of skin surface temperature observed by them when the skin was irradiated so that it absorbed the same quantities of (i) non-penetrating, long infra-red, (ii) penetrating, short infra-red, and (iii) visible radiation.

The more penetrating radiations provoked less elevation of the surface temperature than occurred with non-penetrating.

SURFACE TEMPERATURE OF THE CLOTHED BODY

The rate of heat loss from the body depends on the temperature of its surface as well as on the warmth of the environment. Some years ago, Bedford wished to measure the average surface temperature of the clothed bodies of a large number of industrial workers engaged in sedentary or very light work. It was not practicable to make many measurements on each person, so preliminary observations were made in the laboratory to determine whether measurements of the surface temperatures of a few small areas would give a reliable index of the mean surface temperature of the whole body (Bedford, 1936). Fifty-one sets of observations were made on 41 persons (11 women and 30 men) of varied age and build, and wearing different types of clothing. By means of a Moll radiation thermopile the temperature of small areas at 21 places on each person was measured, and the average taken. It was found that the average of three readings taken in a diagonal line across the back—at the shoulder, centre of back, and on the buttock—gave a close approximation to the mean surface temperature as calculated from the whole 21 readings. On the average the mean surface temperature was underestimated by only 0.4° F, and the root-mean-square error of estimation was 1.4° F. It was concluded that when the warmth conditions were fairly uniform, *i.e.* in the absence of intense localized radiation or of strong air movement, this simple method gave reasonably accurate results, and it was therefore employed in the factory investigation.

In all, 2,558 sets of observations were made in this way, and the mean surface temperatures thus found were correlated with the environmental temperatures. The relation between mean body surface temperature (t_b) and the air temperature (t_a) and mean radiant temperature (t_w) is given by the equation:

$$t_b = 0.399t_a + 0.363t_w + 25.76$$

Thus, if the air and surroundings are at the same temperature a rise of 1°F in the temperature of the environment is accompanied by a rise of 0.76°F in the mean surface temperature of the body. This relationship applies when the air velocity averages about 30 ft/min.

THERMAL EXCHANGES BETWEEN THE BODY AND ITS SURROUNDINGS

So long as the body temperature remains constant the heat produced within the body is equal to the body heat loss; and the body heat production equals the total energy production (metabolism) if the body is at rest, or, when the body is doing work, the energy production less the energy expended in external work. The energy balance is expressed by the equation:

$$M - W = E + R + C + S$$

where M is the metabolic rate;
W is the rate at which energy is expended in mechanical work;
E R, and C are the rates of heat loss by evaporation, radiation and convection, respectively; and
S is the rate at which heat is being stored within the body.

In the equation, $(M - W)$ represents the total heat production, and $(E + R + C)$ the total heat loss.

In studies of the thermal exchanges between the body and its surroundings the metabolism is generally determined by indirect calorimetry. The total heat loss is taken as the rate of energy production, less any energy expended as work, and less any heat stored in the body. Heat storage is indicated by change in body temperature: when the body temperature remains unchanged storage is nil. In earlier work the storage was calculated from the rise in rectal temperature, and the weight and specific heat (0.83) of the body. Heat storage can only be ascertained approximately in this way, for the change in rectal temperature does not measure the change in the mean temperature of the body if there has been any substantial change in the temperature of the surface tissues. For example, when the body is exposed to a cool environment the rectal temperature may fall only very slightly, while the temperature of the skin falls several degrees. The reduction in the heat content of the body (negative storage in our terminology) in such a case is substantially higher than would be indicated by the slight shift in rectal temperature. In the more recent researches (Burton, 1935; Hardy and DuBois, 1938; Winslow *et al.*, 1939), the shift in the mean body temperature has been calculated by taking a weighted mean of the rectal temperature and the average skin temperature.

Under moderate environmental conditions it seems that the mean body temperature can be estimated with fair accuracy by adding one-third of the mean temperature of the skin to two-thirds of the rectal temperature. Yet Nelson and his associates (1947), who perforce used this method for the calculation of heat storage during work at high temperatures, were dubious about its accuracy, and expressed the view that success in partitioning heat losses depends largely on the magnitude of the changes in heat storage involved; and Bazett (1949) pointed out that in cold environments, when heavy clothing is worn, the accuracy obtained may be quite low if a steady state is far from being attained.

The heat loss by evaporation is deduced from the loss of weight, and the difference between the total heat loss and the evaporative loss gives the combined radiation and convection losses. Frequently the body heat loss is only partitioned into these two fractions. When the radiation loss is distinguished from the convection loss the former is calculated either from direct measurements of the radiation intensity from various areas of the body surface, or from measurements of the surface temperature of the body. Some estimate of the effective radiating surface of the body is needed. Winslow and his colleagues (1936) have elaborated a technique of partitional calorimetry for the more accurate determination of the radiation and convection exchanges.

Estimates of the evaporative loss which have been made during the past half-century agree very well, especially when it is borne in mind that physiological reactions are liable to vary considerably. Thus, Rubner (1896) found that the evaporative loss of a man normally clothed in surroundings at 63.5° F was 23 per cent of the total heat loss. Soderstrom and DuBois (1917) put the average figure at about 24 per cent, but the proportion varied between 20 and 34 per cent in normal subjects. Benedict and Root (1926) related water loss to total metabolism, and from their equation it can be calculated that the evaporative loss is about 30 per cent or more of the total; but the equation only holds for subjects with a skin temperature such as is attained after a night's rest in bed. It appears that at ordinary room temperatures the heat loss by evaporation is between 20 and 30 per cent of the total heat lost by the resting subject, and this is confirmed by more recent work (Houghten *et al.*, 1929; Gagge *et al.*, 1938).

The apportionment of the remaining 70 or 80 per cent of the total heat loss into radiation and convection losses has interested physiologists for many years, and conflicting estimates have been recorded. The most widely quoted apportionment is that of Rubner (1896), who calculated the heat balance of a clothed subject in still air at 63.5° F. He found the proportions of the total heat lost by evaporation, convection and radiation to be, respectively, 23, 32 and 45 per cent. A closely similar apportionment was arrived at by Aldrich (1928) as the result of a study of the heat losses from a cloth-covered calorimeter. He put the radiation loss at normal indoor temperatures as 60 per cent of the combined radiation and convection losses. From measurements of the mean surface temperature of the clothed body, I calculated that in still air at temperatures of from 55° to 75° F the radiation loss amounted to from 62 to 69 per cent of the combined loss by radiation and convection (1936). Increase of air velocity reduces the radiation loss by reducing the surface temperature of the body, and convection then accounts for a greater proportion of the combined radiation and convection losses.

As the temperature of the environment rises, the loss by radiation and convection declines and the evaporative loss increases. This is clearly illustrated by the work of Houghten and his colleagues (1929 and 1931). They studied the thermal exchanges between the bodies of clothed subjects and their environment. One series of observations was made on men who were working, and the other on men who were only slightly active and who can conveniently be referred to as 'resting' subjects. The body surface areas of these subjects ranged from 18.4 to 21.4 sq. ft, and the rate of working was 33,000 ft lb/hour. Typical American indoor clothing was worn.

The apportionment of the heat losses for the resting and working subjects is shown in Fig. 27.

For each group of subjects two sets of curves are shown, one for air at a relative humidity of 20 per cent and the other for 95 per cent humidity.

In the resting subjects the total heat loss was practically constant between air temperatures of 70° and 88° F when the humidity was 95 per cent, while with a humidity of 20 per cent approximate constancy of total heat loss was maintained up to a temperature of 110° F. The curves showing the combined convection and radiation losses are almost identical for the two extremes of humidity. At the lower temperatures rather more evaporation took place with dry air than with moist. For a short range of temperature in warm atmospheres the two evaporation curves are congruent, and then at temperatures above 90° F the evaporation increased rapidly with dry air, but declined in moist air. At temperatures higher than 98° F the combined radiation and convection losses were negative, *i.e.* the temperature of the environment was above that of the body surface, and in consequence the body gained heat by radiation and convection

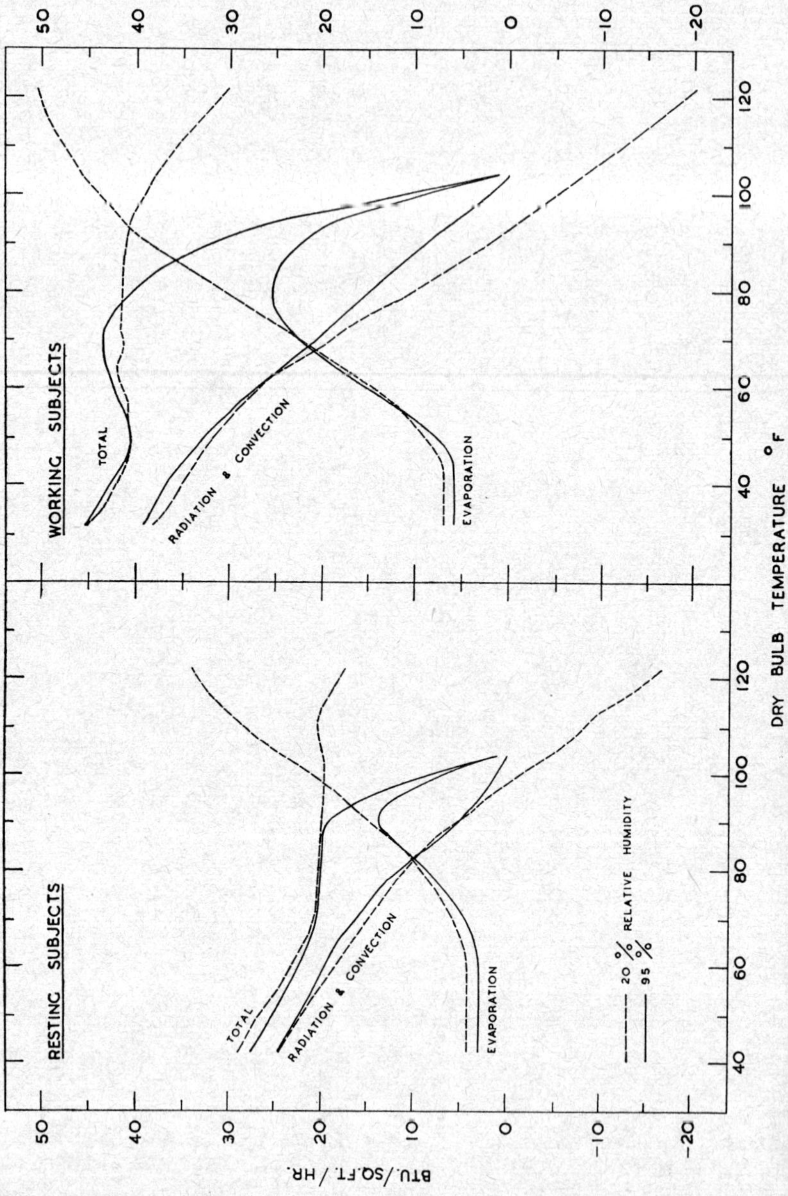

Fig. 27. Body heat losses of clothed subjects; (a) *resting, and* (b) *working. (After Houghten* et al. *(1929 and 1931).)*

instead of losing it. In such conditions, if the total dissipation of heat from the body is to be adequate the evaporative loss must be great enough to compensate for the other heat gains. The curves show that in the resting subjects the loss by evaporation did, in fact, amount to substantially more than the total heat loss at temperatures of 98° F or above, when the air was dry, with the result that until the temperature reached 110° F the total loss of heat remained practically constant.

The curves for working subjects are similar in type to those for resting subjects, but they are at different levels. The total heat loss (and the total heat production) was the higher in the working subjects. At the lower air temperatures the combined radiation and convection losses of the working subjects were considerably higher than those of the resting subjects, which indicates that work raised the temperatures of their body surfaces and thus made easier the dissipation of heat. At the higher temperatures the difference between the radiation and convection losses for the two groups diminished, so that at temperatures of over 80° F there was little to choose between the groups. At these higher temperatures the extra heat production due to work had to be shed by the evaporation of more sweat; sweating began at a lower temperature, and the curves show that at an air temperature of 80° F the evaporative loss of the working subjects was treble that of the resting subjects. In these observations the total heat loss of the working subjects began to decline at lower temperatures than those at which the total losses from the resting subjects became inadequate. In the moist atmosphere the decline began at a temperature of 90° F in the resting subject, but in the working subjects it began before 80° F was reached.

We have seen that, according to the work of Houghten and his colleagues, body heat production is a function of effective temperature. The total heat loss from the body was also found to be constant for a specific effective temperature, although the apportionment between radiation and convection on the one hand and evaporation on the other differed greatly for different combinations of temperature and humidity. An example of this constancy of total heat loss and variation of its apportionment is shown in Fig. 28.

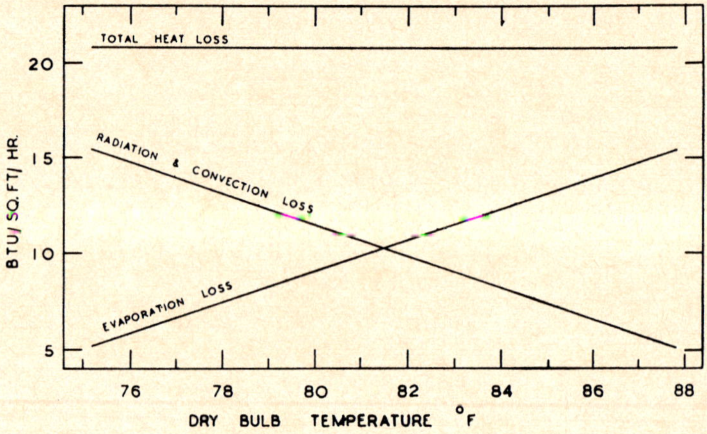

Fig. 28. Body heat losses at various temperatures and humidities yielding an effective temperature of 75° F. (Houghten et al.*)*

Nelson, Eichna, Horvath, Shelley and Hatch (1947) studied the thermal exchanges of men in hot surroundings. For nude, standing persons whose bodies were completely wet (assumed to be when 10 per cent or more of the sweat was unevaporated) the evaporative heat loss was described by the equation

$$E = 1.44V^{0.37}(P_s - P_a)$$

where E is the heat exchange by evaporation from the skin, in kcal/m^2/hour;
V is the air speed in ft/min; and
P_s and P_a are respectively the vapour pressures on the skin and in the air, in mmHg.

When the body surface is not completely wet the rate of evaporation is less than the maximum given by this equation. If the mean skin temperature is known it is possible to estimate the effective evaporating surface relative to the maximum with fair accuracy, and then to compute the evaporative loss; but without direct measurement of the skin temperature, or some independent method of predicting it which has not yet been devised, the calculation cannot be made with satisfactory accuracy (Machle and Hatch, 1947). Until such means of predicting the skin temperature can be found calculations of the upper limit of evaporative capacity are of limited value.

The combined radiative and convective exchanges $(R + C)$ were described by the equation

$$R + C = (5.65 + 0.53V^{0.5})(t_s - t_a)$$

where $(R + C)$ is in kcal/m^2/hour;
t_s and t_a are respectively the mean skin temperature and the air temperature, in °C; and
V is the air speed in ft/min.

The equation applies to conditions in which the mean radiant temperature was substantially the same as the air temperature.

When clothing is worn the calculation of body heat exchanges becomes more difficult. The heat capacity of the clothing causes complications. The emissivity of the surface of the clothing may be distinctly less than unity, even for radiation of long wavelength, although for much ordinary clothing it probably does not differ much from that value. The surface area of the clothed body is greater than that of the nude one, in nine subjects wearing fatigue uniform it was 1.22 to 1.52 times as great.

The average temperature of the clothed surface is difficult to ascertain with accuracy owing to the greater differences between the temperatures of individual areas and the difficulty of calculating a weighted average; and a similar difficulty arises with respect to the mean vapour pressure at the surface. Uneven wetting of different parts of the clothing is yet another source of difficulty.

With these difficulties in mind Nelson and his colleagues studied the heat exchanges of clothed subjects. In spite of the complexities of the problem they were able to obtain gross coefficients of evaporation for a range of rates of sweating. For the combined radiative and convective exchanges $(R + C)$ of clothed standing subjects they obtained the equation

$$R + C = (2.66 + 0.65V^{0.5})(t_s - t_a)$$

where t_s is the mean temperature of the surface of the clothed body.

The convection constant is 1.2 times the coefficient for the nude body, which is compatible with the increase in the surface area of a clothed man. The added constant for radiation is lower than expected, and no reasonable explanation could be offered.

Chilling in Cool Environments

Much emphasis is laid on the heat storage within the body and the consequent rise of body temperature during prolonged exposure to high environmental temperatures. At the other extreme, long exposure to low temperatures results in excessive loss of body heat and consequent chilling. The curves of Houghten and his colleagues have shown that the rate of heat loss from the body may exceed the rate of heat production. Even if the temperature of the deeper tissues, as indicated by the rectal temperature, remains fairly steady, there may be substantial chilling of the superficial tissues.

Superficial chilling is well illustrated by observations made by Vernon, Bedford and Warner (1930) on themselves during a study of the heating and ventilation of schools. Figure 29 shows the average skin and mouth temperatures of the three observers during the course of three-hour experiments when the air temperature averaged 50.4° F and the air velocity 38 ft/min. The mouth temperature fell rapidly, and at the end of the three-hour spell it was only 97.1° F, or 1.1° F lower than at the beginning. The forehead temperature, which is usually very steady, fell 2° F, while the temperatures of the palm, fingers and foot fell 13°, 18° and 7° F respectively. At the start of the experiments the subjective sensations of warmth, which were expressed on a numerical scale for statistical purposes, averaged about half-way between 'comfortable' and 'too cold'. but at the end of the three-hour exposure the subjects felt 'much too cold'. They were, in fact, miserably cold after their long and progressive chilling. These observations give no measure of the total loss of body heat, but such measurements have been made by Winslow, Gagge and Herrington (1939). In one series of experiments the clothed and resting subjects of Winslow and his colleagues were exposed to air at a temperature of 73° F and

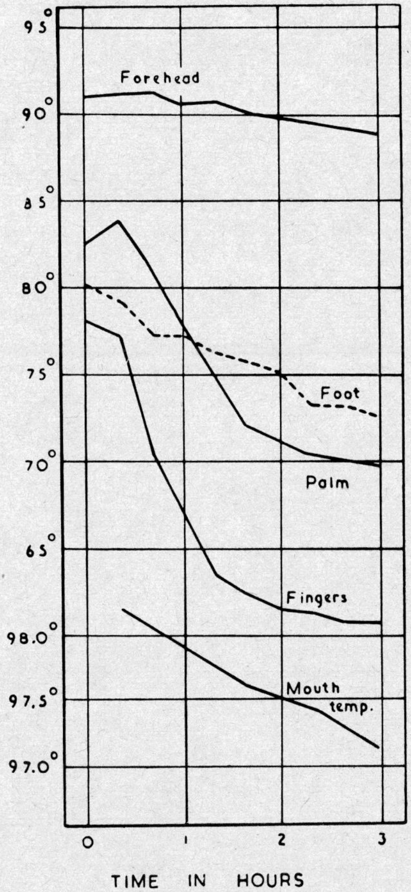

Fig. 29. Chilling during three-hour exposure to temperature of 50° F.

with a velocity of 520 ft/min for periods of over three hours. Some of the data relating to these experiments are plotted in Fig. 30.

During a period of three hours, from the 20th to the 200th min of the experiment, the rectal temperature declined by 0.7° F, the mean temperature of the skin fell by 2.7° F, and the temperature of the clothing surface also fell slightly. Throughout the period the weighted mean body temperature, calculated from the rectal temperature and the skin tempera-

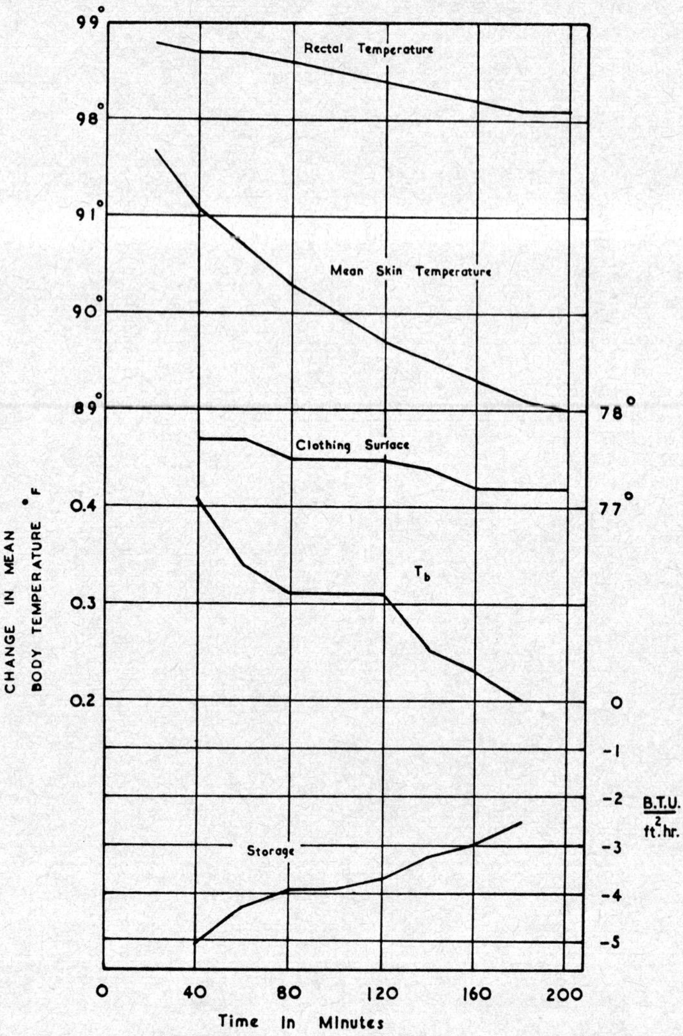

Fig. 30. Progressive chilling of resting clothed subjects in air at 73° F and velocity 520 ft/min. (Windlow et al.*)*

ture, declined. During the whole period the total heat loss from the body exceeded the rate of heat production, but the excess of the rate of heat loss over the rate of heat production diminished as the experiments progressed.

The Effects of Clothing

The efficiency of the protection afforded by clothing clearly depends on the insulating value of the clothing. An example of the protection given by light indoor clothing can be drawn from the work of Gagge, Winslow and Herrington (1938). Their subjects were completely at rest in a semi-reclined position. In one series of experiments they were nude, while in the other they wore the usual American indoor clothing. Figure 31 shows the heat production and heat losses at various operative temperatures.* Heat storage within the body is also shown.

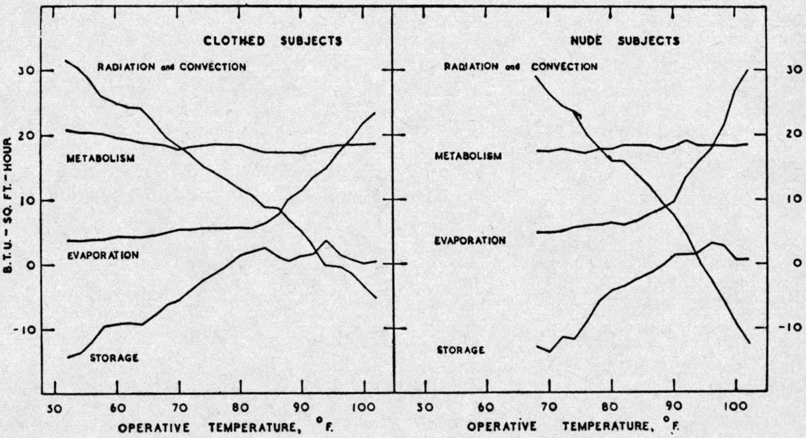

Fig. 31. Heat losses and heat production in clothed and nude subjects. (Gagge et al.*)*

The heat production for both clothed and nude subjects was substantially the same between operative temperatures of 68° and 102° F. At temperatures lower than 68° the heat production increased. The curves for evaporation for both groups were almost identical up to a temperature of 92° F, but at higher

* 'Operative temperature' is a concept introduced by Gagge and his colleagues. It is a function of the air temperature and the mean temperature of the surroundings in which these two variables are weighted in proportion to the radiation and convection constants of the individual subjects. Operative temperature is very similar to globe thermometer temperature.

temperatures evaporation took place more rapidly from the nude subjects than from the clothed. In these experiments clothing had its main influence on the combined radiation and convection losses. The radiation and convection curve for clothed subjects is less steep than that for the nude. Both these curves show zero loss at an operative temperature of 94° F. At temperatures below 94° F the radiation and convection loss from the clothed subjects was less than from the nude ones, while at higher temperatures the clothed subjects gained less heat from the environment than did the nude. Inspection of the curves for heat storage shows that the clothed subjects lost heat faster than they produced it at temperatures below 78° F, while in the nude subjects this excess heat loss began when the temperature fell below 88° F. At temperatures higher than these there was some storage of heat within the bodies of the subjects.

The study of clothing has interested physiologists and others since Count Rumford made measurements of the thermal conductivity of various clothing materials 160 years ago. A century later Rubner did valuable work on conductivity and air permeability, and in recent years many workers have developed more exact methods of measurement and have studied the physical properties of clothing fabrics. During the war of 1939-45 much attention was necessarily given to the design of special clothing for protection against extremes of temperature. The results of this work had to be discussed with non-technical members of various war-time committees and it occurred to a group of workers (Gagge, Burton and Bazett, 1941) that a new unit for expressing the insulating power of clothing would be useful.

The unit was called the 'clo' and its magnitude was chosen so that it was approximately equal to the insulation provided by normal indoor clothing worn by sedentary persons in a comfortable indoor environment. The derivation of the unit is as follows:

As Burton said, most people in towns wear about 1 clo unit of thermal insulation most of the time: Europeans may wear more and Americans less than 1 clo unit but the differences are not great and are due to different indoor temperatures.

If a man is sitting and at rest, he produces heat at the rate of about 50 kcal/m^2 of his body surface per hour.

A normal indoor environment is specified, *i.e.* a dry-bulb air-temperature of 21° C (70° F), an air velocity of 10 cm/sec (20 ft/min) and a relative humidity of less than 50 per cent. For the man to be comfortable under these conditions, his clothing is such that his total heat loss equals his heat production, *viz.* 50 kcal/m^2/hour.

About 12 kcal/m^2/hour is lost by evaporation of moisture leaving 50 – 12 or 38 kcal/m^2/hour transferred by conduction through the clothing. The average temperature of the surface of the body is taken to be 33° C (91.4° F). By analogy with Ohm's law, the resistance to heat flow (r) is equal to the difference in temperature divided by the rate of flow of heat:

$$r = \frac{\Delta t}{\text{rate of heat flow}} = \frac{33\text{-}21}{38} = 0.32°\text{C/kcal/m}^2\text{/hour}$$

But this resistance to the flow of heat is the sum of the resistance of the clothing and the resistance of the air. One must obtain a value of the resistance to heat flow of the surrounding air for the clothed human body when the rate of air velocity is low.

According to Winslow *et al.* (1940), the value for air under such conditions is 0.14° C/kcal/m^2/hour. Therefore the resistance of the clothing is 0.32 – 0.14 or 0.18° C/kcal/m^2/hour. The clo unit is defined as 0.18° C(kcal/m^2/hour).

If R_{cl} denotes the total resistance to heat transferred from the skin to the outer surface of the clothed human body, and I_{cl} denotes the insulation of the clothing, then since insulation is the reciprocal of resistance,

$$I_{\text{cl}} = \frac{R_{\text{cl}}}{0.18} \text{ 'clo' units.}$$

The clo unit has survived criticism and is still used today.

Estimated value of the insulation of various assemblies of clothing expressed in 'clo' units [data from P. O. Fanger (1970)].

Clothing assembly	*Insulation* (I_{cl}) *'clo' units*
Shorts only	0.1
Tropical assembly: shorts, open-neck shirt with short sleeves, lightweight socks and sandals	0.3-0.4
Light summer assembly: long lightweight trousers, open-neck shirt with short sleeves	0.5
Light working clothes: long working trousers, short pants, woollen socks, open-neck cotton shirt	0.6
U.S. Army 'Fatigue', male	0.7
Typical business suit	1.0
Heavy business suit: including trousers, jacket and waistcoat	1.5
U.S. Army 'cold-wet' uniform: cotton-wool mixture vest and pants, wool-nylon shirt, trousers, field coat, coat liner, wool socks	1.5-2.0
Polar weather suit: heavy wool assembly	3-4

Fanger (1970) has set up a heat balance equation for the clothed human body taking into account the following variables which are set out in the table opposite.

Fanger argues that for a long exposure to a moderate thermal environment during which the rate of metabolism is constant, the rate of heat production will equal the rate of heat lost and storage can be neglected so that one may write:

$$H - E_d - E_{sw} - E_{re} - L = K = R + C$$

A wealth of data has accumulated over the years giving the rate of metabolism for a wide variety of activities. The metabolic rate (M), is usually expressed in kcal/m^2/hour of nude body surface, *i.e.* M/A_{Du}.

From the argument given earlier (see p. 84) it can be seen that $H = M - W$. Fanger defines the external mechanical efficiency η as $= W/M$ so that $W = \eta M$ and thus $H = M - \eta M = M(1 - \eta)$ kcal/hour. For many activities, however, H is equal to M for practical purposes, since no external

Variable	*Symbol*	*Units*
Internal heat production of the body	H	kcal/hour
Surface area of the nude human body	A_{Du}	m^2
Heat production per unit area	H/A_{Du}	$kcal/m^2/hour$
Thermal resistance of clothing	I_{cl}	'clo'
Dry-bulb temperature of the ambient air	t_a	°C
Mean radiant temperature	t_{mrt}	°C
Pressure of aqueous vapour in ambient air	p_a	mm. Hg
Relative air velocity	v	m/sec
Mean skin temperature	t_s	°C
Heat loss per unit body surface area through evaporation of secreted sweat	E_{sw}/A_{Du}	
Convective heat transfer coefficient	h_c	$kcal/m^2/hour$ C
External mechanical efficiency	η	
Ratio of the surface areas of the clothed to the nude body	f_{cl}	
Heat loss by diffusion of aqueous vapour through the skin	E_d	
Latent heat loss through respiration	E_{re}	
Sensible heat loss through respiration	L	
Heat transfer from the skin to outer surface of clothed body	K	
Heat transfer by radiation from the outer surface of the clothed body	R	
Heat transfer by convection from the outer surface of the clothed body	C	

mechanical work is being done. He derives expressions for the remainder of the terms in the heat balance equation; these are given below.

$$E_d = 0.35A_{Du}(1.92t_s - 25.3 - p_a)$$
$$E_{re} = 0.0023M(44 - p_a)$$
$$L = 0.0014M(34 - t_a)$$
$$K = A_{Du} \cdot \frac{t_s - t_{cl}}{0.18I_{cl}}$$
$$R = 3.4 \,.\, 10^{-8}A_{Du}f_{cl}[(t_{cl} + 273)^4 - (t_{mrt} + 273)^4]$$
$$C = f_{cl}h_c(t_{cl} - t_a)$$

all heat transfer being expressed in kcal/hour.

Collecting the terms one obtains the following expression:

$$M(1-\eta) - 0.35A_{Du}(1.92t_s - 25.3 - p_a) - E_{sw} - \\ - 0.0023M(44 - p_a) - 0.0014M(34 - t_a)$$

$$= A_{Du} \cdot \frac{t_s - t_{cl}}{0.18I_{cl}}$$

$$= 3.4 \,.\, 10^8 f_{cl}(T_{cl}^4 - T_{mrt}^4) + f_{cl}h_c(t_{cl} - t_a)$$

for the heat balance for the active clothed human body and to obtain the values in terms of heat transfer (in kcal) per unit area of the nude body (m^2) it is necessary to divide all terms by A_{Du}.

The relation of this expression to one which will enable one to predict conditions compatible with thermal comfort has been worked out by Fanger and it will be discussed in the following chapter.

CHAPTER VI

THE WARMTH FACTOR IN COMFORT

INTRODUCTION

OUR feelings of warmth are not dependent solely on the temperature of the environment. Differences in clothing, in muscular activity, in nutrition, age and general bodily build, and, not least, in acclimatization, all tend to influence the feelings of warmth experienced by different persons in the same environment; and in consequence it often happens that in a room where most people are comfortable some feel too warm while others are too cool. The diversity of opinion as to what constitutes a comfortable temperature is sometimes very wide. Interesting examples of this diversity are to be found in the literature. Reid (1844) mentioned that on a certain occasion in the House of Commons one member complained of suffocating warmth while another was shivering with cold. Conversation elicited that some members demanded a temperature as low as 52°F while others required one of 71°F. Billings (1923) referring to the ventilation of the Capitol, Washington, said that some members wanted a temperature of 67° or 68°F while others would have liked 75°F, so that if the temperature were to be kept uniform it would be impossible to please everybody. An amusing squabble about the temperature at the Théâtre Lyrique in Paris was mentioned by Morin (1863). During a warm week in May, when the temperature on the stage was about 70° to 73°F the manager and the prima donna complained of the high temperature, while the director of the opera grumbled that he was cold. Morin remarked that the only way of overcoming such 'prejudices and interested oppositions' lay in the strict observance of well-made regulations.

In my investigation of the comfort of persons engaged in very light industrial work (1936), wide differences in sensations of comfort were apparent. Large numbers of workers were interviewed and questioned as to their feelings of warmth, and the temperature of the environment was measured. The responses of the workers were classified on a seven-point scale:

Sensation of warmth	*Numerical scale*
Much too warm	1
Too warm	2
Comfortably warm	3
Comfortable	4
Comfortably cool	5
Too cool	6
Much too cool	7

For the purpose of statistical treatment it was convenient to assign numerical values to these groups, and a sensation scale of 1 to 7 was thus obtained.* The following table shows these 'comfort votes' classified according to the air temperature measured at a height of 4 ft from the floor.

Comfort votes in relation to dry-bulb temperature

Dry-bulb temperature °F	*Number of observations*	*Percentage of persons who felt:*	
		Comfortable	*Comfortably cool to comfortably warm*
54 to 55.9	39	28	67
56 to 57.9	100	46	76
58 to 59.9	161	58	80
60 to 61.9	478	77	88
62 to 63.9	667	84	91
64 to 65.9	549	78	92
66 to 67.9	527	69	87
68 to 69.9	382	60	88
70 to 71.9	151	45	83
72 to 75.9	31	45	61

With the temperature between 64° and 66° F some persons were too warm while others were too cold; and diversities of opinion were apparent over the whole range of temperatures

* It is sometimes convenient to mark 'comfortable' as zero, with positive numbers indicating increasing warmth and negative numbers increasing chilliness.

studied. An air temperature of 65° F was that preferred by most of the people questioned, but temperatures which differed considerably from that value were tolerated and often found quite comfortable.

It was probably differences of opinion such as have been quoted that led Rubner (1897) to remark, when discussing the suitability of clothing, that it is a mistake to rely on personal feelings of comfort, for, said he, the feeling of comfort is in many people obtuse, undeveloped and variable. The heating and ventilating engineer cannot, however, escape the fact that when persons are uncomfortable they tend to complain, and if he is to succeed in designing satisfactory installations he must pay attention to personal feelings of comfort.

THERMAL COMFORT IN RELATION TO PHYSICAL ENVIRONMENTAL FACTORS

It is of some importance to know how well the various thermal environmental indices discussed in Chapter IV are associated with subjective sensations. I have tested the reliability of various indices in a study of factory workers (Bedford, 1936). The factory buildings in which the investigation was carried out included new and old, single- and multi-storey structures, and the methods of heating and ventilating the buildings were as varied as the buildings themselves. *The observations were confined to the winter months when artificial heating was in use.* The air temperatures encountered ranged from 54° to 76° F, the effective temperatures from 52° to 70° F, and the equivalent temperatures from 46° to 76° F. Measurements were also made of the forehead, hand and foot temperatures of the workers, and of the mean surface temperatures of their clothed bodies. In all, 3,085 sets of observations were made. Ninety-four per cent of the observations were made on women and girls. Nine-tenths of the subjects were seated at their work, but they had to leave their seats from time to time in order to fetch material. Some groups were tested on three occasions, some twice, and others but once. The observations were made on nearly 2,000 different persons.

The reliability of the various scales as indices of comfort was tested by using the method of correlation.*

The coefficients of correlation of thermal sensations with different environmental measures are given in the table.

Correlation between thermal sensations and environmental indices

Sensations correlated with	*Correlation coefficient*
Equivalent temperature	−0.52 ± 0.010
Globe thermometer temperature	−0.51 ± 0.010
Effective temperature	−0.48 ± 0.010
Air temperature (dry bulb)	−0.48 ± 0.010
Dry kata thermometer cooling power	+0.43 ± 0.011

Considering the correlation coefficients set out in the table together with the physical characteristics of the different scales of warmth, we can conclude that up to a temperature of about 75°F the equivalent temperature scale is the best of the measures included in the comparison. It can be used in calm atmospheres or where there is much air movement, and it gives full weight to the effects of radiant heat. It is true that the scale

* Correlation is a statistical measure of the degree of association between two variables. This degree of association is shown numerically by the 'coefficient of correlation' (r), which is a number lying between the limits of 1 and −1. A correlation coefficient of unity indicates that there is perfect association between the two variables, the higher value in the one variable corresponding with the higher value in the other, and the lowest in the one corresponding with the lowest in the other. A correlation of −1 indicates that there is a perfect inverse association between the variables, the highest value of the one variable corresponding to the lowest of the other. Any degree of association falling short of perfection is given as a decimal fraction, either positive or negative. When the correlation is so large that its value is unlikely to have arisen through errors of sampling the coefficient is said to be significant.

The statistical treatment of these data was carried out with the full realization that the sensation scale was an arbitrary one. We cannot say that in one environment we feel twice as hot as in another, nor can it be assumed that the steps in the sensation scale necessarily indicate equal intervals of sensation. Although the use of an arbitrary scale cannot be avoided, it may be argued that a more reasonable scale could have been derived by assuming a normal distribution of the personal feeling of warmth. This point was examined, and it was found that the use of such a scale gave results which did not differ significantly from those obtained by the use of the simple 7-point scale which has been described.

makes no allowance for the effects of atmospheric moisture but, until the temperature approaches 75° F, humidity exerts no great influence on sensations of warmth. It should be noted that the equivalent temperatures used in these calculations were generally measured directly by means of the eupatheoscope. In some other instances, when it was not practicable to use the eupatheoscope, equivalent temperatures were measured by Dufton's method in which two special kata thermometers, one black and one silvered, are used. The equations expressing equivalent temperature in terms of the separate thermal variables (p. 107) were deduced from the partial correlations of warmth sensations with the individual thermal factors, as was also the equation from which the scale of equivalent warmth (Fig. 18) was constructed.

In Chapter IV it was pointed out that in certain conditions the globe thermometer will give a false impression of the thermal environment; but where there is no great amount of air movement, as in the average workroom or office, the reading of the globe thermometer is a good environmental index, since it makes adequate allowance for both air temperature and radiant heat.

When there is no appreciable air movement, and where there is no radiation from hot sources, the ordinary thermometer gives a good indication of the thermal environment.

For the conditions encountered in this investigation the effective temperature scale had no advantage over the reading of the dry-bulb thermometer; but it must be borne in mind that those conditions were neither hot nor humid, nor was there a wide range of air movement. The correlation between these comfort votes and corrected effective temperature was calculated by Dr. Chrenko, and the coefficient is 0.51. Thus, when effective temperature is adjusted to allow for radiation its value as an index of comfort is improved.

As an index of comfort the rate of cooling of the dry kata-thermometer compares unfavourably with the other scales which have been mentioned. The defect of the kata-thermometer in this respect lies in its sensitivity to the effects of wind. This can be illustrated by considering the combinations of air movement and temperature which yield the same cooling power.

Conditions yielding a dry kata cooling power of 8

Air velocity (ft/min)	*Temperature of air and walls (°F)*
10	47.3
100	67.7

Most people would be fairly comfortable with a temperature of 67.7° F and an air velocity of 100 ft/min, but even with stagnant air few would be at all satisfied with a temperature as low as 47.3° F. The ordinary thermometer makes no allowance for the effects of wind, and it has therefore been assumed to be inferior to the kata-thermometer as an index of the thermal environment; but from this comparison it appears that the excessive sensitivity of the kata-thermometer to wind is a more serious defect than the failure of the ordinary thermometer to take wind into account.

THE EFFECTS OF THE SEPARATE THERMAL FACTORS

The correlation of sensations of warmth with various thermal scales has been discussed. These scales differ considerably amongst themselves, and a constant value of one of them may represent very different values of the others. It is therefore desirable that the influence of the separate thermal factors should be determined as accurately as possible.

The cooling effect of air movement depends on the difference between the air temperature and the temperature of the body surface, as well as on the air velocity. This effect can be expressed by the product obtained by multiplying the temperature difference by some function of the air velocity. In the effective temperature scale the 0.75 power of the velocity is used (Missenard, 1933), but over the range of velocities encountered in my observations sufficiently accurate results could be obtained by using the square root of the velocity. It had already been found that the equivalent temperature as indicated by the eupatheoscope could be estimated quite well by an equation in which the square root of the velocity was used (Bedford and Warner, 1934*b*), and in more recent laboratory experiments the convection loss from the human body has been found to vary as the square root of the velocity (Winslow, Gagge and Herrington, 1939). The difference between the body surface temperature and the air temperature can be allowed for by using the value $(100 - t_a)$ where t_a is the air temperature, so that allowance for the effects of air movement can be made by using a term in $\sqrt{v}(100 - t_a)$.

The four separate thermal factors are thus accounted for when the air temperature, the mean temperature of the surroundings, the partial pressure of water vapour in the air, and the product $\sqrt{v}(100 - t_a)$ are taken into account. When each of these in turn was correlated with numbers assigned to various sensations of warmth, the other three factors being kept constant, it was found that in each case the correlation coefficient was statistically significant.

From the partial correlation coefficients so obtained an equation relating sensations of warmth, on the numerical scale given earlier, to the four thermal factors was calculated, *viz.*:

$$S = 11.16 - 0.0556t_a - 0.0538t_w - 0.0372f + 0.00144\sqrt{v}(100 - t_a)$$

where S is the numerical value assigned to the sensation of warmth ('comfortable' = 4);
t_a is the air temperature, in °F;
t_w is the mean temperature of the surroundings (walls, etc.), in °F;
f is the partial pressure of water vapour present in the air, expressed in mmHg; and
v is the air velocity in ft/min.

The total correlation (R) of sensations of warmth with this combination of four thermal factors was 0.51, a value which is not significantly different from the zero-order coefficients for sensations with equivalent temperature and globe thermometer temperature.

Humidity and Comfort

It can be calculated from the above equation that, at the temperatures considered, the effect on sensations of warmth of a moderate change in humidity is very small. This conclusion is in agreement with the findings of other investigators. Miura (1931) found that, in clothed subjects, calm air at a temperature of 69° F with a relative humidity of 50 per cent was equivalent to air at 70°F and 30 per cent relative humidity, thus, at a temperature of 69° to 70° F a change of 20 per cent in the relative humidity affects sensations of warmth only as much as a change of 1° F in the air temperature. From the above equation it can be calculated that at the same temperature a shift of 1° F in the air temperature is equivalent to a change of 18 per cent in the relative humidity. Phelps and Vold (1934) studied the relation between skin temperature and atmospheric conditions, and found the same skin temperature at 69° F and 50 per cent relative humidity as at 70° F and 33 per cent humidity. At these temperatures the effective temperature scale makes rather more allowance for humidity, for a shift of 1° F is shown to be compensated for by a change of only 12 per cent in the relative humidity.

The allowance for humidity made by effective temperature at comfortable levels of temperature seems to be excessive. In a recent study of American subjects, Koch, Jennings and Humphreys (1960) found that the most comfortable condition was associated with an air temperature of 77.6° F when the relative humidity was 35 per cent and a temperature of 76.5° F when the humidity was 85 per cent. The respective effective temperatures of these two combinations are 70° F and about 74.5° F. Rowley, Jordan and Snyder (1947) had already concluded that at air temperatures of 70° to 80° F a change of relative humidity from 35 to 60 per cent had no effect on comfort, and that comfort votes were more closely related to dry-bulb temperature than to effective temperature.

Radiant Heat

Consideration of the constants in the equation shows that in perfectly still air the effect of a given change in the temperature of the surroundings on sensations of warmth is practically the same as that of a similar change in the air temperature, but as the air velocity increases the effect of the mean radiant temperature relative to the air temperature decreases. Thus, at velocities of 0, 20, 50 and 100 ft/min respectively, the change in air temperature equivalent to a change of 10° F in the mean temperature of the surroundings is 9.7°, 8.7°, 8.2° and 7.7° F.

In their experiments on gas fires and electric radiators, Fishenden and Willgress (1925) found that while they felt warm enough at 65° F without extra radiation, at 60° F they needed extra radiation with a mean horizontal component of about 30 Btu/sq. ft/hour to evoke warmth, and at 55° F they needed radiation of an intensity of 75 Btu. It is reasonable to assume that the mean horizontal component of this additional radiation would be received by about one-third of the body surface area. Calculating the mean temperatures of the surroundings represented by the varying combinations of air temperature and radiation found by these authors to give conditions of equal warmth, it can be found that the combinations of radiant heat and air temperature yielded fairly similar equivalent temperatures, according to the equation in Chapter IV. Some of the data are shown in the table below.

Conditions of equal warmth given by Fishenden and Willgress (1925) compared with the calculated equivalent temperatures

Conditions assessed as	*Air temperature °F*	*Mean horizontal component of extra radiation (Btu/ft²/hour)*	*Estimated mean temperature of the surroundings °F*	*Calculated equivalent temperature °F*
Comfortable	65	0	65.0	65.0
	60	30	70.0	64.4
	55	75	80.0	66.1
	50	120	90.0	67.8
	45	170	101.7	70.2

Concerning the high radiation values it should be remarked that in my investigation the mean radiant temperature was never as high as 90° F, and it is not certain that the equation for the estimation of equivalent temperature holds for such large differences between the temperatures of the air and of the surroundings as are shown in the table. With cold air, and with intense radiation on one side of the body, one feels cold on one side and scorched on the other, and it is not possible to make an accurate assessment of one's general state of warmth. It is very interesting to note that with the high radiation values a rather higher equivalent temperature was necessary than with low radiation. Evidence pointing in the same direction comes from the Pierce Laboratory of Hygiene at New Haven. There, Winslow and his colleagues (1938*c*) compared environments giving the same operative temperature, (*a*) when the air and walls were at approximately the same temperature, and (*b*) with cool air and warmer walls. They found that with the cool air-warm wall condition the mean skin temperature was lower than when the air and walls were at the same temperature. Thus for equal skin temperatures a higher operative temperature was required when the mean radiant temperature was substantially higher than the air temperature

The effect of cold walls on sensations of warmth was studied by Houghten and McDermott (1933). Three of the four walls of a small test room could be maintained at the required temperature. It is not easy to calculate accurately the equivalent temperatures corresponding to the values given by these authors, but rough estimates show that the conditions found by them to yield equal comfort give approximately the same equivalent temperatures.

In the expression *'mean temperature of the surroundings'* the term surroundings refers to all surfaces from which radiation reaches the observation point, and this includes not only walls and other boundary surfaces, but also the surfaces of machines and of the occupants of the room. In ordinary circumstances the surface temperature of a person is distinctly higher than that of the walls, etc., of the room which he occupies. Thus the presence of a number of persons in a room raises the mean temperature of the surroundings somewhat above the mean temperature of the walls. The magnitude of this effect clearly depends on the number of persons present and their proximity to the observation point. In workrooms it was found that the increase in the mean temperature of the surroundings due to the presence of the operatives varied from about 1° F in a room which was but sparsely occupied to about 4° F in a room where people sat fairly close together at rows of benches. In a densely occupied room such as a crowded cinema or ballroom the effect would be distinctly greater than the 4° F just mentioned.

Giving evidence to the Committee of the House of Commons on Ventilation, Warming and Transmission of Sound, in 1835, Michael Faraday remarked: 'I think that an individual may find a decided difference in his feelings when making part of a large company, from what he does when one of a small number of persons, and yet the thermometer gives the same indication. When I am one of a large number of persons, I feel an oppressive sensation of closeness, notwithstanding the temperature may be about 60° or 65° F, which I do not feel in a small company at the same temperature, and which I cannot altogether refer to the absorption of oxygen, or the evolution of carbonic acid, and probably depends upon the effluvia from the many present; but with me it is much diminished by a lowering of the temperature, and the sensations become much more like those occurring in a small company.'

Faraday here attributed the unpleasant effects to effluvia from the body. The air temperature mentioned by him is a reasonable one, and would not itself cause any feeling of oppression. Aqueous vapour is an effluvium from the body, and increased humidity probably accounted for part of the effect of crowded assemblies to which Faraday referred. It seems, in view of the state of personal hygiene, or lack of it in those days, likely that the principal cause of the feeling of closeness was body odour.

THE COMFORT ZONE

In view of the diversity of opinion which exists as to environmental conditions conducive to comfortable warmth, it is obviously impossible to specify a thermal environment that will please everybody, but the conditions which are most generally acceptable can be ascertained.

For the persons concerned in my study the most comfortable condition was an equivalent temperature of 62° F, and that is the condition which should be aimed at in rooms where persons are employed on very light tasks. It may be that in offices, where the physical work involved is even less than on those light industrial tasks, an equivalent temperature 1° or 2° F higher would be preferred. In calm air an equivalent temperature of 62° is given by conditions which yield a globe thermometer reading of about 65° F, and where there is no appreciable radiation effect an air temperature of 65° F or an effective temperature of 61° F is satisfactory. A dry kata cooling power of 6 will generally be found to be associated with comfortable warmth.

Unless the temperature of a room is automatically controlled fluctuations will occur, and it is frequently found that the temperature of a workroom or other compartment varies considerably during the day. As the temperature rises above the most comfortable level, or falls below it, an increasing proportion of the occupants will feel too warm or too cool, and it is important that we should know between what limits the temperature should be allowed to vary; in other words, we should define the comfort zone.

The comfort zones described by different workers have been determined according to different standards. Some observers have taken the range between votes of 'slightly too warm' and 'slightly chilly'. Others have taken as their comfort zones those conditions in which not less than 50 per cent of votes ranged from 'comfortably warm' to 'comfortably cool'. In yet another investigation the whole range of temperature in which votes of 'comfortably cool' or 'comfortably warm' were recorded was taken as being the comfort zone.

Leopold (1947) emphasized that close tolerances may be needed to satisfy a group of people, even though individual tolerances may be fairly wide. He urged that attention should be directed to the incidence of discomfort. A range of conditions in which nearly 50 per cent of persons may experience discomfort is too wide to be accepted as a comfort zone.

When one is concerned with the incidence of discomfort over a range of conditions a simple method of examining observational results is to plot the proportion of persons experiencing discomfort against the temperature or other thermal stimulus. For more detailed and accurate information the statistical technique known as 'probit analysis' (Finney, 1947) is of great value. The method was invented by Fechner (1860) and applied for the first time to studies of thermal comfort by Chrenko (1953).

Winter Comfort Zones

For American and Canadian subjects respectively, Yaglou (1926) and Partridge and MacLean (1935) give zones of 63° to 71° F and 61.5° to 71.5° F effective temperature. These are the limits within which not less than 50 per cent of the subjects felt comfortable. In environments warmer or cooler than the comfort zone the average person was uncomfortably warm or uncomfortably cool. It is desirable that conditions should be maintained which will be found comfortable by as many persons as possible, and there is, therefore advantage in adopting a rather narrower zone for practical purposes. I suggested that the comfort zone should be taken as that in which not less than 70 per cent of the subjects were 'comfortable', and in this range at least 86 per cent of the votes recorded ranged from 'comfortably cool' to 'comfortably warm', *i.e.* never more than 14 per cent were uncomfortably cool or warm. The zone is set out below in terms of different measures of warmth.

Scale of warmth	*Temperatures (°F) between which not less than 70 per cent of comfort votes were 'comfortable'*
Equivalent temperature	58° to 66°
Globe thermometer temperature	62° to 68°
Effective temperature	57° to 63°
Air temperature (dry bulb)	60° to 68°

People who are engaged in wholly sedentary work are likely to need temperatures a degree or two higher than those shown above. Sitting in an assembly hall, for instance, entails less energy expenditure than is required for repetition tasks that are performed while seated, so that while an air temperature of 65° F may be the optimum for light industrial work it would be rather low for persons sitting at rest.

Some years ago the author was asked to investigate complaints of discomfort in the Debating Chamber of the House of Commons. From considerations such as the foregoing it was recommended that with a reasonable degree of air movement a temperature of 67° to 68° F should be maintained during ordinary sitting, that when the House was crowded the temperature should be reduced by 2° or 3° F to compensate for the extra heat radiated by the occupants, and that during late sitting the temperature should be raised slightly.

Those doing work calling for greater energy expenditure than is required for the light industrial tasks that have been considered need somewhat cooler conditions. With light air movement and moderate humidity those doing light but active work will probably be comfortable with air temperatures of 60° to 65° F during winter, while for those doing heavier work 55° to 60° F will probably be suitable. These temperatures are suggested in the light of experience, but they are not based on the questioning of large numbers of people.

The range of effective temperature recommended for British subjects represents distinctly lower temperatures than those found desirable for American and Canadian subjects. In part this difference is due to the fact that in Great Britain we are acclimatized to lower indoor temperatures than are general in American and Canadian buildings, but the heavier indoor clothing normally worn in Great Britain is another contributory cause.

Interesting experiments on the influence of clothing were made by Yaglou and Messer (1941). Women commonly like temperatures rather higher than those which suit men, so that a temperature that is acceptable to women is often too warm for men. Experiments were carried out in an air-conditioned room, and in a special series of tests women were dressed in men's clothes and men in women's. When the subjects wore their customary indoor clothing, in a room at 72° F with a relative humidity of 30 per cent and an air velocity of 20 ft/min, the mean skin temperature of the women was about 2° F lower than that of the men, while the mean temperature of their clothing surface was 2° F higher. The hands and feet of the women were 5° F cooler than those of the men. Most of the men were comfortable, while the women were cold. When the air and wall temperatures were raised to 76° F the women were comfortable but the men were too warm. Under conditions of comfort (71.5° F for the men and 76° F for the women) the mean skin temperature was the same in the men and the women (92° F), but the mean clothing temperature of the women was about 5° higher. When the men wore their own summer clothes weighing about 3 lb, they wanted a temperature of 76° F, and when women wore men's winter clothes, weighing 6.2 lb instead of their own 1.5 lb, they wanted a temperature of 72° F, or virtually the same as did the men when wearing these clothes. These experiments show that the differences between standards of temperature for men and women are almost entirely due to differences in clothing. Moreover, Fanger (1970) found that there were no differences between the temperature preferred by male and female subjects when they wore standardized clothing.

Summer Comfort Zones

During hot weather we become acclimatized to higher temperatures and go more lightly clad than in winter. Temperatures which would be found uncomfortably warm in winter are accepted as pleasant in summer. In the variable summer climate of Great Britain we experience substantial variations of temperature, but the periods during which the cooling of ordinary buildings would appreciably enhance comfort are usually short and infrequent. It is largely on that account that air-conditioning has not hitherto been widely used for summer cooling in Great Britain. By ordinary ventilation methods it has generally been possible to maintain a reasonable degree of comfort.

Nevertheless, people at work may experience considerable discomfort during the summer months. Hickish (1955) studied the comfort of factory workers engaged on light manual tasks during warm weather in the London area. The incidence of discomfort due to excessive warmth is shown below.

Incidence of discomfort due to excessive warmth at various air temperatures

Air temperature °F	No. of persons questioned	Percentage of persons who felt Too warm or much too warm	Much too warm
63-	29	0	0
65-	141	2.1	0
67-	216	4.6	0
69-	387	8.8	0.3
71-	365	10.4	1.1
73-	124	19.5	3.2
75-	71	50.7	19.7
77-	138	76.8	46.4
79-	26	96.2	69.2
81 & over	40	100.0	75.0

The upper limit of the comfort zone was taken to be 75° F, which in these observations represented a corrected effective temperature of 71° F. There was a sharp increase in the incidence of discomfort as the air temperature exceeded 75° F. It might be expected that in these clothed, working subjects sweating would occur at about that temperature, and it was suggested that the sudden increase in discomfort was probably due to the onset of sweating.

In the more trying summer conditions which are experienced in the United States, however, air-conditioning is often necessary for the maintenance of comfort and efficiency, and so that satisfactory temperatures can be maintained it is necessary to know the summer comfort zone. Yaglou and Drinker (1928) gave a zone of 64° to 79° F effective temperature, but they apparently took the whole range of effective temperature in which any votes of 'comfortably cool' or 'comfortably warm' were recorded. On the 50 per cent basis the zone would be approximately 66.5° to 75° F effective temperature. From numerous studies sponsored by the American Society of Heating and Ventilating Engineers it appears that the optimum effective temperature indoors during summer varies from 69° to 73° F. The most desirable temperature varies slightly with geographical location, but the variations between the requirements of individuals in any one location are greater than the variations in the optima for different localities.

SUMMER COOLING

It appears that the optimum indoor effective temperature for summer is only about 3° or 4° F higher than that for winter, but it does not follow that in air-conditioned buildings the air should always be cooled so that its temperature is so near the winter figure. In winter one may leave a room warmed to 65° or 70° F and pass immediately into the external air at a temperature below freezing point without suffering chilling, for one would normally don more clothing before passing outside. But in summer, on entering a cooled building from the heat outside, one does not put on more clothing. The chilling which may thus be experienced is likely to be the greater because the clothing is damp with sweat. An environment that would be found quite pleasant by a person with a dry skin and clothing can be very chilly to one who is perspiring freely. Adaptation to summer heat appears to make people more sensitive to temperature changes than they are in winter. People vary greatly in their reaction to 'cold shock'. In one American study it was found that most of the 275 subjects found the change to cool conditions pleasant, but others suffered unpleasant or even harmful chills (Newton *et al.*, 1938).

Yaglou (1937*b*) investigated the disturbances of body heat-loss consequent on exposure to contrasting temperatures. Some of his subjects withstood the sudden change to a cooler environment very well, while others were badly chilled. In one series of experiments subjects were exposed for an hour or more to a temperature of 95° F with 40 per cent relative humidity, and then passed to a cooler room in which the temperature was 80° F and the relative humidity 45 per cent, and there they remained for 2 hours. At the end of that time they returned to the hot room for a further hour. As Yaglou points out, these are the air conditions frequently used in the design of cooling systems for theatres, restaurants and railway cars. Two of the subjects were young men who seldom complained of cold, and two were women who always complained when the temperature was lowered below 85° F in summer. The two groups are respectively described as cold resistant and cold intolerant. In the hot room the subjects were flushed, and nearly 80 per cent of their total heat loss took place by evaporation. When the cooler room was entered the heat loss was increased by 40 to 50 per cent, because the loss by radiation and convection was immediately increased, while evaporation from the damp clothing still continued at approximately the same rate as in the hot room. There was no immediate change in the rate of heat production. After an hour or so in the cooler room the men subjects had adapted themselves to the new environment, but the women felt cold throughout the 2-h period. When the hot room was re-entered there was a further disturbance of the heat balance, this time in the reverse direction. Heat loss by radiation and convection was immediately decreased, but the evaporative loss remained low for some minutes.

In further experiments the effects of a change from an environment at 95° F and 40 per cent humidity to one at 88° F and 27 per cent humidity were studied. The disturbance of body heat-loss was much smaller, and the subjects were able to adjust themselves without experiencing untoward symptoms. These experiments were made on only a few subjects, and cannot therefore be regarded as conclusive but they support the idea that, in places which persons enter for only short periods, air-conditioning by dehumidification, with but little reduction of air temperature, is desirable.

Generally excessive cooling should be avoided. Yaglou (1941) shows that exposure to a sudden temperature drop of 10° F or more is harmful to a considerable proportion of persons in hot weather when the skin and clothing are wet with perspiration. The initial chilling, which lasts only a few minutes, suppresses sweating and may reduce one's ability to withstand heat for some time after returning to the hot atmosphere.

Although experience in America teaches that excessive cooling may lead to cold shock, it appears that in suitable conditions substantial cooling can be tolerated. Thus, in certain dry localities where the shade temperature may exceed even 120° F, rooms have been cooled to 70° F with only 50 per cent relative humidity, to the great satisfaction of their occupants (Grocott, 1948). The explanation of this may be that in such a hot dry atmosphere, although sweating is profuse in the open air, the sweat is evaporated very rapidly so that the garments are scarcely damp, and when persons pass to the cool environment indoors they experience no chill from clammy clothing.

Formerly it was accepted in the United States of America that the comfort zone for summer was a few degrees higher than that for winter, but it is now recommended that in air-conditioned buildings the air temperature should be maintained all the year round at 73° to 77° F, with a suitable relative humidity and moderate air movement.

COMFORT IN THE TROPICS

Recent investigations have provided information as to the degree of warmth that is found comfortable by persons living in the tropics, acclimatized to heat and clad in the light garments that are customarily worn in the tropics.

From observations made in the hot, arid zones of Australia, Drysdale (1950) concluded that while people may have a preference for an air temperature of about 75° F they become accustomed to a feeling of heat without being uncomfortable until the temperature approaches 84° F, when real discomfort is liable to be experienced. Drysdale did not regard effective temperatures as good index of comfort for hot, dry conditions.

Webb (1952) made observations on people living in Singapore. The most comfortable condition for his subjects was an effective temperature of about 78° F, and the comfort zone was about 77° to 80° F effective temperature. From Webb's observations Ho (1952) derived an equation expressing comfort votes in terms of the three separate variables: air temperature, vapour pressure and air speed. The total correlation of comfort votes with these variables was 0.67, as compared with the coefficient of 0.55 obtained by correlating comfort votes with effective temperature, so that it does not appear that effective temperature was the best possible index over the narrow range of conditions encountered.

Ellis (1952) studied the thermal comfort of naval ratings serving in a warship in the tropics. More than 80 per cent felt comfortable, comfortably cool or comfortably warm (*i.e.* they were not uncomfortable) when the effective temperature did not exceed 79° F, and less than 20 per cent reported that they were sweating when the effective temperature was below 78° F (air temperature 84° F and wet-bulb temperature 76° F). It was concluded that on mess decks in the tropics the effective temperature should be kept below 78° F, and that for optimum comfort it should not exceed 74° F.

Ellis (1953) also made a study amongst men and women, European and Asian, living in Singapore. In terms of effective temperature the comfort zones were found to be 73° to 79° F for European men, 72° to 78° F for European women, 71° to 81° F for Asian men, and 76° to 80° F for Asian women. Within these zones at least 80 per cent of the comfort votes represented reasonable comfort.

The upper desirable limit of effective temperature derived from Webb's observations is in good agreement with the limits found by Ellis. Likewise, the maximum desirable dry-bulb temperatures of 82° to 85° F for civilians and 84° F for naval ratings agree well with Drysdale's figure of 84° F as the upper acceptable limit for indoor conditions in the arid regions of Australia.

Ambler (1955), in an investigation carried out at Port Harcourt and Lagos, Nigeria, concluded that effective temperature was a fairly good index of comfort. For Europeans not engaged in heavy work the threshold value for the onset of discomfort was an effective temperature of 80° F, while 83° F was very uncomfortable for untrained persons.

SKIN TEMPERATURE AS AN INDICATION OF SENSATIONS OF WARMTH

The skin temperature varies with the temperature of the environment, and it is therefore to be expected that skin temperatures will be correlated with sensations of warmth. Various observers have recorded significant correlations between sensations of warmth and skin temperatures and, indeed, some of the correlations are quite high. Generally, the site chosen for the skin temperature measurements has been the forehead, but the temperatures of other areas of skin have also been studied in this connection.

Ward (1930) recorded the sensations of warmth of herself and four other subjects, and at the same time measured their skin temperatures (*a*) on the forehead, (*b*) on the cheek, (*c*) over the carotid artery, and (*d*) on the sternum beneath the clothing. The association between skin temperature and sensations was quite close. The coefficients of correlation between comfort votes and the skin temperatures of the forehead and over the carotid in the four subjects were respectively 0.88 ± 0.02 and 0.96 ± 0.01; while the corresponding correlations for the observations made on herself were 0.88 ± 0.02 and 0.91 ± 0.01. Maximum comfort was experienced when the skin temperature of the forehead or over the carotid was 92.3° F. The subjects were uncomfortably cool when the skin temperature in these parts fell below about 89° F and too warm when it rose above 96° F. These high correlations led Ward to conclude that the temperature of the skin over these areas might be used as an index of the subjective sensations of warmth of the individual. Phelps and Vold (1934), discussing these data, commented that in equilibrium conditions they found Ward's data especially significant; they agreed with her conclusion.

These and other studies of the relation between warmth sensations and skin temperature were carried out on very small groups of subjects, and the results obtained cannot safely be assumed to be of general application. From Ward's data it seems that as the result of a series of observations on a single subject in equilibrium conditions, the subject may be 'calibrated' so that it is possible to estimate his subjective feeling of warmth from a knowledge of his skin temperature with fair accuracy. Yet it was possible that from the results of observations on a large population it might be quite impossible to estimate the feeling of warmth of a particular individual with any useful degree of accuracy. This point was explored in my investigation among factory workers (1936).

The warmth-sensations of the workers were correlated against measurements of the skin temperatures of their foreheads, hands and feet. The correlation coefficients computed from the main series of observations are shown in the table.

The Correlation between Sensations of Warmth and Skin Temperature

Sensation correlated with	*Coefficient of correlation (r)*	*Number of observations*
Forehead temperature	−0.23 ± 0.01	2,567
Hand temperature	−0.36 ± 0.01	2,541
Foot temperature	−0.33 ± 0.01	2,449

Each of these correlation coefficients is significant and shows a definite association between skin temperature and warmth sensations, but all are far lower than those reported by Ward for her small group of subjects.

The average sensation was 'comfortable' when the forehead temperature was 93.7° F, the hand temperature 84.9° F and the foot temperature 76.7° F. To change the average sensation from 'comfortable' to 'comfortably cool' or 'comfortably warm', or from 'comfortably warm' to 'too warm', a change of 8° F in the forehead temperature was required, and in the hand and foot temperatures changes of 16° and 21° F respectively were needed. Although there was a good linear relationship between the average sensations and skin temperature there were wide differences in the sensations recorded with any particular skin temperature. Thus, although the average sensation was 'comfortable' when the forehead temperature was 93.7° F, at forehead temperatures of 93.0° to 93.9° F sensations over the whole range were recorded. Four per cent of the people with forehead temperatures within these limits were too warm or much too warm, and 8 per cent were too cool or much too cool.

When the skin temperature is high one tends to feel warmer than when the skin is cool; and the feeling of warmth is, of course, also related to the temperature of the environment. At any particular air temperature there was considerable variation in the skin temperatures observed, and the question therefore arose whether or not with the same air temperature those people with the cooler skins actually felt cooler than those whose skins were warmer. Examination of the data showed that with air temperature held constant there was still a distinct, although slight, relation between the temperature of the skin and the feeling of warmth: those with the cooler skins tended to feel cooler. Variations in skin temperature may, therefore, account partly for the differences in opinion as to what constitutes a comfortably warm environment.

When the partial correlations of warmth sensations with forehead temperature, hand temperature and foot temperature were calculated, it was found that the total correlation (R) of sensations with the three skin temperatures was 0.44. This total correlation is lower than the correlation of warmth sensations with equivalent temperature or air temperature or effective temperature. The correlation of 0.44 represents a root-mean-square error of estimation of sensation from skin temperature of 0.77 of a unit on the sensation scale. The probable range of errors of estimation is, therefore, about 1.5 units on the scale. That is to say that if from the measurements of the skin temperatures of a person's forehead, hand, and foot it is estimated that he is quite comfortable, he may in fact be feeling too cool or too warm. Clearly, for the prediction of the comfort of individual subjects out of a large population skin temperature measurements are of no practical value. On the whole, the comfort of the individual occupants of a room can be predicted with slightly greater accuracy from a knowledge of the warmth of the room by assuming that every occupant is experiencing the average sensation that would be expected at that room temperature, than by estimating the state of warmth of each individual from a knowledge of his skin temperature.

Results confirming this conclusion were obtained by Glickman, Inouye, Keeton and Fahnestock (1950), who used the mean skin temperature computed from measurements made at many points. Sensations of warmth were predicted with significantly greater accuracy from a knowledge of the effective temperature, or of the simple dry-bulb temperature, than from a knowledge of the mean skin temperatures of the individual occupants.

Fanger (1967) investigated the relations between human metabolic heat production per unit of nude body surface area and (*a*) mean skin temperature and (*b*) evaporative heat loss amongst North American male and female students who were thermally comfortable. The results showed that with increasing heat production, the mean skin temperature falls if comfort is to be unaffected, and evaporative heat loss rises.

In his recent monograph (1970) Fanger uses the data from these experiments to obtain values for t_s and E_{sw} to be used in his heat balance equation; they are given by the following expressions:

$$t_s = 35.7 - 0.032 \frac{H}{A_{Du}} \text{ expressed in } {}^\circ\text{C}$$

and

$$E_{sw} = 0.42 A_{Du} \left(\frac{H}{A_{Du}} - 50 \right) \text{ kcal/hour.}$$

This enables t_s to be eliminated from the expression for E_d and an expression to be used instead of E_{sw} in terms of heat transfer.

The heat balance equation now becomes, if one writes H' for M/A_{Du} and puts η equal to zero,

$$\begin{aligned} H' - 0.35(43 - 0.061H' - p_a) - 0.42(H' - 50) \\ - 0.0023H'(44 - p_a) - 0.0014H'(34 - t_a) \\ = \frac{35.7 - 0.032H' - t_{cl}}{0.18 I_{cl}} \\ = 3.4 \,.\, 10^{-8} f_{cl}(T_{cl}^4 - T_{mrt}^4) + f_{cl} h_c (t_{cl} - t_a) \end{aligned}$$

Fanger states that the basic comfort equation is given by equating the heat production function with the thermal environmental function as follows:

$$H' - 0.35(43 - 0.061H' - p_a) - 0.42(H' - 50) - 0.0023H'(44 - p_a) - 0.0014H'(34 - t_a) = 3.4 \,.\, 10^{-8} f_{cl}(T_{cl}^4 - T_{mrt}^4) + f_{cl}h_c(t_{cl} - t_a).$$

Of course, one now has to obtain an expression for the mean surface temperature of the clothing t_{cl} and this is derived from equating the heat production function with the heat transfer through the clothing expression, and solving for t_{cl}:

$$t_{cl} = 35.7 - 0.032H' - 0.18I_{cl}[(H' - 0.35(43 - 0.061H' - p_a) - 0.42(H' - 50) - 0.0023H'(44 - p_a) - 0.0014H'(34 - t_a)]$$

This leaves the problem of determining values for the convective heat transfer coefficient h_c.

The coefficient will vary according to whether the heat transfer is by free convection or by forced convection. Nielsen and Pedersen (1953) give, for free convection, $h_c = 2.05(t_{cl} - t_a)^{0.25}$ and Winslow, Gagge and Herrington (1939) give, for forced convection, $h_c = 10.4v^{0.5}$ where v is less than 2.6 m/sec. For most practical cases, Fanger states that the free convection formula should be used when the relative velocity (v) is less than 0.10 m/sec and the forced convection formula for v greater than 0.10 m/sec.

It is now possible, using these equations to calculate the optimum combinations of dry-bulb air-temperature, aqueous vapour pressure of ambient air, mean radiant temperature and relative air velocity which will evoke sensations of thermal comfort amongst persons in thermal equilibrium with the thermal environment, for various activities and types of clothing.

Since heat transfer by three different routes from the body to the environment is involved, the equation finally developed is complicated so that, for practical purposes, a computer should be used for obtaining solutions. This has been done by Fanger and he has given the results in the form of a series of tables and diagrams for direct practical applications in which it is assumed that the mechanical efficiency η is zero which simplifies the computations.

A further simplification of the expression for the temperature of the surface of clothing is:

$$t_{cl} = 35.7 - 0.032H' - 0.18I_{cl}(H' + 6)$$

For a person sitting quietly, H' may be taken as 50 kcal/m²/hour and if ordinary clothing is worn I_{cl} is about 1 'clo' so that, for comfort, and it must be recalled that this equation applies to comfortable conditions, the surface temperature of the clothing works out to be 24° C or about 75° F. It is of some interest to note that this result confirms the assumption made by Bedford in 1936.

For the same rate of metabolism, *i.e.* 50 kcal/m²/hour one finds that the term in the 'comfort equation' which represents the heat lost by the evaporation of sweat, *i.e.* $0.42(H' - 50)$ vanishes.

Students may also be surprised to find that relative humidity does not appear in the 'comfort equation', especially, as Hill (1919) pointed out, 'among meteorologists, physicists, hygienists and ventilating engineers there exists an unaccountable affection and reverence for "relative humidity".' In Hill's day it was seriously suggested that heating winter air which lowered its relative humidity would increase the rate of evaporation of moisture from the human body. In evidence given before the Committee of the House of Commons on the Ventilation of this House, 1914, an engineer said that where the temperature of the air coming into the House was only 40° F at a relative humidity of 80 per cent it would be raised to 63° F and the relative humidity would fall to 50 per cent and to offset the increased evaporation from the body, the air would have to be humidified. Hill's colleague, the physicist O. W. Griffith, showed by a simple calculation that, in the case cited, humidification was unnecessary since the rate of evaporation would actually *decrease* by some 4.5 per cent. Griffith also pointed out that, so far as *heat losses* are concerned, changes in the humidity at ordinary cool room temperatures are immaterial and that convection, even that which is set up by the body alone, completely swamps the effects.

Fanger's equation confirms this and it shows that the change from air which is absolutely dry to air saturated with moisture is equivalent to lowering the temperature by about 1.5° to 3° C provided that the persons concerned are thermally comfortable. There is also another very important reservation: the moderate effects of humidity only apply to steady-state conditions.

The deservedly famous 'effective temperature' scale appears to overestimate the effects of humidity on comfort at ordinary temperatures. The experiments upon which the scale was based were conducted under the following conditions: people walked from one room maintained at a certain temperature and a very low humidity to another at a lower temperature and high humidity. The experiments were repeated at various levels of temperature and humidity in order to establish combinations of temperature and humidity which felt equally warm or cool. Appraisals of warmth by clothed subjects depended upon sensations evoked from stimuli of the face and hands. First impressions were noted because, even when the period of exposure was increased to a few minutes, subjects were incapable of detecting quite large changes in humidity. This was attributed by the investigators to adaptation and they proceeded on the assumption that all combinations of temperature and humidity which felt alike in warmth on first impression would probably result in the same degree of adaptation after prolonged exposure. This assumption is now known to be false and the effects of humidity were due, as Yaglou himself said, to the phenomena of adsorption and desorption of moisture.

When subjects passed from a relatively warm and dry atmosphere to a moist cooler one, the exposed skin and clothing adsorbed moisture and the heat of adsorption imparted a transient sensation of warmth. When the subject returned to a drier atmosphere, the excess moisture quickly evaporated and produced a transient cooling effect.

Hill (1919) in discussing the effects of mist and damp air noted that dew may be deposited on the face and clothes of a man coming into a warm moist atmosphere from a biting cold day and the dew, on giving up its latent heat of vaporization, *i.e.* 580 cal/g, markedly increases the sensation of warmth and closeness.

The effects of moisture on clothing are very complicated indeed. As Sir Napier Shaw (1933) put it, 'the fact that humidity is itself important in weaving sheds may be accepted as an intimation that there is something peculiar about fibres that is worth attention and should find a place when we are thinking of the hygrometer and its use. . . . Exposed to the air they take up moisture at temperatures below saturation and alter their shape and volume by doing so. And curiously enough the condition of the fibre depends not on the amount of water contained in the air which forms its environment but on the fraction by which the actual quantity is related to the quantity at saturation at that temperature. This curious property depends upon the capacity of the fibres to condense water under specified conditions in the pores and the force exerted in doing so is remarkable, as every yachtsman knows, for a rope which is made taut when dry will tear something down if it is moistened, and cordage is as sensible to the dampness of the air in the cold Antarctic as in the torrid tropic. The amount of water involved is very different but the fibre is always satisfied with its fractional share and not otherwise. . . . It is very remarkable that so long as it retains its structure every substance which has formed part of a living organism–wool, hair, leather, whalebone, cotton, hempen rope–is similarly sensitive to the humidity of the air.'

Practical experience teaches that the power of clothing to insulate the body from cold is reduced when the fabric becomes damp but, as Burton pointed out, many laboratory studies of the thermal insulation of fabrics fail to confirm this. He said that this was because the laboratory studies were conducted under steady-state conditions whereas in the wearing of clothing in practice a steady state of temperature and humidity is never realized. The passage of moisture through materials used for thermal insulation in buildings has posed similar problems.

However, Fanger has given evidence of the applicability of his comfort equation to the investigations of other workers including those conducted in tropical regions and optimum temperatures calculated from the equation have been in good agreement with temperatures obtained from field studies. For example, Ellis (1953) studied 152 European and Asian residents in Singapore and found an optimal temperature of 27° C whilst the temperature given by Fanger's equation was 27.4° C.

This basic comfort equation was first published in 1967 and it was pointed out at the time that whilst the equation was useful for engineers in the future, it was based on physiological data obtained at ambient temperatures of 60°, 66° and 72° F or 16°, 19° and 22° C, *i.e.* over a range of 6° C, and that the extent to which one could extrapolate to wide ranges of environmental temperatures could only be judged by experiment. Whilst evidence since obtained has tended to validate the original equation one cannot use it to determine the *degree* of thermal comfort since only the optimal thermal comfort point is defined. It is comparable to the situation which would result if the number 4, which represents 'comfortable' on Bedford's scale of warmth, were inserted in the equation which he derived, *i.e.*

$$4 = 11.16 - 0.0556t_a - 0.0538t_w - 0.0372f + 0.00144\sqrt{v}(100 - t_a)$$

or

$$0.0556t_a + 0.0538t_w + 0.0372f - 0.00144\sqrt{v}(100 - t_a) = 7.16$$

Equations of this type would not enable one to estimate the thermal comfort of the occupants of existing buildings.

Fanger has accepted this point and, with his comfort equation as a basis, he has derived an index from which it is possible to predict the thermal sensation for any given combination of metabolic heat production, clothing insulation and the four physical parameters of the thermal environment.

It is at this stage that a psycho-physical scale is used. The scale which is now most commonly used in America is that adopted by the American engineers (ASHRAE).

Sensation	*Numbers assigned by* *ASHRAE*	*Fanger*
Cold	1	−3
Cool	2	−2
Slightly cool	3	−1
Neutral	4	0
Slightly warm	5	+1
Warm	6	+2
Hot	7	+3

Fanger here agrees with Bedford who pointed out that it would have been better to have marked 'comfortable' as zero with positive numbers indicating increasing warmth and negative numbers increasing chilliness. It is true that Bedford's scale and that of the ASHRAE are not identical but it is tacitly assumed that 'neutral' in this context means 'comfortable'.

And it is a curious fact that the word 'comfort' does not even appear in the American scale. Nevertheless, there are more similarities than differences when one compares the British and the American scales: both are semantic differentials with seven points between the polar terms 'cold' and 'hot' for one scale, and 'much too warm' and 'much too cool' for the other scale.

If one assumes that thermal sensations for a given level of activity depend upon the heat load on the human body one may express this as follows, which also implies that the thermal load might also vary with the rate of heat production of the body:

$$Y = f\left(L \,.\, \frac{H}{A_{Du}}\right)$$

In this expression Y represents the number assigned to sensation on the scale of warmth, L represents the thermal load on the body which is defined as the difference between the internal heat production and the heat lost to the environment, *i.e.* $H - E_d - E_{sw} - E_{re} - R - C$ for a man 'hypothetically kept at the comfort values of the mean skin temperature and the sweat secretion at the actual activity level'. By pooling the results obtained from various experiments—the original monograph should be consulted for the details—Fanger obtains the following equation:

$$Y = (0.352e^{-0.042M'} + 0.032)L$$

where M' represents the level of activity expressed in kcal/m^2/ hour of nude body surface. It is quite unnecessary to write out the formula in full: it is complex and unsuitable for calculations by hand. A comprehensive table is given in the Appendix of values of Y which are referred to as the 'predicted mean vote' or PMV for about 3,500 combinations of the variables which were obtained from a computer. This table is reproduced by kind permission of Professor Fanger and the Danish Technical Press.

Such equations and the resulting mean votes present problems of interpretation. For instance, Bedford showed, like others who came after him, that the effects on thermal comfort of moderate changes in humidity were small and that if this factor were disregarded his equation relating comfort votes to environmental conditions became:

$$S = 10.89 - 0.0574t_a - 0.0525t_w + 0.00162\sqrt{v}(100 - t_a)$$

Assuming that the mean radiant temperature is equal to the air temperature and that in calm air one can neglect the term for air velocity, $S = 10.89 - 0.11t$ and if t is equal to 62.6° F (17° C) S is equal to 4 on Bedford's scale, *i.e.* 'comfortable'. But what is the value of S if the temperature rose to 70° F (21.1° C)? According to the equation, the value of S would be 3.19, which lies between 3 for 'comfortably warm' and 'comfortable' for which 4 is assigned. Bedford did not pose the question of the interpretation of this result or others like it because at the time he was concerned with formulating an equation for 'equivalent temperature' as measured by a modified eupatheoscope. Referring back to standard conditions of an enclosure of perfectly still air in which the air and the surroundings are at the same temperature, $S = 10.89 - 0.1099t_e$ where t_e is the 'equivalent temperature'. By putting $10.89 - 0.1099t = 10.89 - 0.0574$, *etc.* and solving for t_e Bedford obtained the equation:

$$\text{Equivalent temperature} = 0.522t_a + 0.478t_w - 0.01474\sqrt{v}(100 - t_a)$$

Three years after the publication of this result, Great Britain was at war. During the second World War, Bedford was concerned amongst other matters, with environmental problems in ships of the Royal Navy especially those associated with high air temperatures and radiant temperatures. He knew that it was unwise to extrapolate from equations for equivalent temperature and chose to adapt the American scales of effective temperature since they were based on experiments conducted over a very wide range of temperatures. But effective temperature did not take thermal radiation into account, and as this was an important factor inside warships, Bedford adapted the effective temperature nomographs, extending the range of air velocities and using the globe thermometer reading instead of the dry-bulb temperature of the air. The resulting index was known as 'corrected effective temperature'. It was later found to be very well correlated with the wet-bulb globe thermometer index (WBGT) of Yaglou and Minard.

In his experiments on the effects of ceiling heating on comfort, Chrenko (1953) found that discomfort was experienced by his subjects even though the equivalent temperature as calculated from Bedford's equation was within an accepted zone of comfort. During this investigation, attention was given to the problem of interpretation of the results obtained from correlating numbers on a psychophysical scale with environmental temperatures. Chrenko (1955) showed that when the percentage incidence of discomfort, whether from being too warm or too cold, was plotted against the mean numerical values of a scale of warmth, the resulting curve tended to be sigmoidal in shape.

One would expect that for a mean comfort vote of zero (or 4 on Bedford's scale), the percentage incidence of discomfort would be zero, at +3 (or 1 on Bedford's scale) there would be 100 per cent incidence of discomfort from the heat. On the other hand, for a mean vote of −3 (or 7 on Bedford's scale) one would expect 100 per cent incidence of discomfort from the cold. Midway between 0 and +3, that is, at +1.5, one would expect a 50 per cent incidence of discomfort from heat and at −1.5 a 50 per cent incidence of discomfort from cold. To verify this hypothesis, Bedford's data were examined but of the recorded 2571 comfort votes only about 12 per cent were too warm or too cold, and there were no data available for mean comfort votes at the midway points of 5.5 and 2.5 (which correspond to −1.5 and +1.5). Confirmation of the hypothesis, so far as warmth was concerned, was obtained from a consideration of the results obtained by Ellis (1952).

Interpolation of results from a sigmoidal curve is inconvenient and extrapolation is hazardous so that it would be much better if a sigmoidal curve were to be represented by a straight line. This may be done by transformation of percentages to probits and plotting probits against mean comfort votes. From the resulting straight line it is simple to obtain probits of the percentage incidence of discomfort for any given value of the mean comfort vote.

A numerical example should help to make this clear. From his data, Bedford (1936) obtained the following equation:

$$S = 11.25 - 0.112t_a$$

where S is the number assigned to sensation on Bedford's scale of warmth, and t_a is the dry-bulb temperature of air in °F.

Suppose that the temperature is 71°F (21.7°C); substitution in the above equation gives the mean comfort vote S as equal to 3.30. The problem is to assign a meaning to this vote. Even if the answer had been 3.0, all that one could have said was that the *average* vote was 'comfortably warm'. But one knows that some people in the group studied would be uncomfortably warm—a glance at the frequency distribution of the comfort votes tells us that. The question is, how many?

The following equation was obtained from Bedford's data as a first approximation:

$$Y = 9.23 - 1.52X$$

where Y is the probit of the percentage incidence of discomfort from heat and X is the mean comfort vote. By substitution of $X = 3.3$ one obtains a value of 4.21 for Y and entering a table of probits gives a percentage of 21.5. Had the temperature been 63°F (17.2°C), the predicted mean vote would be 4.19 and substitution of this value in the probit equation gives Y equal to 2.87 which corresponds to a percentage of 1.6 per cent. Thus, as one expected, as the mean comfort vote moves towards the neutral point which on Bedford's scale is represented by 4, the incidence of discomfort is reduced to almost negligible value.

But the argument so far has only taken into consideration discomfort from too much heat. When the ambient temperature falls and the risk of discomfort from heat is reduced, the risk of discomfort from cold increases. But this can be evaluated in a manner similar to that employed for the risk of discomfort from heat. Therefore a comfort zone can be delineated with some precision: the matter was discussed in detail by Chrenko (1955).

Two years later, Chrenko (1957) investigated the effects of heated floors on sensations of warmth, and as in the experiments on ceiling heating the following scale of thermal sensation was used:

Thermal sensation	*Numerical mark assigned*
Much too warm	+3
Too warm	+2
Comfortably warm	+1
Comfortable	0
Comfortably cool	−1
Too cool	−2
Much too cool	−3

The warmth or otherwise of the sole of the foot was examined by the use of this scale and when the sensations of warmth on the sole of the foot were plotted against the percentage incidence of discomfort—due to overheating of the sole of the foot—a sigmoidal curve was obtained.

The percentages were converted to probits and the following equation was obtained:

$$Y = 2.71X + 0.85$$

where Y is the probit of the percentage incidence of discomfort and X is the mean thermal sensation on the scale above. On substitution of the mid-point of this scale, *i.e.* +1.5 in the equation one obtains a value of Y of 4.92 which on entering a table of probits gives a percentage of 47 per cent.

Throughout the preceding discussion discomfort has been

considered as a quantal response to the thermal environment, that is, a response which occurs or does not occur. The question arose as to where the line should be drawn across the scale of sensation and this did not present any difficulty, since by definition one cannot be thermally uncomfortable in the range from 'comfortably cool' to 'comfortably warm'. Outside this range a given individual subject must be thermally uncomfortable. If a number of persons are exposed to a given thermal environment or thermal stimulus, those who are comfortable will experience sensations to which the numbers –1, 0 or +1 may be assigned whilst those who are uncomfortable thermally will experience sensations to which –3, –2, or +2 or +3 may be assigned. As the temperature rises one expects the numbers of +2 and +3 votes to increase and when one obtains the proportion of such votes to the total obtained from a group of subjects one obtains the incidence, conveniently expressed as a percentage, of discomfort for the given environmental stimulus. Similarly as the temperature falls, one expects the numbers of votes of –2 and –3 to increase and a progressively increasing proportion of total votes indicate the incidence of thermal discomfort from cold.

Fanger does not use the term 'thermal discomfort' in his analysis of a large number of American and Danish experiments on thermal comfort; instead he uses the expression 'predicted percentage of dissatisfied'. He defines as 'dissatisfied' those who voted –2 (cool), +3 (cold), +2 (warm) or +3 (hot) on the grounds that Gagge, Stolwijk and Hardy (1967) provided evidence that 'real discomfort' is first expressed by those voting higher than +2 or lower than –2. He describes as 'dissatisfied' those who felt 'decided discomfort' according to these criteria. Therefore it follows that the term 'predicted percentage of dissatisfied' is really the same as 'predicted percentage of decided discomfort' and as such analogous to the expression which we have been using, *i.e.* the 'percentage incidence of discomfort'.

The percentage incidences of those 'dissatisfied' were plotted against ambient temperature, which since it ranged from about 19° to 32°C enabled two probit lines to be drawn, one for those experiencing discomfort from being too warm and another for those who experienced discomfort from being too cold.

From the data obtained the predicted percentage incidences of those dissatisfied, *i.e.* experiencing discomfort, were plotted against the predicted mean vote.

As one would expect from our previous discussion, the predicted percentage of 'dissatisfaction' or incidence of discomfort for votes of -1.5 or $+1.5$ were respectively 52 per cent and 51.4 per cent, *i.e.* about 50 per cent. As a first approximation, an equations for the 'warm dissatisfied' line is:

$$Y = 3.04 + 1.33X$$

where Y is the probit of the percentage of those dissatisfied and X is the mean comfort vote.

It is instructive to compare this equation with that derived from Bedford's data, *i.e.*

$$Y = 9.23 - 1.52X$$

in which Y represents the probit of the percentage incidence of discomfort from warmth and X is the mean warmth vote on Bedford's scale: this is equal to $(4 - X')$ where X' represents the numbers which I assigned to the Bedford scale of sensation.* The Bedford equation may be transformed thus:

$$Y = 9.23 - 1.52(4 - X') = 3.15 + 1.52X'$$

In Fanger's equation for $X = 0$, *i,e*, 'neutral', the probit of the expected incidence of 'dissatisfied', *i.e.* those experiencing discomfort from being too warm, is 3.04 from which one obtains a percentage of only 2.5. In the Bedford equation, for $X' = O$ one obtains a probit of the expected incidence of discomfort from warmth of 3.15 which gives a percentage of 3.2.

So far as discomfort from cold is concerned, Fanger's data yield as a first approximation:

$$Y = 3.04 - 1.34X$$

* "By the present writer".

whilst for Bedford's data one obtains

$$Y = 3.30 - 1.42X'$$

One is tempted to conclude that there is very little real difference between the ASHRAE and Bedford scales of warmth in practice.

Fanger's results also show once again that no matter what the environmental temperature, it is impossible to satisfy everybody in a large group. At the optimal temperature for Fanger's data, *i.e.* about 25.6° C (77° F), there is a risk of discomfort of 5 per cent. In Bedford's investigation the optimal temperature was about 17.2° C (65° F) and the risk of discomfort was 8 per cent.

Fanger's equations for comfort are a considerable advance on previous work in this field since the equations take account of the activity of the subject and the clothing worn. Bedford's equations are valid for the subjects of his investigation only and Bedford did not specify the clothing worn or the activity. However, from the unpublished records one ascertains that the work was that associated with light electrical industries and, with the metabolic data now available, it is reasonable to assume that the metabolic rate of Bedford's subjects was of the order of 100 kcal/m^2/hour. Entering Fanger's tables at this level of activity, one finds that the predicted mean vote of 0.03, *i.e.* about 'comfortable', is found when the temperature is 18° C and the air velocity 0.15 m/sec which is near enough to Bedford's optimal temperature and average air velocity. This combination of variables is associated with a 'clo' value of 0.75. So it is fair to assume that the insulating value of clothing worn by women in factories in the England of the thirties was of this order.

To understand why the optimal temperature of 25.6° C was found for the American subjects, it should be remembered that they were sitting around for three hours in clothing with an insulation of 0.6 clo which is the insulation given by the following clothing assembly: athletic shorts, woollen socks, open-neck cotton shirt and cotton trousers, with the shirt tail hanging out.

SOME CAUSES OF DISCOMFORT

Temperature Gradients

Methods of heating or ventilating rooms which give distinctly higher temperatures about the head than near the floor are undesirable, for such conditions tend to produce a stuffy feeling in the head, while the feet are cold.

A hundred years ago, Reid (1844) remarked that one often encountered great differences of temperature within the same compartment so that the feet might be exceedingly cold while the head was oppressed with heat. He said that when travelling in coaches in cold weather he had repeatedly noticed a difference of 12° to 16° F between the temperature of the air around the head and that at the feet. He urged that coach-makers should admit the air in such a manner that, in cold weather, it might receive some heat from the upper part of the coach before it reached the floor, and he suggested that the floor of the coach should be heated.

The Commissioners who, in 1857, reported to the General Board of Health on the heating and ventilation of dwellings, made large numbers of measurements of the temperature gradients in rooms. In their report the Commissioners said that in a comfortable and healthy apartment the floor should be at the highest temperature in the room; while the temperature should gradually decrease towards the ceiling. Little attention seems to have been paid to this recommendation, for systems of heating which create very considerable temperature gradients have been widely used, both in industrial buildings and in dwellings.

An illustration of the effects of such a system can be drawn from my own observations. In one factory, women and girls were engaged in soldering metal containers. The shop was heated by closed stoves, and the air temperature at head level averaged 4° F higher than that near the floor, while the difference was sometimes as great as 10° or 11° F. In addition to these air temperature gradients there was a local radiation effect, for the hands and faces of the workers were exposed to the radiation from hot soldering irons and often also from high-pressure gas burners used for lighting, while the lower parts of their bodies were screened from this radiation. When questioned as to their feelings of warmth, many of these persons said that they were generally comfortable, but that their feet were cold, and one remarked that she was too warm, but that her feet felt chilly.

When the skin temperatures observed in this factory were compared with those noted at similar room temperatures in other factories it was found that in the soldering factory the forehead temperature was about 2° F higher, and the temperature of the hands about 5° F higher, than in other factories, while, on account of the colder air near the floor, the feet tended to be cooler. Complaints of cold feet were twice as frequent in the soldering factory as in the other factories. The floor of the soldering factory was of wood blocks, so the complaints of cold feet were not due to very low temperature of the floor itself.

Cold Feet

Much serious discomfort is caused by cold feet, for it is impossible to be really comfortable when the feet are chilly. Discomfort from this cause is not peculiar to modern times. We have noted the comment on steep temperature gradients made by Reid a century ago, and Bernan (1845) says that a certain Welsh prince had an officer in his court called his footbearer, whose duty it was, at meal times, to sit with his back to the fire and keep his master's feet warm by cherishing them in his bosom. Cold feet must have been common in the draughty castles of Saxon and Norman times.

In my (1936) investigation, all complaints of cold feet were noted. In the main series of observations complaints were volunteered by 5 per cent of the persons questioned. As would be expected, complaints were more frequent in cool rooms than in warmer ones. When the temperature was between 54° and 58° F, 9 per cent of the subjects made complaint of cold feet, but a number of persons complained even when the air temperature was above 66° F.

The temperature of the feet is fairly closely related to the temperature of the air near the floor, and comparing foot temperatures with air temperatures measured 6 in from the floor it was found that the foot temperatures of the complainants were, indeed, slightly lower than the average for all the people tested, but the difference was only 1° F or rather more, and such a small difference is not likely to have been enough to cause complaint. On the other hand, those who complained felt distinctly cool when their feet were at the temperature which was found comfortable by the average person. When sensations of warmth were compared with equivalent temperatures it appeared that for comfort those who complained needed distinctly higher environmental temperatures than those which were found satisfactory by the average person.

Excessive Local Heat

With some applications of radiant heating local discomfort may be caused if the temperature of the heating surfaces is too high. When overhead sources of radiant heat are used care is necessary to ensure that the radiation to the heads of the occupants is not excessive. When floor heating is used unduly high floor-surface temperatures can be very uncomfortable especially to those who must stand or walk on such floors for prolonged periods. Further consideration will be given to these matters when various methods of heating are discussed.

Draughts

The prevention of draughts is a major problem for ventilation engineers. The entire elimination of draughts is a difficult matter, and complaints tend to arise even when much care has been devoted to the design of the ventilating system. The problem is complicated by great divergencies of opinion as to what constitutes a draught. One man's meat is another man's poison, and an air current that is found pleasant and refreshing by some people will provoke others to angry complaint. How many wordy battles have been fought about the opening of windows?

The speed at which an air current becomes noticeable depends on its temperature. Baetjer (1924) found that at temperatures of 60° to 65° F currents impinging on the cheek with speeds below about 40 ft/min were not perceptible. At 54° F a current of 30 ft/min was just noticeable, while at 86° F currents with velocities lower than 120 ft/min were unnoticed.

In my investigation strong air currents were not encountered. The maximum air speed was about 100 ft/min, and in the majority of observations the speed was below 50 ft/min. With these low air velocities it was to be expected that complaints of draughts would be few, and in fact only between 1 and 2 per cent of the subjects volunteered complaints. There was no consistent relationship between the frequency of complaints and the air velocity measured at head level, but complaints were more frequent when the currents about the feet exceeded 40 or 50 ft/min. Yet it does not appear that the cause of complaint was always the actual speed of the air currents, for complaints were made when the air velocity was below 20 ft/min. Those who complained had skin temperatures rather below the average, and probably partly on that account they had a distinct preference for a warm environment. For comfort these people required a temperature 4° F higher than that found most comfortable by the average person.

Exposure to the stream of air from a ventilating jet may cause one to feel a draught. If the stream of air is at or below the temperature of the surrounding air the convective heat loss from the skin will be increased and the local effective temperature will be lowered. On the other hand, if the ventilating air is admitted at a temperature higher than the general temperature of the room sensations of chilliness due to draught may be prevented, and according to its temperature and speed the air stream may even cause local sensations of warmth.

Studying convective stimuli, Turner (1955) found that with a sudden increase of air speed to 50 ft/min it was necessary to raise the temperature of the air stream by 8.8° F if a sensation of cooling of the forehead was to be avoided, while with a speed of 100 ft/min the temperature had to be raised by 14.4° F. On the other hand, if with speeds of 50 and 100 ft/min respectively the temperature was raised by more than 12.4° or 16.7° F local sensations of warmth were caused.

Rydberg and Norbäak (1949) considered the relation between the speed of an air stream and the difference between the temperatures of the stream and of the surrounding air. They concluded that freedom from draught will be achieved if the ratio between the excess of the temperature of the stream over that of the room, in ° F, and the air speed, in ft/min, is equal to 0.07:

$$(t_s - t_r)/V = 0.07$$

When cold air is being discharged from ventilation inlets, as in summer air conditioning, it is not possible to ensure freedom from draughts, but Ryberg and Norbäak conclude that with a given constant refrigerating load the feeling of draught will be at a minimum when in the above equation the temperature difference is the excess of the room temperature over the temperature of the air discharged from the inlets, and the velocity is that of the air leaving the inlets.

Further Comments

It is sometimes assumed that the perfect indoor environment is characterized by completely uniform conditions in the enclosed space. Such conditions are rarely, if ever, achieved outside special climatic chambers and even here it was found that, for a group of people all wearing similar standardized clothing and with the same level of metabolism, 5 per cent of the group experienced some thermal discomfort. It is further assumed that deviations from these supposedly ideal conditions must necessarily be associated with more discomfort and the examples previously discussed tend to confirm this assumption.

Consider the example of Reid in his coach: his feet were 'exceedingly cold' whilst his head was 'oppressed with heat'. Using the scales of warmth to assess the sensations of the heat and feet *separately*, one could assign –3 to the feet and +3 to the head. So far as thermal balance of the body was concerned equilibrium might have been achieved inasmuch as the heat lost by the feet would be compensated by the heat gained by the head. The same would not occur with sensations so that +3 would be added to −3 giving zero and so comfort. Common experience indicates that discomfort remains. The human body can differentiate between different thermal stimuli in different parts of the body. Integration takes place at a different level: Reid criticized the design of the coach!

Since it is possible to have cold or hot feet and yet otherwise be thermally comfortable one may be misled if in an environment which is not uniformly heated one attempts to make an assessment of the general or overall bodily warmth. It would be advantageous to make assessments of at least three areas of the body, *i.e.* the head, the feet and the rest of the body. In this way one avoids making the assumption that the optimal ambient temperature is the same for each region of the body which is implicit in the calculations based on the overall heat loss from the body which have led to comfort equations in the past. A reasonable assumption would be that for warmth of the head there exists an optimal temperature of the ambient air at head level and so far as the feet were concerned there could likewise be an optimal temperature of the ambient air at or near the floor.

Bedford (1936) found that as the temperature of the air measured only 6 in from the floor increased the incidence of complaints of cold feet decreased until at between 70° and 73.9° F they ceased. Yet for these subjects the optimal temperature measured 4 ft from the floor was 65° F. Therefore a vertical gradient of air temperature of some nearly 9° F was associated with thermal comfort. One might expect this from a consideration of the way the skin temperature of the foot varies with ambient temperature; according to Bedford's data a change in ambient temperature of 10° F results in a change of 8° F in the temperature of the foot but only 1.4° F in the forehead.

Even at comparatively high ambient temperatures, there is something like 10° F difference between the peripheral temperatures of the head and the feet.

The hypothesis that different parts of the body require

different ambient temperatures for thermal comfort was tested by Chrenko who found, during his investigation of ceiling heating, that *so far as the head of the subject was concerned,* the optimum air temperature was 60° F.* If the optimum air temperature at floor level so far as the feet are concerned is 72° F then this confirms the observations of Bedford and the hypothesis mentioned above. It also follows from this that one does not need to specify an absolutely uniform thermal environment for thermal comfort for sedentary subjects. This is of some importance to environmental engineers concerned with the design of internal environments. They need not attempt to simulate the carefully controlled uniform conditions obtaining in the experimental climatic chamber in the interests of the thermal comfort of the occupants of a building. On the contrary, there is evidence, to be discussed in a later chapter, that a non-uniform thermal environment is to be preferred.

* This finding, hitherto unpublished, appears to confirm the recommendation of Yarnold (1947).

CHAPTER VII

SENSATIONS OF WARMTH AND COLD

MODE OF STIMULATION

GENERAL sensations of warmth are not governed directly by the rate of heat production within the body nor by the rate of heat loss from it, for the latter may remain practically constant over a range of temperature within which different sensations may be experienced. Neither do variations in the proportions of heat lost by radiation, convection, and evaporation govern impressions of warmth, for the proportions of heat dissipated in these ways may vary widely with the same effective temperature, that is, when the same feeling of warmth is experienced. Thermal sensations do not depend on the temperature of the deeper tissues, as indicated by the rectal temperature, for one may feel warm when the body temperature is low, or chilly during a fever.

We experience sensations when our sense-organs are stimulated. These sense-organs, or receptors, consist in specialized groups of cells which react to changes in their environment. In the skin there are various types of sense-organs, some of which respond to changes in pressure and give rise to sensations of touch, while others are stimulated by temperature changes and are responsible for our sensations of warmth and cold. Sensation depends on a changing environment. If the body remains perfectly in equilibrium with its surroundings so that no change occurs in the temperature of the skin or of the deeper tissues, the thermal receptors are not stimulated and no sensation of warmth or cold is evoked. Then the subject will feel 'comfortable' according to the scale of sensations set out at the beginning of Chapter VI.

Although when individuals or small groups of subjects are studied a close association is found to exist between general sensations of warmth and skin temperature, the association is not perfect. Sensations do not depend only on the temperature of the skin surface, nor are they necessarily governed by the direction of change of the skin temperature, for, as Bazett and McGlone (1927) point out, although sensations of cold usually accompany a falling skin temperature, one may feel very cold when the skin temperature is rising rapidly, as after sea-bathing.

Bazett and McGlone studied temperature sensations in relation to the temperature gradients in the superficial tissues. They inserted thermojunctions to various depths and also measured the temperature at the surface of the skin. When rapid changes of sensation occurred there was always some layer of the tissues investigated in which the temperature was falling when the subject was feeling cold, or rising when he felt warm. Sensations of cold were associated with a steep thermal gradient in the subcutaneous tissues. Comfortable conditions, when temperature sensations were not perceived, were accompanied by a moderate gradient, and sensations of warmth by a low gradient.

Later, these workers and their colleagues (1932) identified the end-organs in the skin from which temperature sensations probably arise. One type of sense-organ, the end-bulb of Krause, 0.1 mm below the skin surface, is responsible for sensations of cold, and another type, the Ruffini ending, at a depth of 0.3 mm, for sensations of warmth. From further studies (Bazett and McGlone, 1932) it appeared that stimulation of these sense-organs might arise from temperature gradients between the end-organs and the circulating blood, for when sensations were induced by variations in the external temperature all the sensations experienced could be explained as the result of cooling or warming of the blood in the immediate neighbourhood of the end-organs. It was thought possible, however, that such changes in the blood temperature might only act indirectly through the production of chemical changes or gradients in the neighbourhood of the end-organs.

Recent work has shown that it is no longer clear that each sense modality has its own specialized nerve ending, especially in the case of the cutaneous receptors. The encapsulated highly organized sensory receptors have a remarkably sparse distribution. Bazett himself tried to determine as completely as possible the number of sensory experiences from a piece of foreskin (prepuce) afterwards sacrificed for histological study. No more than four types of sensation, *i.e.* heat, cold, touch and pain were evoked, but the piece of tissue contained *seven* clearly differentiated types of nerve ending.

The specialized cutaneous sense-organs are practically restricted to regions of the body which are devoid of hair. Not surprisingly, the hairless regions like the sole of the foot, the palm of the hand and various mucous membranes are highly sensitive areas and as such chosen for study by the early histologists. There is some evidence that the complicated encapsulated endings have been produced by the constant mechanical deformation to which nerve terminals in hairless regions are exposed.

One may conclude that, whatever the function of the specialized endings in the hairless regions, they are not the only structures that will transmit information in their alleged modalities.

Other studies of the thermal responses of the skin have been made by Hardy and Oppel (1937, 1938), who irradiated the skin and measured the skin temperature with a sensitive thermopile. They also concluded that stimulation depends on change in the normal thermal gradient in the skin (Oppel and Hardy, 1937*a*) and that separate end-organs respond to increases and to decreases of temperature (1938).

INFLUENCE OF QUALITY OF RADIATION ON SENSATIONS

Oppel and Hardy (1937*a*) used three kinds of radiation in their earlier experiments: (i) visible radiation of 0.4 to 0.7 microns; (ii) penetrating infra-red radiation in a waveband of from 0.8 to 3.0 microns; and (iii) non-penetrating infra-red radiation of longer wavelengths than 3 microns. The minimum amounts of incident radiation required to evoke a sensation of warmth, when the skin of the forehead was irradiated for 3 sec, were in the ratio of 3 : 2 : 1, respectively, for the visible, penetrating infra-red, and non-penetrating infra-red radiations. Proportions of the incident radiation which varied with the wavelength were reflected at the surface of the skin. Allowing for reflection, the minimal amounts of absorbed radiation which were required to give rise to sensations were in the ratio of 2.2 : 1.5 : 1. The more penetrating the rays, the less sensitive one is to them. Hartley (1934) described experiments with various incandescent heaters in which he showed that the threshold intensity for pain was higher with the sources which emitted the more penetrating radiation, and that exposure to radiation of a given intensity could be borne for longer when the radiation was of a more penetrating kind.

The minimum stimulating rates for the various radiations used by Oppel and Hardy did not bring about the same rise in the temperature of the skin (Fig. 32).

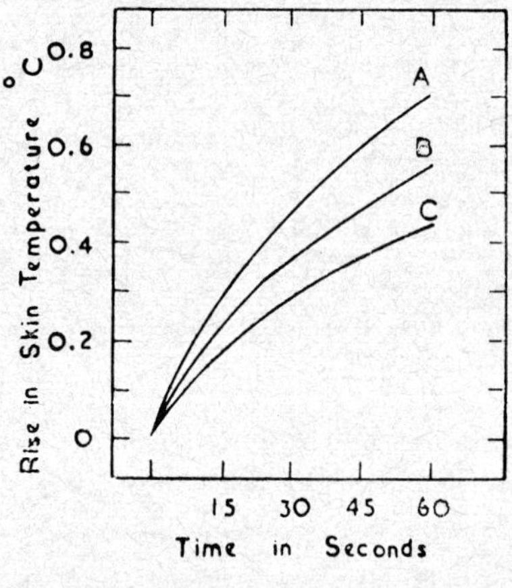

Fig. 32. Rise of skin temperature produced by minimum stimulating rates of (a) *visible,* (b) *Penetrating infra-red, and* (c) *non-penetrating infra-red radiation. (Oppel and Hardy.)*

A relatively large amount of visible radiation was necessary to evoke a sensation and it caused considerable rises in the temperatures at the skin surface and beneath it. The amount of penetrating infra-red required was less and the resulting temperature elevations were less. Still less non-penetrating infra-red radiation was needed, and it caused the least changes in the temperatures of the surface and of the deeper tissues. It is estimated that these three types of radiation were absorbed by the skin in such a way that with each of them a similar disturbance of the normal thermal gradient in the tissues was caused.

VARIATION OF MINIMAL STIMULUS WITH AREA STIMULATED

The sensation produced by a heat stimulus of fixed intensity increases as the area stimulated increases. Hardy and Oppel (1937) studied this spatial summation of the responses of thermal receptors. They investigated the relationship between the area stimulated and the minimum strength of stimulus needed to evoke a sensation after irradiation for 3 sec. Non-penetrating infra-red radiation was generally used, but some comparisons were made with visible and penetrating infra-red radiations.

The smallest intensity of radiation which could be perceived as warmth was 0.00015 g cal/sq. cm/sec, when the area of skin irradiated was not less than 200 sq. cm. Further increase in the area made no sensible difference to sensitivity. The rise of skin temperature produced by this radiation was at the rate of 0.001°C/sec, *i.e.* a temperature rise of 0.003°C in 3 sec will evoke a sensation of warmth. This represents the threshold of excitation of the end-organs.

As the area irradiated decreased below 200 sq. cm the strength of stimulus per unit area required to cause a sensation increased, but the total amount of radiation necessary, *i.e.* the product of intensity and area, decreased as the exposed area decreased. The intensity of radiation required to evoke a sensation by the stimulation of a single end-organ was calculated to be between 0.031 and 0.025 cal/sq. cm/sec, corresponding with a rise in the temperature of the skin surface of 0.48°C in 3 sec.

The sensitivity of the skin to heat varies in different parts of the body surface, but over the forehead the sensitivity is uniform. Observations made on the forehead, when areas of skin varying from 1 to 40 sq. cm were stimulated, showed a logarithmic relation between the area and the intensity of stimulus required to evoke a sensation. The data are represented by the equation

$$I \times A^{0.78} = 0.00813$$

where I is the intensity of stimulation, in cal/sq. cm/sec; and
A is the area stimulated, in sq. cm.

Turner (1955) measured the threshold values of radiative stimuli in two subjects using the limiting method. The area of skin exposed to the stimulus was varied between 15 and 200 sq. cm, and the period of exposure was 5 sec. The individual responses were variable. The threshold was taken as the median effective stimulus, *i.e.* that intensity of radiation which evoked a 50 per cent response.

Threshold intensities of radiative stimuli in relation to the area of skin exposed (Turner, 1955)

Area of skin exposed (sq. cm)	*Threshold intensity of radiation (cal/sq. cm/sec)*	
	Subject I	*Subject II*
15	0.0014	–
30	0.00054	–
70	0.00023	0.00012
200	0.000123	0.000058

Hardy and Oppel exposed their subjects for only 3 sec whereas Turner's subjects were exposed for 5 sec, moreover, as Chrenko (1962) pointed out Hardy and Oppel appear to have taken the minimal effective stimulus as the threshold, whereas Turner took the median effective stimulus, so that any agreement between the results of Hardy and Oppel and those for Turner's Subject I is fortuitous.

Chrenko (1964) also used the method of limits in an investigation of the threshold intensities of thermal radiation that evoked a sensation of warmth on the forehead. The subjects were 23 men and 22 women. The radiation was that emerging through a circular aperture of 4 in diameter in a spherical furnace of 12 in internal diameter.

In different experiments the furnace temperature ranged from 590° F (310° C) to 1,490° F (810° C), and the peak wavelengths of the radiation emitted ranged from 5.0 to 2.7 microns. The intensity of the radiation reaching the subject was varied by altering the distance between the subject and the furnace, and the area of the skin exposed to the radiation was kept at 40 sq. cm by placing a screen close to, but not touching, the forehead. The period of exposure was 3 sec in all experiments. The temperature of the skin on the forehead was measured.

There was much variation between subjects. Thus, with the furnace temperature at about 1,000° F the average thresholds of individual subjects ranged from 0.0009 to 0.0029 cal/sq. cm/sec.

There was evidently a tendency for the threshold to decline as the temperature of the furnace fell and the peak wavelength increased; and when the threshold intensities were plotted against the skin temperature of the forehead, disregarding the wavelength of radiation, it was clear that as the skin temperature rose the sensitivity to radiation increased. The pooled results from all the subjects were analysed statistically, and a partial regression equation expressing threshold intensity in terms of furnace temperature and skin temperature was derived. According to that equation an increase of 100° F in the temperature of the furnace, with constant skin temperature, was associated with a rise in the threshold of 0.000045 cal/sq. cm/sec; while with constant furnace temperature a rise of 1° F in the skin temperature was associated with a fall in threshold of 0.000133 cal/sq. cm/sec.

The peak wavelength of the radiation in the experiments of Hardy and Oppel was 3.5 microns, and the minimal intensity of radiation evoking a sensation of warmth was 0.0004 cal/sq. cm/sec when the area of skin exposed was 40 sq. cm as in Chrenko's experiments. For a similar wavelength the average threshold of Chrenko's women subjects was 0.0014 cal/sq. cm/sec. In order to estimate the minimal effective stimulus Chrenko subtracted twice the standard deviation of 0.0005 from the mean and thus arrived at the identical figure of 0.0004 cal/sq. cm/sec reported by Hardy and Oppel.

SENSATIONS OF COLD

Hardy and Oppel (1938) continued their work by investigating the stimulation of sensations of cold by radiation. As their source of 'cold radiation' they used a block of solid carbon dioxide, and the cold stimulus was the net loss of heat from the skin as the result of the radiation exchange between it and the carbon dioxide block.

The minimum rate of change in the skin temperature for the production of a sensation of cold, when the area stimulated was 200 sq. cm or more, was 0.004° C/sec, whereas the rate for sensations of warmth was 0.001° C. For small areas, on the other hand, the sensory threshold for warmth was higher than that for cold. Heat is conducted into the skin more readily when the skin is exposed to a hot source of radiation than it is abstracted when the skin is radiating to a cold surface, so that a cold radiation stimulus produces a greater change in the surface temperature than that caused by a stimulus of the same intensity from a hot source. Thus, an increase of radiation loss to the solid carbon dioxide of 0.00033 cal/sq. cm/sec from an area of 197 sq. cm caused the skin temperature to fall at the rate of 0.004° C/sec, while exposure of the same area to the non-penetrating infra-red radiation from an electric hot plate with an intensity of 0.000021 cal/sq. cm/sec caused a rise in the skin temperature of only 0.004° C in 3 sec, or rather more than 0.001° C/sec. 'Cold' radiation produces about twice the rate of change in skin temperature, calorie for calorie, as does radiation from hot sources.

The results of the work of Hardy and Oppel are summarized in Fig. 33.

The diagram above clearly shows that, whatever the area of the skin stimulated, if the surface temperature does not rise at the rate of 0.001° C/sec, or fall at the rate of 0.004° C/sec, no thermal sensation is evoked. The relation between the area

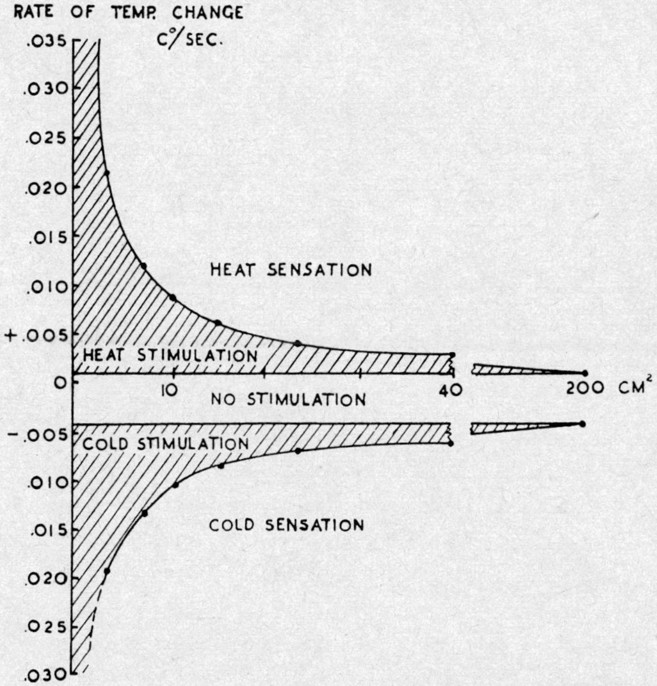

Fig. 33. Sensory thresholds for heat and cold in relation to area stimulated and rate of change of skin temperature. (Hardy and Oppel.)

stimulated and the minimal rate of change of skin temperature for sensation is also shown.

SENSITIVITY TO THERMAL CHANGES

We have seen that sensations are evoked by quite small thermal stimuli and small changes in the temperature of the skin. The minimum stimulus for a sensation of warmth is given as a change of 0.00015 cal/sq. cm/sec in the radiation intensity over a period of 3 sec. Such a change would be caused if the temperature of the walls of a room were suddenly raised by 2° F, the temperature, velocity and humidity of the air remaining unchanged. Changes in these atmospheric conditions which would bring about corresponding shifts in the convection exchange or in the evaporative cooling would also cause sensations.

Hardy and Oppel remark that the body's heat-regulating mechanism may be brought into operation by even smaller thermal stimuli, which are insufficient to evoke sensations. It may be that stimuli which are below the sensory threshold can excite vasomotor adjustments reflexly. Thus the surface temperature of the body may be adjusted to slowly changing conditions without thermal sensations obtruding themselves on our consciousness.

Our sensitivity to changes in the thermal environment may lead to false impressions concerning the warmth of our surroundings. If on a cold day one enters a warmed room immediately on going indoors a sensation of warmth is experienced. It may be, however, that although the room is distinctly warmer than the external air, it is yet not warm enough for sedentary persons to remain comfortable over a long period. After the heat generated by outdoor exercise has been dissipated the sedentary body may not produce heat at a rate sufficient to maintain thermal comfort, and sensations of cold may be experienced. First impressions may be very misleading. On a cold day a stay of at least half an hour is necessary before the warmth of a room can be adequately judged from sensations alone. It is unfair, and indeed foolish, to walk into a factory or classroom after a brisk walk out-of-doors, and then proceed immediately to pass judgment on the adequacy of the heating.

CHAPTER VIII

THE FRESHNESS FACTOR IN THERMAL COMFORT

DURING the first World War of 1914-18, the Health of Munition Workers Committee defined an invigorating atmosphere as being: (i) cool rather than hot; (ii) dry rather than damp; (iii) diverse in its temperature in different parts and at different times rather than uniform and monotonous; and (iv) moving rather than still. Sir Leonard Hill had already shown the enervating effects of excessive temperature, humidity and of stagnation of air. He also said '*for air to feel fresh it must have such cooling and evaporative power as suffices to keep the head pleasantly cool and the skin free from sensible perspiration*'.

Whilst the temperature of the interior of the body remains comparatively constant varying from about 36.4° to 37.5°C, the organ demanding this close range of temperature being the brain, the temperature of the surface of the body varies considerably. Bedford's data, illustrated in Fig. 21, show this very clearly. There can be as much as a 20°F difference between the surface temperatures of the forehead and foot at low ambient air temperatures.

But whereas very little heat production occurs in the foot, the nerve cells of the brain have a very high rate of metabolism and about 15 per cent of the total volume of blood which the heart supplies to the body as a whole, flows through the brain. Furthermore, there is a lavish blood supply to the superficial tissues of the scalp and face. This is why the temperatures of the forehead and face vary comparatively little with ambient temperature.

Whereas sensations of warmth may be referred to any part of the surface of the body, impressions of freshness or the converse, stuffiness, are referred mainly to the head. It often happens that one may feel comfortably warm in a given room and yet one may also feel fresh or the room may seem to be stuffy or oppressive. Impressions of freshness or stuffiness are determined in the main by the thermal environment inasmuch as low environmental temperatures tend to evoke impressions of freshness and high temperatures evoke impressions of stuffiness. But one can feel cool in a room which is definitely stuffy or warm in a room which seems fresh.

Vernon, Bedford and Warner (1926*b*), and later Bedford and Warner (1939) studied the effects of the thermal environment on impressions of freshness. They measured the temperature, humidity and average velocity of the air, and also determined the variability of air velocity and of air temperature.

The earlier workers were very much concerned with sensations of air movement and they considered that impressions of freshness were due to local stimulation of the skin which was partly thermal and partly tactile. They used the following scale for the assessment of sensations of air movement:

Sensation of air movement	*Number assigned*
Very stagnant (or stuffy)	0
Very stagnant to stagnant	1
Stagnant or stuffy	2
Medium to stagnant	3
Medium	4
Medium to fresh	5
Fresh	6
Fresh to very fresh	7
Very fresh	8

Bedford and Warner (1939) used the same scale, but instead of ‘sensations of air movement’ they substituted the term ‘impressions of freshness’, and this scale was also used by Munro and Chrenko (1949).

THE IMPORTANCE OF AIR MOVEMENT

Experience teaches, and observations show, that cool air is more bracing than hot air. Freshness increased at the lower temperatures. A fall of 5° F in summer or of 3.5°F in winter was enough to increase freshness by one unit on the arbitrary scale of sensations. When the air temperature was held constant there was found to be a clear relationship between freshness impressions and air velocity ($r = 0.62$ and 0.49 for the summer and winter observations respectively). A given change in air velocity had more effect on freshness impressions in summer than in winter. In the winter observations, in which the air temperature averaged 61° F, a change of about 40 ft/min in the velocity increased the freshness by one unit of the scale of sensations, say from 'medium' to 'fresh'. In summer, when the temperature averaged 10° F higher than in winter, an increase in freshness equivalent to one unit on the scale accompanied an increase of only about 20 ft/min in the velocity. Thus, in these observations, the invigorating effect of air movement was more noticeable at high temperatures than at low ones, whereas when general sensations of warmth are considered a given change in the amount of air movement has a greater effect when the temperature is low than when it is high.

One gathers from the foregoing results that the effect of air velocity on impressions of freshness varies with the temperature of the air but there is the possibility that some form of seasonal adaptation may also be involved as the observations at the lower air temperature were made in winter (*i.e.* those at 61° F) and those observations made at about 70° F were in summer.

Experiments on the threshold air velocity which would evoke the sensation of air movement were conducted recently by Chrenko* and it was found that (*a*) thresholds varied with air temperature, and (*b*) for a given air temperature the threshold was considerably lower in summer than in winter. Moreover, in both seasons, the threshold increased in linear fashion with air temperature. The results tended to confirm the observations of Baetjer (1924) but one should bear in mind that Baetjer exposed only the cheek whereas in Chrenko's experiments the whole head of the subject was exposed to the stream of moving air.

The relation between these thresholds and the temperature of the air is given by the following equations:

$$V_s = 1.48t_a - 78.5 \qquad \text{(observations in summer)}$$
$$V_w = 2.87t_a - 120.9 \qquad \text{(observations in winter)}$$

* Hitherto unpublished observations.

in which V_s and V_w represent the respective median threshold air velocities, expressed in ft/min, for summer and winter seasons and t_a is the air temperature expressed in °F. The equations are valid for a range of temperature from 64° to 80°F in summer and from 62° to 74°F in winter.

The predicted threshold at a temperature of 61°F in winter is 55 ft/min; in summer at a temperature of 71°F it would be 27 ft/min. It is important to notice that for the same temperature (71°F) in winter the threshold would be 83 ft/min.

These results also indicate how sensitive people are to air movement: the thresholds reported here are median thresholds, *i.e.* the air velocity which evokes the response on 50 per cent of the occasions when the stimulus is presented. Minimal values of air velocity evoking the response were much lower. It should also be mentioned that in these experiments the temperature in the laboratory was kept constant at a given level and only the air velocity was altered during a given experiment.

But these experiments do not enable one to make assessments of freshness or otherwise in a given environment unless a scale is used. The use of the early nine-point scale of Bedford *et al.* was considered but it was decided that it was more convenient to use a seven-point scale comparable with the seven-point scale of warmth so far as statistical treatment of the results was concerned and also in view of the use of similar seven-point scales by other workers. The scale used in the threshold experiments was as follows:

Impression	*Number assigned*
Much too stuffy	1
Too stuffy	2
Slightly stuffy	3
Neither fresh nor stuffy, *i.e.* neutral	4
Slightly fresh	5
Too fresh	6
Much too fresh	7

The votes 3, 4, and 5 are associated with comfort and votes 1, 2, 6 and 7 are associated with discomfort.

One may talk about a 'sensation of warmth' or a 'sensation of air movement' but it is better to use the word 'impression' when one considers 'freshness'. Freshness, for example, is not synonymous with a sensation of air movement although air velocity at head level is a very important physical characteristic of a given environment so far as impressions of freshness or otherwise are concerned. Impressions of freshness were correlated with the air velocity at head level and the air temperature at the same level and a regression equation was obtained by computer:

$$F = 0.0128v - 0.0946t_a + 9.954$$

where F represents the numbers assigned to impressions of freshness, v is the air velocity expressed in ft/min and t_a is the air temperature expressed in °F.

This particular equation applies to observations made during winter. The lowest temperature of the air during the experiments was 62°F and if one assumes completely calm air, when v approaches zero, it can be calculated from the above equation that $F = 4.06$, *i.e.* the environment seems to be neither fresh nor stuffy.

Suppose that the air velocity is 56 ft/min and the temperature is kept at 62°F, the predicted freshness vote becomes 4.78 which is very nearly equal to 5, *i.e.* 'slightly fresh'. This is the air velocity equal to the median threshold of perception of moving air. When the velocity is raised to 73 ft/min, the predicted vote becomes 5.

These results show that the value of air velocity of 50 ft/min often chosen by heating and ventilating engineers in this country for design purposes is too low. It would be of interest to ascertain the effects of the various temperatures and air velocities mentioned above on the warmth of the head of the subjects which was also assessed during the experiments by using the 7-point scale.

Accordingly, a computer was used to obtain a regression equation relating warmth to air temperature and air velocity: the equation is

$$W = 0.1421t_a - 0.00879v - 8.799$$

in which W represents the numbers on the scale of warmth, *i.e.*

+3 stands for 'much too warm' and so on down the scale to –3 or 'much too cool', t_a and v representing air temperature and velocity as before.

Suppose that the air velocity is so low that it can be neglected, then the equation shows that, if the air temperature is 62°F, the predicted mean warmth vote will be 0.01, *i.e.* zero for all practical purposes, *i.e.* exactly comfortable.

It will be recalled that on the basis of experiments on ceiling heating Chrenko found that the optimal air temperature for warmth of the head was 60°F and the present results tend to confirm his earlier work in this respect.

For the environment to feel slightly fresh ($F = 5$) the air velocity must be 73 ft/min and if this value is substituted in the warmth equation with t_a still at 62° F, the predicted warmth vote for the head becomes –0.63 which is on the cool side of comfortable. For experiments conducted in summer, the equation for predicting the warmth of the head of the subjects is

$$W = 0.2017t_a - 0.0087v - 14.35$$

In very calm air, this equation indicates that the optimal temperature for warmth of the head is 71°F.

This result indicates that some adaptation may probably have taken place and it provides clear-cut evidence in support of Bedford's statement made when he was discussing summer comfort zones. Hickish (1955) took 75°F as the upper limit of the comfort zone in his observations in the field: Bedford (1936) suggested a temperature of 68°F for a winter comfort zone. The equation for predicting the mean freshness vote amongst Chrenko's subjects in summer was

$$F = 0.0156v - 0.1093t_a + 11.48$$

When the equations for winter and summer observations are compared it can be shown that, if temperature is held constant, in winter a change of 77 ft/min in air velocity is required to change the freshness vote by one unit whereas for the summer observations a similar change requires 64 ft/min.

The coefficients for air temperature in both equations are

similar. In the equations for predicting warmth of the head this is not the case: in calm air at a temperature of 70°F (21°C) the mean warmth vote in winter would be +1.2, whilst in summer it would be −0.4 but, as the air temperature rises, the differences become smaller and by extrapolation both lines meet at a temperature of about 87°F at which level, as might be expected, all subjects would be very uncomfortable. But the discomfort would be relieved by increasing the air velocity, *e.g.* the equation predicts that in summer at a temperature of 80°F increasing the air velocity at head level to 145 ft/min will lower the mean warmth vote from just under +2 (uncomfortably warm) to +0.6, midway between 'comfortable' and 'comfortably warm', and, at the same time, bring the freshness vote up to 5, *i.e.* slightly fresh.

Variable Air Currents

Variability of air movement was measured with the aid of a specially constructed hot-wire anemometer (Vernon and Manley, 1926). Velocities were determined at intervals of 2 or 2.5 sec over periods of 5 min, and the variability of air movement was determined by calculating the mean variation of velocity and expressing it as a percentage of the mean velocity over the period of observation. Figure 34 shows the results of two sets of observations. The data plotted in curve *A* were obtained in a workroom with good cross-ventilation, and those in curve *B* in a poorly-ventilated office. In the first example the average air velocity was 72 ft/min, with a mean variation of 47 per cent, and in the second the velocity averaged 15 ft/min and the mean variation was 16 per cent.

In the summer observations, when air temperature and average air velocity were both held constant, there was a highly significant partial correlation between freshness and variability of air velocity ($r = 0.43 \pm 0.08$). With the same air temperature and average velocity, an increase of 14 per cent in the mean variation of the air velocity was accompanied by an increase in freshness equal to one unit of the sensation scale. In winter the residual correlation between freshness and variability of air

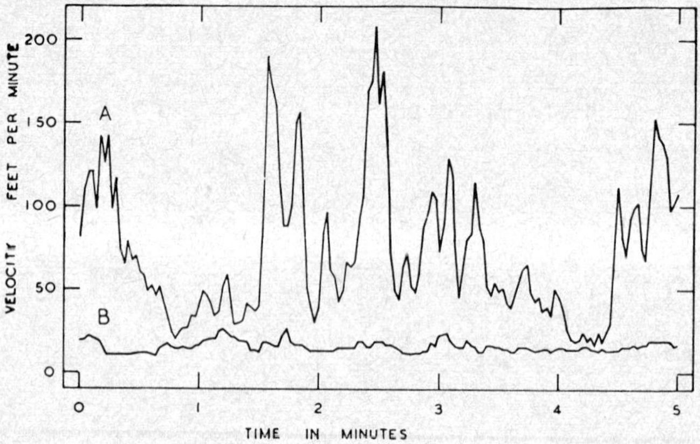

Fig. 34. Variability of air velocity.

movement, with constant temperature and average velocity, was insignificant. If variability of air movement had any influence in winter it was of far less importance than in summer. We have already noted that the mean air velocity had less influence in winter than in summer, while, on the other hand, a given temperature change had about 40 per cent more effect on freshness impressions in winter than in summer.

TEMPERATURE VARIATIONS

Synchronously with the observations on variability of air movement, other observations were made with a delicate thermopile so constructed as to measure rapid fluctuations in air temperature. The temperature changes were expressed as the mean variation about the average temperature, in °F. The temperature variations, which were measured only during the winter observations, were but slight. In 73 per cent of the observations the mean variation of temperature was less than 0.4°F and in only 4 per cent was it 1°F or more. When air velocity and mean air temperature were held constant the partial correlation coefficient for freshness and temperature variations was insignificant. The slight temperature variations which did occur had no appreciable influence on sensations.

THE INFLUENCE OF HUMIDITY

We have seen that at the room temperatures commonly encountered in Great Britain during the winter months the effect of changes of humidity on feelings of warmth is small; but there is evidence that changes in humidity which are insufficient appreciably to affect sensations of warmth may yet have distinct effect on comfort. At a temperature of 70°F an uncomfortable feeling of wetness is experienced at relative humidities of 70 per cent or over (Miura, 1931). Out-of-doors, when allowance is made for the effects of sunshine, barometric change, and the ion-content of the atmosphere, days of low humidity are significantly pleasanter than humid days (Winslow and Herrington, 1935). Markham (1937) expressed the opinion that mental alertness was associated with the lower humidity ranges, and lethargy with the higher. He suggested that for mental activity a relative humidity of under 55 per cent was ideal, while for rest and recuperation a higher humidity, possibly in the region of 70 per cent, was desirable.

Bedford and Warner found a significant correlation between freshness and humidity when temperature and air velocity were kept constant. From the regression constants it appears that under the conditions covered by the observations, a rise of 8.5 per cent in the relative humidity could be compensated for by a fall of 1°F in the air temperature. This effect of humidity is roughly three times as great as would be expected from the results of studies of the influence of environmental warmth on general thermal sensations. Thus a change of humidity which would be insignificant from the standpoint of warmth may have a distinct effect on one's impression of freshness.

FRESHNESS DUE TO LOCAL STIMULATION

Impressions of freshness or of stuffiness, it has already been remarked, depend on the stimulation or lack of stimulation of areas of skin. A sudden increase in the velocity of the air, say from 20 to 300 or 400 ft/min, causes an increase in the pressure on exposed areas which is sufficient to stimulate the tactile endings in the skin and evoke touch sensations. Even if the speed of a rapid air current is practically constant, any variation in the direction of the wind, or any slight bodily movement, will cause the current to impinge on fresh areas of skin and thus arouse tactile sensations. Air currents moving at velocities which are much too low to stimulate the tactile receptors in a sedentary person may excite touch sensations in persons who are in motion, for the air velocity which matters from this standpoint is the velocity relative to the person, and not the absolute velocity. A room is less likely to appear stuffy to a person who is moving about than to one who is sitting still.

The air velocities encountered by Vernon, Bedford and Warner during their observations were not usually high enough to evoke sensations recognizable as those of touch. The freshness impressions were almost certainly due to stimulation of the thermal receptors. The changes in the average temperature, and in the average air velocity, which were found to correspond with a change of one unit in the freshness scale were such as would be expected, if they occurred suddenly, to alter the skin temperature at a rate high enough to excite a thermal sensation. Changing direction of air currents and slight body movements expose fresh areas of skin and thus ensure the sudden changes necessary for stimulation, even when the air velocity is fairly constant. Any variability in the velocity of the air currents causes further stimulation, and it is worthy of note that this variability, even when it is expressed as a percentage of the mean velocity, tends to increase as the velocity increases. With constant air velocity and average air temperature, variability of air temperature had no significant influence on impressions of freshness. This is what would be expected from the work of Hardy and Oppel, for the mean variation of air temperature was small and insufficient to cause shifts in the skin temperature such as would stimulate the thermal receptors.

RADIANT HEAT

Although in the observations of Vernon, Bedford and Warner attention was paid only to the temperature, humidity and rate of movement of the air, there are other variables which exert their influence on comfort and which tend to evoke impressions of stuffiness or of freshness.

Intensity of Radiation

Commonly the mean radiant temperature and the air temperature are very similar, but at times they vary considerably. In thin-walled structures in severe winter weather the mean radiant temperature may be several degrees below air temperature, while in hot weather, or with certain methods of heating in winter, the air may be distinctly cooler than the surrounding surfaces.

Different combinations of air temperature and mean radiant temperature may yield the same overall degree of warmth, say the same equivalent temperature or the same operative temperature; for instance, cold walls can be compensated for by an increased air temperature. It does not follow, however, that all combinations of air temperature and mean radiant temperature which give the same equivalent temperature are equally comfortable. In the open we find the combination of cool air and the warmth of the sun's rays most pleasurable; and indoors, too, our sensations are influenced by the balance between the air temperature and the amount of radiant heat. As long ago as 1857 the Commissioners of the General Board of Health advocated as one of the requirements for comfort that the walls of a room be at least as high in temperature as the general temperature of the room, while they included cold walls or floors amongst the conditions which make for discomfort. Although they had no statistical evidence to offer on this point, Bedford and Warner (1939) held the same view. In field investigations they had experienced feelings of stuffiness at comfortable equivalent temperatures when they were unable to find any satisfactory explanation other than the fact that the mean radiant temperature was 6°F or more below the air temperature. Yaglou (1937*b*), also states that a combination of warm walls and cool air is unquestionably more comfortable and pleasant than one of cold walls and warm air.

Winslow and his colleagues (1938*c*) found that when their subjects were exposed to environments (*a*) with air and walls at approximately the same temperature, and (*b*) cooler air and warmer walls, different sensations were experienced when the same operative temperatures were maintained. In hot environments discomfort was less, and in cool ones greater, when the walls were warmer than the air. That is to say, with the same operative temperatures the subjects felt cooler when the walls were warmer than the air than when air and walls were at the same temperature. It was thought probable that local or temporary stimulation by the cold air might have produced the observed effects. There may have been local chilling of the respiratory membranes or of the skin.

Munro and Chrenko (1949) studied the effects of radiation on the reactions of a group of 145 persons, each of whom was exposed three times to a given equivalent temperature (60°, 65° or 70°F), once with the boundary surfaces of the room at air temperature, once with the surface temperature 5°F below air temperature, and once with it 3°F above air temperature.

Of those who were exposed to the comfortable equivalent temperatures of 65° and 70°F, 84 per cent found the warm-wall situation more pleasant than the cold-wall condition. The warm-wall condition felt distinctly fresher than either of the others, and the cold-wall condition felt rather less fresh than that in which the walls were at air temperature.

In the fresher conditions, in which the walls were warmer than the air, the air temperature was lower than in the comparable experiments with cold walls, and that lower air temperature was associated with a lower vapour pressure. When the results of these experiments were subjected to statistical analysis, the following equation was obtained:

$$F = 5.45 - 0.216f + 0.047(t_w - t_a)$$

in which F represents the number assigned to the impression of freshness on Bedford and Warner's scale, f is the pressure of aqueous vapour expressed in mmHg and t_w and t_a are respectively the mean radiant and air temperatures expressed in °F. This equation applies to sitting subjects who were thermally comfortable and as all the experiments were conducted in calm air the effects of bodily warmth and air velocity need not be considered. The constants in the equation show that changes in atmospheric humidity have a much greater effect on impressions of freshness than the differential between the temperatures of the walls and the air. A change in humidity of 5 mmHg will exert a definite effect on freshness–about one unit on the arbitary scale–but will have no significant effect on sensations of bodily warmth.

The pleasant effects of the 'warm-wall' environment were not so much due to the difference between the wall temperature and the air temperature–the equation shows that this factor is insignificant–but rather due to the low air temperature together with the low humidity. The effect of the latter factor is in all probability due to intra-nasal evaporative cooling. This depends upon a vapour pressure gradient between the moist surface of the nasal mucous membranes and the vapour pressure of the inspired air. It has nothing to do with relative humidity.

The experiments, in fact, confirm the findings of Bedford and Warner, with respect to the effects of humidity.

The explanation of Bedford and Warner's impressions of stuffiness at comfortable equivalent temperatures in the field in which the mean radiant temperature was some 6°F lower than the air temperature is not the low radiant temperature but the high air temperature. Suppose that the comfortable equivalent temperature was 70° F, then this would be approximately equal to the arithmetic mean of the air temperature and the mean radiant temperature. Thus

$$70 = t_a + t_w/2 = 73 + 67/2$$

and a temperature of 73° F of the air at head level might well have evoked impressions of stuffiness.

Quality of Radiation

We have seen that radiation of the shorter wavelengths can penetrate the skin to some extent, whereas there is no appreciable penetration when the wavelength is 3 microns or more. To produce the same heating effect on the skin surface, and to evoke a sensation of warmth, a lower intensity of radiation is necessary with the non-penetrating wavelengths than with the shorter radiations. When one sits before a fire on a cold day the radiant heat from the fire compensates for the usually cool atmosphere. The radiation from the fire includes visible and penetrating infra-red rays as well as non-penetrating infra-red, and it may be on that account that under such conditions one prefers an equivalent temperature rather higher than would be necessary for comfort if the air temperature were higher and the radiation less intense. Such an effect of penetrating radiation may also account partly for the observation of Winslow and his colleagues noted in the preceding paragraph, for although reference is made to conditions of cool air and warm walls, the sources of radiation were three 2.5 kW electric heaters, the radiation from these being reflected from the copper walls of the experimental booth.

By reflex action the warming of an area of skin can cause vasodilation in the mucous membranes of the nose, and cooling of the skin brings about constriction of the vessels in these membranes (Mudd, Goldman and Grant, 1921). Some years ago, Sir Leonard Hill (1931-34) found that irradiation of the skin by long (non-penetrating) infra-red rays could cause such vasodilation of the mucous membranes that the nasal airways were appreciably narrowed, so that in a suitable subject breathing through the nose became difficult, and an unpleasant feeling of stuffiness resulted. He found, also, that irradiation with short infra-red rays produced a reverse effect, so that one's nose could become blocked by exposure to one type of radiation, and the stuffiness could be relieved by subsequent exposure to another type. Nose-closing radiation was said to be emitted from dark or dull-red sources, such as some electric heaters, while brighter sources such as electric lamps, coal fires, and modern gas fires were 'nose-openers'. Bright sunlight, it was averred, did not cause any nasal congestion. These effects were by no means rare, for Hill observed them in 50 per cent of the subjects tested by him. The nose-closing effect was produced up to a distance of 40 ft from a 300-W gas-filled incandescent lamp with a screen of cellophane interposed. A screen of desquamated horny layer of skin, 2 in^2, set in a sheet of cardboard, kept the affected side of the nose shut when held in front of the face up to a distance of 27 ft away from the lamp. Hill remarked that the sensitivity of the skin to these rays is very great. The intensity of the radiant heat from the lamp is much less than that due to the presence of a clothed man 10 ft away, or to an increase of 0.1° F in the temperature of the walls of the room. This change in the temperature of the walls is only about one-twentieth of that which Hardy and Oppel found necessary to evoke a sensation of warmth.

The attribution of such divergent properties to the kinds of radiation employed in warming buildings was a matter of no small importance, and investigations were undertaken, in Great Britain by Dufton and Bedford (1933), and in the United States by Winslow, Greenburg and Herrington (1934).

Experiments made by Dufton and Bedford on a group of thirty blinded ex-soldiers yielded no evidence that differences in the quality of radiation exerted any effect. Further experiments were carried out on other subjects, and nose-closing was found to occur not only with exposure to a dull electric fire, but also when the subjects were exposed to brighter radiation as from the clinical gas lamp which had been described as a nose-opening source. Exposure to unobstructed sunlight of a subject who was stripped to the waist also caused nose-closing. It appeared that any type of radiant heat could be made to produce this nasal congestion. Convective heating could also produce the effect. When subjects were stripped to the waist and exposed to a rapid increase in air temperature the inspiratory and expiratory pressures rose, and breathing became difficult. Direct conduction also evoked the response. The nasal airway became blocked when a hot fomentation was applied to the arm. Any kind of sudden warming of the skin could bring about nose-closing. The congestion could be relieved by sudden chilling of some area of skin. The nose could be obstructed by irradiation of the face or by the application of a fomentation to the arm, and instant relief could be given by sponging the leg, or the other arm, with cold water. No relief could be obtained by switching on an electric lamp.

Very similar results were obtained by Winslow and his

colleagues. They produced constriction of the nasal airways by exposure to direct radiation, by immersion of the arm in hot water, and by exposure to high air temperatures. Both groups of workers attributed the nose-closing reaction to a warming of the skin, and scouted the idea that it was due to qualitative effects related to specific wavelengths. It is true that on some occasions Winslow and his associates found that switching on a light was followed by a reduction in the nasal congestion, but it was later found that the clicking of a dummy switch also produced similar results. Hence it appeared that the apparent light-antagonism was due to a psychological rather than a physiological reaction.

Since these refutations were published, Hill (1935, 1939) has pursued his claims, and van Dishoeck (1935) and Lehmann (1939) also claim to have demonstrated the nose-opening effects of short infra-red radiation.

Whatever may be the ultimate truth about these alleged nose-opening rays, it does not appear that the nose-closing effects which have been demonstrated are of any great practical significance. Sudden warming is necessary to produce them, and normally both nostrils are available for breathing. It is only in especially sensitive persons, such as those with deflected septa, that any difficulty is likely to arise.

In experiments in which people were exposed to radiant heat and their impressions of freshness or of stuffiness noted, Yarnold and Fitzpatrick (1940) found that these impressions varied with the wavelength of maximum radiant energy. Stuffiness was associated with wavelengths of about 3, 4.1 and 4.8 microns, with regions of freshness in between. In 1947, Yarnold published a paper giving further details of his experiments and he stated that 'providing a subject is comfortable as regards warmth, the feelings of stuffiness are caused mainly by excessive air temperature, which, in the absence of radiation of suitable wavelength, should not exceed 61°F. Those wavelengths which are not absorbed by the skin, *i.e.* 2, 3.5 and 4.5 microns, induce sensations of freshness, while those which are absorbed, 3, 4.1 and 4.7 microns, cause feelings of stuffiness.' But these interesting findings have yet to be confirmed.

The various results obtained by different investigators could be explained by considering some of the possible physiological mechanisms involved when the skin is stimulated by thermal radiation. Under normal circumstances, the temperature of the surface of the skin is lower than the temperature of the deeper tissues and this gradient of temperature may reflexly cause vasoconstriction of the nasal blood vessels which tends to widen the nasal airways. When the gradient of temperature is reduced by heating the outer surface of the skin, vasoconstriction is reduced and is followed by vasodilatation which tends to close the airways. This effect would be produced most readily by radiation of long wavelengths which are mostly absorbed at the outer layers of the skin. If, in this situation, short-wave radiation impinges upon and penetrates the skin, heating will take place below the surface and the original gradient of temperature would be re-established, reflex vasoconstriction occurs and the nasal airways are widened. If the surface of the skin is already cool and there is maximum vasoconstriction, no amount of radiation either with long-wave or short-wave will cause nose-opening; the only effect will be heating of the skin and nasal congestion. This interpretation appears to be supported by the experiments of Lehmann (1939).

To illustrate the uncomfortable conditions and respiratory congestion produced by the combination of a warm head and cold feet Hill (1919) gave the following example. A small greenhouse in the garden on the east coast of England was used for sedentary work and on a fine sunny September day with 'white floating clouds and a strong W. breeze' conditions were, with windows and door open, agreeable so long as the sun was clouded. But discomfort in the form of unpleasant heat of the head, stuffiness and headache occurred when the sun was out, in spite of the breeze blowing in. The radiant heat from the sun heated up the head and the feet became relatively cold. The nasal air-ways became partially obstructed and sneezing resulted. But 'on fetching a hot-water bottle and rug and making the feet cosy and warm, comfort returned and the nose became fully open. With the feet very warm and the breeze blowing in through the door, the sun's heat on the head no longer produced stuffy headachy feelings.'

Unfortunately, Hill did not give any figures for the intensity of thermal radiation in his greenhouse, but at head-level in sunlit offices the present writer has measured mean radiant temperatures of over 100° F (38° C) which produced decidedly uncomfortable conditions. However, his experience led Hill to state, not for the first time, that the ideal method of warming and ventilating rooms would give radiant heat, a warm floor and agreeable movement of cool air—the conditions, as he said, of a sunny spring day out of doors.

So far as the environmental engineer is concerned with designing indoor environments these criteria of Hill should be interpreted with a certain amount of caution, even though as Hill wrote, 'the skin can stand, without being scorched, twice as much radiant energy per unit of time in the form of visible sun rays as it can of dark heat rays, because the visible rays largely penetrate, and, warming the blood locally, are carried away as heat to cooler parts of the body'. One can predict, admittedly by extrapolation, from data given by Chrenko (1964) that the threshold intensity of solar radiation evoking a sensations of warmth on the forehead is very much higher than the threshold for infra-red radiation. For example, if the temperature of the skin of the forehead is kept constant,

$$Y = 10.723/L_{max} + 4.27$$

where Y is the threshold expressed in mW/cm^2 and L_{max} is the wavelength of maximum intensity of the thermal radiation which may be taken for sunlight as about 0.5 microns; the threshold is thus about 26 mW/cm^{-2} which is more than twice the threshold for infra-red radiation, for the average person.

One must also remember that a considerable amount of solar radiation is reflected by the living human skin whereas virtually all infra-red radiation is absorbed at the surface of the skin. It follows from this that, *for a given intensity of thermal radiation,* infra-red radiation will raise the surface temperature of the skin of the head to a higher level than radiation with a distribution of energy amongst the shorter and visible wavelengths, *e.g.* solar radiation, and since heating of the head is associated with feelings of stuffiness such feelings are more likely to occur with 'dark' radiation.

The spectral reflectance of human skin with respect to thermal radiation is given in the following table:

Wavelength of radiation expressed in microns		*Percentage reflectance of living human skin*	*Solar radiation energy**
0.25		7	
0.30	Ultraviolet	6	5
0.35		20	
0.40		23	18
0.45		32	29
0.50		38	31
0.55	Visible	38	27
0.60		56	25
0.65		65	22
0.70		68	20
0.80		60	16
1.0		50	10
1.2		35	7
1.4		12	4
1.6	Infra-red	18	3
1.8		13	2
2.0		8	2
2.2		7	
2.4		5	
2.6		5	

Data from Jacquez *et al.* (1955) and Dimitroff *et al.* (1956).
* Data for solar radiation from Moon (1940): units in kcal/m^2/min/μm.

The skin of hairless animals behaves similarly to human skin so far as reflectance to thermal radiation is concerned. The wavelength of maximum energy for solar radiation is about 0.5 micron and the maximum reflectance of human skin appears to occur between the wavelengths of 0.65 to 0.80 microns where nearly 70 per cent of the incident energy impinging on the skin is reflected from it. These physical measurements thus confirm practical experience and previous experiment discussed earlier.

The calculation of the actual amount of solar radiation absorbed by the human body is very complicated but it has been shown that, for a clear sky, the average intensity of solar radiation incident on a man is about 230 kcal/m^2/hour which represents a heat input roughly five times the resting heat production by metabolism.

ATMOSPHERIC ELECTRICITY

A hundred years ago writers on ventilation discussed the possible bearing of atmospheric electricity on bodily comfort, but those discussions could only be in the nature of speculation, since it is only in comparatively recent years that exact observation of the electrical state of the atmosphere has been possible.

Gases contain particles–atoms or molecules, and groups of molecules–which carry positive or negative electrical charges. The velocity of these particles in an electrical field varies with the intensity of that field and with the size and charge of the charged particles, or ions. In air, the small ions have a mobility of 1 to 2 cm/sec/V/cm, whereas the large ions in a field of unit intensity have a mobility of only 0.01 to 0.0005 cm/sec. The large ions are formed by the aggregation of small ions with condensation nuclei, which are present in great numbers in the air of urban localities. Ions of intermediate size, with a mobility of 0.1 to 0.01 cm/sec in unit field, also occur.

Gagge and Moriyama (1935) made counts of the ions present in the external air at New Haven, Connecticut. There was a rise in the number of positive small ions during the summer months, while the negative ions, usually fewer in number than the positive, declined in number during the summer. The heavy ions were twice as numerous in the winter as in summer, and the intermediate ions increased slightly in winter. Heavy negative ions averaged about 10,000/ml and heavy positive ions 8,000 to 9,000/ml in the winter months, while heavy ions of both signs each declined to about 4,000/ml in the summer. The intermediate positive ions numbered 2,000 to 4,000/ml during the course of the year, and the negative ions of intermediate size were somewhat smaller in number. The light ions of either sign numbered about 100/ml, with the positive ions rising to over 200/ml in the summer.

Yaglou, Benjamin and Choate (1932) showed that in unoccupied rooms the ionic content of the air is much the same as out-of-doors, but in occupied rooms there is a striking decrease in both positive and negative ions. They counted the small ions with mobilities from the maximum possible down to 0.2 cm/sec. Figure 35 shows the effects of occupation on the ionic content of the room, as found by Yaglou and his associates. The counts were taken in a small unventilated chamber of 550 ft^3 capacity in which there were seven persons. There was a sharp drop in the number of ions of both signs directly after the occupants entered the room, and the low ion counts continued until the room was vacated.

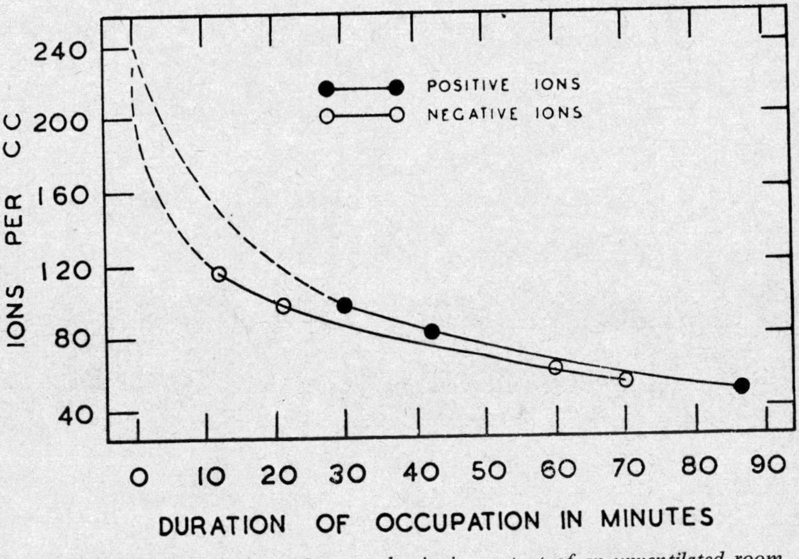

Fig. 35. Influence of occupation on the ionic content of an unventilated room. (Yaglou et al.*)*

The minimum outdoor air supply required to maintain the normal ion content of the air in a crowded room was about 160 ft^3 per person per min. With the more usual air supply of 30 ft^3 per min the number of ions was not appreciably greater than when the room was unventilated. Owing to diffusion and adsorption by metal conductors, mechanical ventilation reduced the ion content of the air by amounts varying up to 30 per cent.

Heating the air by means of a central fan system increased the ion content, and cooling it by a similar system decreased the ion count. The use of water sprays for washing, humidifying, and dehumidifying the air produced many large negative ions and deprived the air of all small ions.

This interest in ionization in relation to ventilation was aroused by the findings of Dessauer (1931) and his associates, who described the effects of artificially ionized air on normal and abnormal subjects. It was claimed that ions of opposite sign exerted opposite physiological and subjective effects. Positive ions produced a rise in metabolic rate, in the pulse rate, and in the blood pressure, and caused unpleasant symptoms such as headache, nausea and general malaise. On the other hand, negative ions produced a feeling of well-being, and caused a decrease in the physiological reactions which were increased by positive ions. Patients with high blood-pressure gained benefit from exposure to negative ions. Following this, Küster and Janitzky (1932) found a fall in the metabolic rate when the air contained an excess of negative ions, and a rise with an excess of positive ions.

American research workers devoted much energy to checking these German claims. Yaglou, Brandt and Benjamin (1933) exposed a group of subjects to air which was artificially ionized. Large numbers of small ions were added to the air so that the ion count was 5,000 to 1,500,000/ml, whereas in normal air counts of 50 to 800 ions were obtained. Some of the subjects in these experiments showed a strong liking for negative ionization but all of this group were persons with high blood-pressure, rapid pulse, or high metabolic rate. Their feeling of relaxation was generally accompanied by a fall in blood pressure, pulse, or metabolism. There seemed to be a tendency for the air to feel fresher when it had a high ion count, especially with negative ionization, but the effect was not enough for it to be said that there was a definite improvement. Yaglou and his colleagues concluded that: 'Under the conditions of the present experiments nothing definite was found to justify the use of artificial ionization in general ventilation'. In other experiments, Brandt (1933) could not find that de-ionized air had any effect.

Herrington (1935) also studied the effect of ionized air on normal subjects. Subjective sensations revealed no effects specifically related to the administration of ionized air. He concurred in Yaglou's opinion that there was no justification for the use of ionization in general ventilation, but he made the reservation that the observations did not close the question as to the cumulative effects of ionization of the air on health and comfort. There still remained, also, the problem of the reaction to ionized air of subjects suffering from hypertension. Inconclusive results were obtained by Yaglou (1935) in a study of the effects of high concentrations of small negative ions on cardiovascular efficiency.

Later work by Herrington and Kuh (1938) failed to show any significant effect of high concentrations of heavy ions on patients with hypertension. Blood pressure, pulse rate and metabolism were unaffected by exposure to either positive or negative ions. In some cases patients were under treatment for two years, for most of which time only negative ions were used, but no subject showed clinical improvement.

Winslow and Herrington (1935) examined the relation between subjective impressions of pleasantness and the ionic content of outdoor air. Those days on which the total ion content was high were judged to be unpleasant, irrespective of sunshine and relative humidity. Since the ions in city air are presumably largely made up of vapour particles from heating plants, automobile exhausts and the like, it was believed that the observed correlation was most probably due to the effects of such substances, and perhaps indicated a subtle influence of odour.

From the available evidence it appeared that the differences which normally occur in the ion content of the air of occupied rooms do not exert any significant influence on subjective sensations.

In more recent years there has been renewed interest in the possible medical and physiological effects of atmospheric ions. In some instances there appears to have been a lack of controlled experimentation, and some of the claims that have been made do not appear to be well founded.

In experiments on isolated strips of the trachea from anaesthetized animals it has been found that positive ions slow or stop ciliary movement, and the mucosa dries with the muscle contracting. On the other hand, negative ions speed up ciliary movement, relax the muscle and increase secretion; they counteract the effect of positive ions (Krueger and Smith, 1957). It also appears that hay-fever patients obtain relief from symptoms while they are exposed to high concentrations of negative ions, whereas positive ions give no relief (Kornblueh, Piersol and Speicher, 1958).

These findings again raised the question whether, in spite of the earlier American investigations, ions might exert some effect on comfort that would warrant the use of ionizers in ventilating installations. With the aim of answering that question a study was carried out by the American Society of Heating, Refrigerating and Air-conditioning Engineers (Humphreys and Jennings, 1962).

Tests were made under three ionic conditions: (*a*) background, when the ionizers were not operating, (*b*) positive ionization, and (*c*) negative ionization. Two of these conditions were used in each 3-hour test, and all possible permutations were used. Tests were made at the comfortable temperature of 78°F, and at the higher temperature of 87°F. At both temperatures the relative humidity was controlled at 50 per cent. The ion counter determined the concentration of ions having a mobility of 0.05 cm/sec/V/cm or greater. With positive ionization the average counts were about 12,000 and 15,000 ions/ml in the tests at the two temperatures. With negative ionization the mean counts were about 4,800 and 5,500 ions/ml. In the background condition the average counts were about 200 to 400 positive, and 100 to 200 negative, ions/ml.

At intervals during the tests the subjects recorded seven subjective impressions on predetermined scales: (1) the general feeling of warmth, (2) the feeling of dampness or wetness due to perspiration, (3) the pleasantness or otherwise of the environment, (4) the feeling of fatigue, (5) the impression of the dryness or humidity of the air, (6) the impression of movement or of stagnation of the air, and (7) the personal feeling or mood. Pulse rates and oral temperatures were also measured.

The results obtained were subjected to rigorous statistical analysis. There was no clear-cut evidence that the ionic state of the atmosphere had any effect on sensations. It was concluded that, at the temperatures and ion concentrations that prevailed in the experiments, if there is any effect it is likely to be too small to be of engineering significance. If there is any effect on thermal sensations the results suggest that it will be in the direction of making one feel cooler with both polarities.

The same experimental results have been used as the basis for another paper (Koch, 1962). Koch found that in the tests made at 87° F five of the seven subjective responses were influenced to a statistically significant degree when the air was enriched with either positive or negative ions. On the whole, the subjects felt somewhat less warm, perspired less, found the air pleasanter, were in a better mood, and felt less fatigued than when they were exposed to untreated air. Yet, analysis of the responses of the 15 individual subjects showed that eight had no statistically significant preference, one felt best without ionization, four preferred ionized air (positive or negative) to the background condition, and two preferred positively ionized air to the background condition. The apparent effect on the thermal sensation was such as might have been caused by lowering the air temperature by about 1° F in these tests at 87° F. At the more common temperature of 78° F, Koch found no significant change with change in the state of ionization. Hence the findings seem to support the conclusion that if there is any effect it is too small to be of engineering significance.

According to Krueger (1962) the physical characteristics of air ions might lead one to suppose that they had no biological significance, for, as he says, an air ion is only a group of from four to twelve gaseous molecules randomly arranged round an electrically charged gaseous atom or molecule. The air ion moves slowly amongst an enormous number of non-ionized molecules: for every ion there are about 10^{12} non-ionized molecules. Nevertheless experiments showed that physiological changes could be induced by gaseous ions.

In his earlier experiments on excised strips of tracheal mucosa and on exposed tracheas of anaesthetized animals, Krueger noted that all the gaseous components of the atmosphere were subjected to ionization and it was therefore impossible to determine which ion produced the observed effect. Subsequent experiments showed that negatively charged oxygen (O^-) and positively charged carbon dioxide (CO_2^+) mediate the physiological effects on the trachea.

It is suggested that the underlying mechanism is that gaseous ions may react with cell fluid to form active radicals which evoke cellular responses. The effect of the essential positive air ion (CO_2^+) is to release a powerful neurohormone 5-hydroxytryptamine (5-HT) which causes all the effects on the trachea which follow exposure to positive air ions. The counter-action of negative air ions depends on the (O^-) radical which accelerates the destruction of 5-HT.

CHAPTER IX

THE NEED FOR VENTILATION

THEORIES OF VENTILATION

THE importance of proper ventilation was emphasized by D. B. Reid (1844) who, a century ago, wrote: 'Mental anxiety may, perhaps, be considered the most powerful enemy to the duration of human life, and, next to it, defective nutriment, whether in quantity or quality. But after these, no other cause, at least in modern times, appears to have inflicted so great an amount of evil upon the human race as defective Ventilation. . . .'

Ventilation was defined by Reid as connoting the due appropriate supply of air to any apartment. He was mainly concerned with the quantities of fresh air that were necessary and with the methods by which that air could be supplied; but he had much to say about the chemical and physical properties of the air. He emphasized the need for maintaining adequate warmth, and made pertinent observations concerning the importance of taking air movement into account. Many of the principles clearly enunciated by Reid have been verified or rediscovered by experimenters during the past thirty years or so.

Various theories have been advanced to account for the harmful effects of bad ventilation. Over two hundred years ago, Gauger (1714) remarked that it was not the warmth of a room but its inequality of temperature and want of ventilation that caused numerous maladies. Bad air was held responsible for the spread of disease, and for the unpleasant sensations that are experienced in ill-ventilated rooms. Modern bacteriological investigations have shown that some infections are airborne and that their spread may be influenced by ventilation, but the composition of the air of such rooms is no longer regarded as the main cause of the headaches and malaise which occur.

Crowded rooms tend to be overheated, and during the eighteenth century it was apparently considered that the discomfort in such rooms was due to excessive warmth. But towards the close of that century, Lavoisier, who had examined the composition of the air of occupied rooms, 'did not hesitate to attribute to carbonic acid the malaise often experienced in crowded assembly halls, the malaise generally attributed to warmth alone' (Leblanc, 1842). Thereafter, for more than a hundred years, excess of carbon dioxide was popularly held accountable for that discomfort.

This belief persisted, although as early as 1862 Pettenkofer had written that the unpleasant symptoms were not due merely to warmth or humidity, or to carbon dioxide excess or oxygen deficiency, but rather to the presence of very small quantities of organic material exhaled from the skin and lungs. In Pettenkofer's view carbon dioxide was not important of itself, but only as an index of the amount of those other noxious substances.

During the next twenty years or so various writers claimed to have shown the presence of toxic organic substances in expired air, and then, in the years 1887 to 1889, Brown-Séquard and d'Arsonval reported animal experiments which seemed to carry conviction. Water condensed from expired air was injected into animals, which quickly died. In other experiments they linked together in series vessels containing animals, the air supply being drawn from vessel to vessel so that the animal in the last vessel breathed the air which had passed through the preceding vessel. The animals in the last two vessels died, while those in the first two remained healthy. In later years these experiments were repeated by many investigators, in each case with negative results. Exhaustive tests failed to corroborate the claims which had been made. Especially convincing work by Haldane and Smith (1892-93), by Billing, Mitchell and Bergey (1895), and by Hill and his colleagues (1913) exploded the anthropotoxin theory, and explained the faulty experimental procedures which had led Brown-Séquard and d'Arsonval to erroneous conclusions.

While the supporters and opponents of the anthropotoxin theory waged their controversy evidence was accumulating which confirmed that even relatively high concentrations of carbon dioxide caused no discomfort, and indicated that excessive warmth was the chief source of trouble. Hermans, in 1883, attributed the effects of bad ventilation to heat and high humidity, and Flügge, twenty years later, made confirmatory experiments.

Early in the present century two Home Office Committees, the Departmental Committee on the Ventilation of Factories, in its Second Report, and the Departmental Committee on Humidity and Ventilation in Cotton Weaving Sheds, accepted the view that overheating was the chief problem raised by bad ventilation, and later experiments by Sir Leonard Hill and his colleagues gave the final proof.

For these experiments a small airtight chamber of about 3 cubic metres capacity was constructed, with provision for heating and cooling, and with electric fans for creating air movement. Seven or eight students were shut in the chamber until the carbon dioxide reached 3 to 4 per cent and the oxygen had fallen to 16 or 17 per cent. The wet-bulb temperature rose meanwhile to 82° or 85° F, with the dry-bulb temperature a degree or two higher. 'The students went in chatting and laughing, but by and by, as the temperature rose, they ceased to talk, and their faces became flushed and moist. Some tried to light a cigarette (to relieve the monotony of the experiment), and were puzzled by their matches going out. They had not sensed the diminution in oxygen, which fell below 17 per cent. Their breathing was slightly deepened by the high percentage of CO_2, but no headache occurred in any of them from the short exposure to from 3 to 4 per cent CO_2. Their discomfort was relieved to an astonishing extent by putting on the electric fans placed in the roof. Whilst the air was kept stirred the students were much less affected by the oppressive atmosphere. They begged for the fans to be put on when they were cut off. . . . The air of the chamber breathed by other students outside caused no discomfort, *whereas a student inside breathing the outside air felt no relief*' (Hill, 1914).

These experiments gave renewed impetus to the study of

ventilation. They showed that the speed of movement of the air, as well as its temperature and humidity, exerts a profound effect on comfort. Although the purity of the air of occupied rooms is of no small importance, the prime consideration from the standpoint of comfort is that the physical environment shall be suitable.

Thus, ventilation requirements are linked with considerations of heating or of cooling. The amount of fresh air that can be endured indoors during cold weather is dependent on the adequacy of the heating installation. If inadequate provision is made for warming a building ventilation will be restricted during cold weather. As Reid remarked: 'Ventilation need not be expected where food, fuel and clothing are deficient. Heat is still more important to the human frame than fresh air'. Man can tolerate an impure atmosphere for long periods, but prolonged exposure to extremes of temperature is likely to lead to acute ill-effects. *Heating and ventilation are, indeed, indissolubly connected, and they should always be considered together.*

COMPOSITION OF NORMAL AIR

Air is a mixture of gases. It consists mainly of nitrogen and oxygen, with a small quantity of carbon dioxide and traces of various inert gases. Ordinarily, it also contains a variable quantity of water vapour.

The composition of outdoor air is remarkably constant, except for its water content. Carpenter (1937) reports on a series of more than 1,100 analyses of the outside air in New Hampshire, Boston (Mass.) and Baltimore. In each place the average concentration of carbon dioxide was 0.03 per cent by volume, and that of oxygen 20.94 per cent. In a single observation the carbon dioxide percentage was 0.060, but in all others the values fell between 0.026 and 0.035 per cent. During the period of investigation the average concentrations did not vary from month to month, or from year to year. The observations in New Hampshire were made not far from a heavily wooded locality, while those in Boston were made only about 200 ft from a heating, lighting and power plant which consumed 35 to 95 tons of coal/day. Although the local conditions of the three observation stations varied so widely there was no evidence that season, or the proximity of large consumers of fuel, caused any measurable differences in the carbon dioxide or the oxygen content of atmospheric air. It can therefore be taken that the composition of normal dry air is:

Oxygen	20.94 per cent by volume
Carbon dioxide	0.03 per cent by volume
Nitrogen and other inert gases	79.03 per cent by volume

However, whereas the concentrations of oxygen and nitrogen vary very little there are variations in the concentration of carbon dioxide which takes part in the biological carbon cycles. Moreover, there have been proposals to consider the concentration of carbon dioxide as an index of combustion and natural ventilation in an urban environment. Therefore it is important to take note of variations in background concentrations of the gas in air. Even in places unaffected by vegetation and urban activities, the concentration of carbon dioxide varies from 303 parts/10^6 to 320 parts/10^6. Much larger diurnal variations have been found in areas where plants are abundant. But measurements made since the end of the last century show that the carbon dioxide content of the atmosphere has been steadily increasing from about 290 parts/10^6 in the 1890-99 decade to 320 parts/10^6 in 1969.

It is reasonable to suppose that this increase is due to burning coal and oil and as more fossil fuels are burnt, more carbon dioxide will be produced until a level is reached which will measurably increase the surface-temperature of the earth.

At present, it is not possible to determine whether such a rise in temperature, which is due to the 'greenhouse' effect of carbon dioxide, will be significant. More basic research on the relation between carbon dioxide content of air and transmission of thermal radiation through the atmosphere will be needed.

POLLUTED AIR

From 5th December to 9th December 1952 the greater part of Great Britain and especially London was covered by a fog accompanied by a temperature inversion. During this period and up to about 27th December there was a death rate of between 3500 to 4000 in excess of the average for the previous decade and post-mortem examinations revealed evidence of irritation of the respiratory tract. Measurements were made of the quantity of smoke suspended in the atmosphere and the concentration of sulphur dioxide and the maximum values were 4.46 mg/m^3 of smoke and 3.75 mg/m^3 or 1.34 $parts/10^6$ of sulphur dioxide. An episode of acute pollution of the atmosphere such as this focused attention upon the pollution of air as a health problem in Great Britain.

The first Clean Air Act became law in 1956 and there has been a steady reduction in the emission of smoke and sulphur dioxide in the air over Britain notwithstanding a 10 per cent increase in population and a 17 per cent increase in annual gross energy consumption. A striking example of the resulting benefit is that in Central London there has been an increase in winter sunshine of 50 per cent to a value almost the same as that at Kew. However, some 14 years were to elapse before a standing Royal Commission was appointed on 20th February 1970 'to advise on matters, both national and international, concerning the pollution of the environment; on the adequacy of research in this field; and the future possibilities of danger to the environment'. The Commission, under the Chairmanship of Sir Eric Ashby, published its first report in February 1971.

In the introduction to the Report a very useful distinction is drawn between the 'internal environment', that is, the environment in which individual men and women live and work, and the 'external environment'. The authors of the Report limited their considerations to the external environment because the internal environment is already the subject of 'a vast complex of legislation and public administration as well as the activities by employers, unions and others' which is being studied by a Committee set up by the Department of Employment under the chairmanship of Lord Robens to review the provision made for the safety and health of persons at work*. Yet the word

* Published in July 1972. HMSO Cmnd. 5034.

'environment' in the public mind is associated with the external environment. A similar impression might also be gained from a perusal of some official reports. For example, the annual Report of the Chief Medical Officer of the Ministry of Health for the year 1966 refers to 'environmental hygiene' as an activity confined to the 'external' environment. The Report records a range of activity and expertise which disposes of the view of some that environmental hygiene is a public health activity of limited interest and even of diminishing medical importance in an age of rapidly advancing medical science.

The Chief Medical Officer's Report stated that 'the clear association of pollution with bronchitis alone should make it imperative that authorities should do their utmost to reduce what is almost certainly one of the greatest environmental hazards in this country at the present time. It is unfortunate that many of the backward authorities are those with high rates of bronchitis.' This was written in 1966. The Royal Commission's Report states that the improvements due to the Clean Air Act 'are not uniformly spread over the country. There are regions in the north of England where there is urgent need for a more vigorous application of the Clean Air Acts.'

The C.M.O.'s Report also notes that many people fail to understand why so much importance is attached to the control of domestic smoke with apparent neglect of other forms of atmospheric pollution.

Nearly forty years ago, Vernon (1934) pointed out that domestic fireplaces alone discharged 2.5 million tons of soot into the atmosphere every year. Owens (1927) noted that the total weight of deposit in various London districts in 1927 varied from 187 to 475 tons/sq. mile whilst the deposit at Rochdale in 1922 was nearly 700 tons/sq. mile—ten times the quantity deposited at Malvern. About forty years later, measurements still indicated that the smoke and fumes from domestic fires were the principal offenders in the pollution of air in urban communities. According to the Asby Report, in 1953 little more than half the smoke contaminating the atmosphere of Britain came from domestic chimneys, a third came from industry and a sixth came from the railways. But the change from steam to diesel and electric locomotives practically eliminated smoke from the railways. By 1968, domestic chimneys were responsible for 80 per cent of the pollution due to smoke. The total smoke emitted in the United Kingdom in 1953 was 2.31 million metric tons; in 1968 it was 0.93 million tons and it is predicted that this will fall to 0.6 million tons in 1975.

The average concentration of smoke near ground level in the U.K. fell from about 175 $\mu g/m^3$ in 1958 to about 60 $\mu g/m^3$ in 1968. During the same decade there has also been a fall in the average concentration of sulphur dioxide near ground level from 175 $\mu g/m^3$ to just over 100 $\mu g/m^3$. Although smoke emitted by diesel engines may be very unpleasant and offensive there is no conclusive evidence that pollutants from road vehicles, *viz.*, carbon monoxide, hydrocarbons, oxides of nitrogen and lead are hazards to health *but sufficient is not known of the long-term effects of the relatively low concentrations at which these pollutants occur.* It is worth bearing in mind that pollutants together may exert a synergistic effect, that is, an effect greater than the sum of the effects of individual pollutants which also must include tobacco smoke. Reid (1964) concludes that 'although air pollution is certainly not the only cause, nor perhaps even the major initiating cause, it is almost certainly a promoting or aggravating factor of serious chronic lung disease'.

Other forms of pollution of the external environment are outside the scope of this book. But those interested should not fail to read the story of the use of toxic chemicals in the countryside and the widespread destruction of wild life in America as told by the late Rachel Carson in *Silent Spring.*

Great Britain has not escaped and in the spring of 1961 'tens of thousands of birds were found littering the countryside, dead or dying in agony'. A special committee of the House of Commons investigated the matter and the hearings convinced the committee of the alarming threat to wild life. The committee went on to recommend that 'the Minister of Agriculture and the Secretary of State for Scotland should secure the immediate prohibition for the use as seed dressings of compounds containing dieldrin, aldrin, or heptachlor, or chemicals of comparable toxicity'.

But these dangerous chemicals are manufactured in factories and workers have been exposed to the risk of poisoning for years, and as far back as 1948 under the Factories Act the Chief Inspector of Factories had to be informed of poisoning or disease in the following cases: lead, phosphorus, manganese, arsenic, mercury, carbon bisulphide, aniline, benzene poisoning, ulceration due to poisoning with chromium, cancerous ulceration of the skin, toxic anaemia, toxic jaundice, compressed air illness and anthrax. The list of prescribed industrial diseases included 20 types of poisoning, four types of infections, and numerous inflammations, ulcerations and new growths including cancer. The insecticides containing organic compounds of phosphorus are just as poisonous to man as they are to animals. The point is that today, exposure to dangerous chemicals is not confined to the occupational environment.

THE AIR OF OCCUPIED ROOMS

The atmosphere of occupied rooms is modified by the presence of the occupants, and as the result of their activities. The air exhaled from the lungs is richer in carbon dioxide, and poorer in oxygen, than that which is inspired, and it contains more water vapour. Organic substances are given off from the body and from the clothing, especially when personal hygiene is deficient, and these give rise to the characteristic body odour or crowd smell. Heat is emitted by the living body, and the ionic content of the air is reduced. Further, bacteria are discharged by the occupants, especially during coughing, sneezing and talking, and these may include the organisms which are responsible for respiratory diseases. With some forms of heating apparatus thc products of combustion may be liberated into the atmosphere of a room, and the activities of the occupants generally raise the dust content of the air well above that of the outside air. Many industrial processes cause various gases, vapours, fumes and dusts to be discharged into the air, and some of these are dangerous to health.

We may say that the primary object of the ventilation of a confined space is to provide a respirable atmosphere: enough fresh air must be supplied to maintain an adequate concentration of oxygen, and to remove the products of respiration at such a rate that no ill-effects arise from excess of carbon dioxide. But nowadays ventilation connotes something more than the introduction of the meagre quantity of fresh air which would suffice to maintain a respirable atmosphere. Satisfactory ventilation provides a healthy, refreshing and comfortable atmosphere.

Carbon Dioxide and Oxygen

Contrary to what was thought at one time, the changes in the concentration of carbon dioxide which are ordinarily encountered in the air of occupied rooms exert no injurious physiological effects. Figure 36 summarizes the results of experiments by Haldane and Priestley (1905) in which the subjects breathed

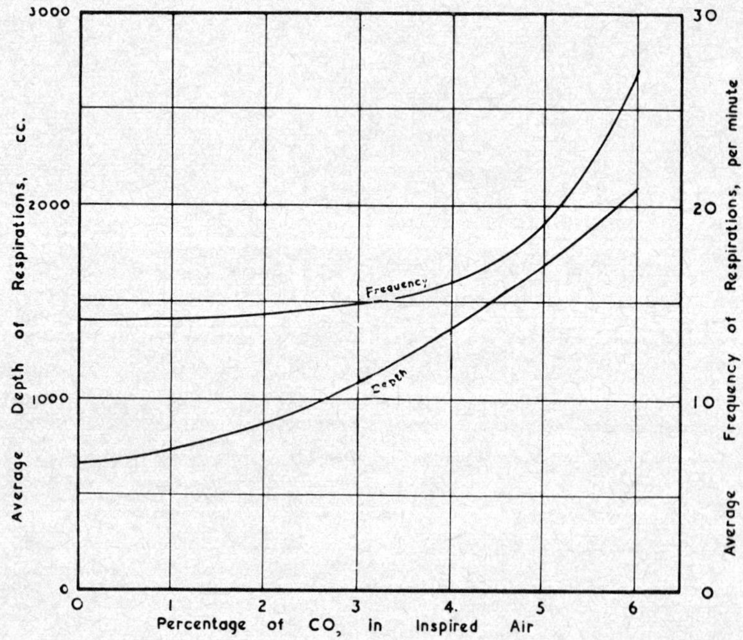

Fig. 36. Influence of carbon dioxide on depth and frequency of respiration.

normal air or air charged with varying percentages of carbon dioxide. A slight increase in the carbon dioxide concentration caused some increase in the depth of respiration, but with a concentration of 2 per cent the depth of respiration increased by only 30 per cent, while the number of respirations/min rose from 14 to 15. With 3 per cent carbon dioxide the depth of respiration was nearly doubled, but there was no further increase in frequency until still higher concentrations were inhaled. With concentrations of up to 3 per cent the carbon dioxide concentration in the alveolar air was still the same as when normal air was breathed, hence the carbon dioxide tension in the blood was maintained constant. The slight increase in respiration which accompanies an increase in carbon dioxide up to, at any rate, 2 per cent is highly unlikely to cause the slightest discomfort, and produces no ill effects.

It was found that the concentration of oxygen could be reduced to 13 per cent without changing the depth or rate of respiration, and that such diminution was not noticed by the subject. However, in practice the oxygen concentration should never be allowed to fall to that level. It should be maintained at not less than 18 per cent unless that is quite impossible.

In gastight chambers, such as submarines or sealed air-raid shelters, distress may arise from excess of carbon dioxide or deficiency of oxygen, but in ordinary buildings no such troubles will be met with. A room is badly ventilated indeed if the carbon dioxide content reaches 1 per cent, or the oxygen concentration falls by a similar amount.

Collier (1940) remarks that a relative or partial anoxaemia may be produced with abnormal ease in ill-ventilated shops among persons who for any reason are already anaemic or are suffering from defective circulation, and quotes Haldane and Priestley (1935) who suspected that in certain neurasthenic patients shallow breathing was somehow causing anoxaemia. Collier adds that under-nourished, anaemic and nervous workers often suffer from a mild, chronic oxygen-lack; but he goes on to say that the environmental cause of this is not a lowered concentration of oxygen in the atmosphere. In so far as the condition can be attributed to defective ventilation it is due, he says, to the absence of the stimulation provided by passage of cool moving air over the skin.

Effects of Body Odours

The odours which are perceptible in occupied rooms arise mainly from the occupants themselves. These body odours are especially noticeable by persons entering direct from the fresh air outside. The sense of smell is quickly fatigued, and on that account odours which are readily noticeable, or even unpleasant, to a newcomer, may be unnoticed by those who have been exposed to them for some time. People vary in their response to body odours. Some do not seem to object to them, while others find them most objectionable, and even nauseating. Odours which are not strong enough to evoke violent complaint may yet impart a staleness to the air which is hardly compatible with good ventilation. The fresh-air supply to an occupied room should be enough to reduce body odours to an imperceptible or barely perceptible level.

It is sometimes urged that ventilation rates as high as those which are required to achieve this end are not necessary. It is said that many people like a 'frowst' and do not object to odours, and that in the interests of fuel economy lower ventilation rates should be permitted. This attitude seems unacceptable. Great care is devoted to ensuring that we have a pure water supply, and no one would suggest that in the interests of economy we should be doomed to drink polluted water. On aesthetic grounds alone it should be one's right to be allowed to live and work in a clean atmosphere which is free from objectionable odours.

Flügge (1905) wrote that the objection to an evil-smelling atmosphere was to be supported not on account of its poisonous properties, which had never been proved to exist, but on account of the resulting feeling of nausea. There is evidence, however, that body odours are undesirable for other reasons. Winslow and Palmer (1915) studied the effects of lack of ventilation upon the appetite for food, in experiments in which the temperature, humidity and air movement were controlled. They concluded that there were substances present in the air of an unventilated occupied room which in some way, and without producing conscious discomfort or detectable physiological symptoms, diminished the appetite. They further expressed the view that the effect of such influence might in time be very important. In later work, Winslow and Herrington (1936) used as their source of odours heated house-dust taken from vacuum cleaners. Such odours, even when not consciously observed, had a clearly demonstrable effect in reducing appetite.

In ventilation the term 'freshness' generally refers to the invigorating properties of an atmosphere, which depend wholly or mainly on its physical characteristics. Dufton (1939) suggests that the differentiation between freshness and fragrance should be preserved. A stale atmosphere would be the reverse of fragrant.

Reid (1844) tells an amusing story to illustrate the effects of ventilation and fragrance on the appetite. About 50 members of one of the Royal Society clubs in Edinburgh dined in an apartment which he had constructed. The room was copiously ventilated, and from time to time the air was lightly perfumed. He says:

> 'Nothing very special was noticed during the time of the dinner by the members; but Mr. Barry of the British Hotel, who provided the dinner, and who, from the members of the club being frequently in the habit of dining at his rooms, was familiar with their constitutions, showed the Committee that three times the amount of wines had been taken than was usually consumed by the same party in a room lighted by gas, but not ventilated–that he had been surprised to observe that gentlemen whose usual allowance was two glasses, took, without hesitation, as much as half a bottle–that those who were in the habit of taking half a bottle took a bottle and a half, and that, in short, he had been compelled twice to send hackney coaches for additional supplies during dinner, though he had provided a larger supply than usual, considering the circumstances under which the members met.
>
> Minute inquiries afterwards assured me that no headache nor any injurious consequence followed this meeting, nor were any of the members aware at the moment that they had partaken more heartily than usual, till Mr. Barry showed them what had taken place.'

Removal of Body Odours

If a foreign gas such as hydrogen is liberated slowly and continuously into a room which is ventilated at a uniform rate, the concentration of the gas in the atmosphere increases exponentially, rapidly at first and then ever more slowly, until an equilibrium concentration is reached. The equilibrium level depends on the rate at which the gas is liberated, and on the ventilation rate; while the speed with which equilibrium is approached depends also on the size of the room. Similarly, if the ventilation is continued and the supply of gas is cut off, the concentration of the gas will follow the course of an exponential die-away curve. It will decline rapidly at first and then more and more slowly.

Yaglou and his colleagues (1936) showed that the concentration of odoriferous substances is affected by some other factors besides the simple diluting effect of the ventilating air stream. Their experiments lasted for 3.5 hours, and an equilibrium odour strength was reached in 1 to 3 hours, according to the ventilation rate; thus, in each experiment equilibrium was reached. It might, therefore, have been expected that the strength of odour reached would have depended on the fresh air supply per person, regardless of the cubic space available for each. If the removal of the odorous substances were affected only by the ventilating air evenly distributed, the influence of the amount of cubic space available would make itself felt during the period before equilibrium was reached, but thereafter the concentration of odours would depend on the air supply per person, and would be independent of the allocation of room space per person. But it was, in fact, found that with a constant air supply per person the strength of odour diminished as the cubic space per person increased.

It seemed probable that this result was due to some extent to absorption of odours by the walls and other surfaces within the room. Later work, however, showed that such absorption was of little or no significance (Yaglou and Witheridge, 1937). Observations were made of the rate of disappearance of body odours from rooms which were virtually airtight, immediately after the occupants had left. The odours disappeared with remarkable rapidity—at a rate far greater than could be accounted for by the exceedingly slight leakage—until a very low intensity was reached, and then the disappearance rate became very slow. It seems possible, say Yaglou and Witheridge, that some constituents of the body-odour complex are extremely unstable, and that these are oxidized and leave behind other and less odorous, but more stable, constituents. Loss of moisture may have something to do with the loss of odour.

It also appeared from this work that as the fresh air supply per person increased the efficiency of ventilation and odour removal decreased. Judged by the rate of removal of the carbon dioxide due to respiration, the ventilation efficiency was about 100 per cent when the air flow was below 300 cu. ft/person/hour, and about 75 per cent with a supply of 1,800 cu. ft/person/hour. With rapid ventilation rates diffusion is less complete, and some of the fresh air supplied to the room leaves it without carrying off its quota of body odours, or of the products of respiration.

FRESH AIR REQUIREMENTS FOR ODOUR CONTROL

The fixing of a standard of air supply is a matter of much difficulty. It is generally recognized that life in the open air is more conducive to vigorous health than an indoor life. Much of the benefit of an outdoor life is undoubtedly attributable to exercise and the effects of sun and wind, but there will be general agreement that the greatest amount of ventilation which can be achieved with comfort and reasonable regard to economy is desirable.

Most of the standards of ventilation which have been proposed for nearly a century past have been based on the efficiency of ventilation in removing body odours. Obviously, the ventilation required to keep the atmosphere of a room free from perceptible odours will depend partly on the personal cleanliness of the occupants, and partly on the keenness of the investigator's sense of smell. It is not surprising, therefore, that widely divergent standards have been suggested from time to time. Different workers have advocated standards for the minimal fresh air supply ranging from 5 to 50 cu. ft/min/person.

In Great Britain several studies have been made of the amount of fresh air required in barracks. Roscoe (1857) found that in soldiers' sleeping rooms a supply of 10 cu. ft/min/person was sufficient to remove odours, but he suggested that the air supply should be at least 20 cu. ft/min. Four years later, the Barracks and Hospitals Commission (1861) also came to the conclusion that a minimum of 20 cu. ft/min should be supplied.

Some idea of the conditions prevailing in barracks at that time may be obtained from the following comment made by Fairbairn, Glaisher and Wheatstone (1857): 'What adds greatly to the discomfort of the men in barracks, and what tends to increase the nuisance and vitiate the atmosphere of the barrack room is, the abominable practice of having an open wooden vessel or tub standing in each room as an urinal. Why cannot that nuisance be removed and what difficulty can possibly exist in forming in the corner of each room a basin and flow pipes, concealed and well ventilated, to carry off these pestilential odours?'

The odour evoked by fresh human urine has been described as 'mawkish': a description of decomposing urine may be left to the imagination.

Some years later, De Chaumont (1875) made many observations in barracks, relating his impressions of smell to the carbon dioxide content of the air. He found that so long as the amount of carbon dioxide due to respiration was not more than 2 parts in 10,000 of air, the air of the rooms seemed fresh and did not differ sensibly from the outer air.

As a working figure it is generally assumed that the average person produces 0.6 cu. ft of carbon dioxide/hour, and on that basis a concentration of 2 parts in 10,000 of respiratory carbon dioxide represents a fresh air supply of 50 cu. ft/person/min. According to De Chaumont, odours were just perceptible when the carbon dioxide due to respiration amounted to 4 parts in 10,000, or the air supply per person to about 25 cu. ft/min.

This last figure agrees very well with the standards earlier recommended for barrack rooms, but De Chaumont advocated the higher standard of 50 cu. ft which, if he experienced the same conditions as Fairbairn *et al.*, is not surprising.

Such a standard is a high one, and in practice it has not been maintained. For example, the London County Council required fresh air to be supplied to cinemas and theatres at the rate of 1,000 cu. ft/person/hour, or about 17 cu. ft/min.

Odours in relation to ventilation and concentration of carbon dioxide. (De Chaumont, 1875)

	Fresh; not sensibly different from outer air	*Rather close; organic matter becoming perceptible*	*Close organic matter disagree-able*	*Very Close; organic matter offensive*
Carbon dioxide due to respiration, parts in 10,000 of air	2	4	7	9
Total carbon dioxide, parts in 10,000	6	8	11	13
Air supply/person/min (assuming production of 0.6 cu. ft CO_2/person/hour)	50	25	14	11

In recent years further research has been carried out by Yaglou, Riley and Coggins (1936), with the object of revising existing standards; and in this work again the absence of disagreeable odours was taken as the criterion of a satisfactory ventilation rate. The fresh air requirements were determined on two bases: (*a*) the primary impression on entering the room from relatively fresh, clean air; and (*b*) the impressions of the occupants. The standard of odour was that assessed as 'moderate', and it is said that the odour at that strength was neither pleasant nor disagreeable; there was little objection to it, but it was regarded as the maximum allowable limit in occupied rooms.

The intensity of odour in a room was found to vary with the number of occupants as well as with the fresh air supply per person; thus the greater the number of occupants the greater the amount of air which must be supplied for each of them if odours are to be avoided.

The fresh air requirement varies with the personal habits of the occupants. Thus, it was found that when the subjects of the experiments went without baths for a week the fresh air requirements increased day by day. Presumably because of the difference in personal habits, and in the bathing facilities available to them, persons of the poorer social grades required more fresh air than did their more fortunate fellows; and the fresh air requirements of school children were commonly greater than those of adults.

The minimum amounts of fresh air required to keep the atmosphere so that the impression given to persons entering from the outer air is not disagreeable are summarized in the table.

The requirements vary for different types of occupant, but as a broad generalization it can be said that for most ordinary

Minimum values of outdoor air requirements
(Yaglou et al., *1936)*

Type of occupants	*Air space/person (cu. ft)*	*Fresh air supply required (cu. ft/min/person)*
Sedentary adults of average socio-economic status	100	25
	200	16
	300	12
	500	7
Labourers	200	23
School children of average class	100	29
	200	21
	300	17
	500	11
School children of poor class	200	38
School children of better class	200	18

conditions an air supply of 17 to 20 cu. ft/min/person will give satisfactory results.

In buildings where much of the air is washed and then recirculated the amount of fresh air can be reduced considerably, but then care must be taken that the water used in the air washer is frequently renewed, otherwise the water may actually impart odours to the air.

At the risk of repetition it should be emphasized that the fresh air supplies mentioned represent minimum requirements. The maximum ventilation that can be achieved with comfort, when regard is paid to necessary limitations such as those imposed by shortage or cost of fuel, is desirable.

AIRBORNE BACTERIA

In sneezing and coughing, bacteria-laden droplets are discharged and may be spread over a distance of 15 ft. The droplets vary in size. The larger ones will fall to the ground fairly quickly. The smaller ones fall less quickly, and as they become yet smaller by evaporation many of them will rapidly become so small that they can float in air—they become airborne. Organisms that fall on the floor, or on fabrics such as blankets, may survive on dust which may later become suspended in the air.

For many years until recently the transmission of infection by airborne organisms has tended to be discounted, and direct droplet infection has been held mainly responsible for the spread of respiratory infections. In recent years, however, renewed attention had been given to the question of airborne infection. Wells and Wells (1936) stressed the need for what they called sanitary air control. Wells (1935) had found that organisms sprayed into the air of a basement room could be recovered on every corridor up to the top floor of a three-storey building, where the concentration reached approximately 1 per cent of that in a basement room. In the National Institute of Health, Washington, psittacosis virus appeared to have been conveyed by air from the basement to the upper floors (McCoy, 1934). Wells and Brown (1936) sprayed influenza virus in the air, and by animal inoculation confirmed its recovery up to 1 h after suspension. Dalla Valle and Hollaender (1940) showed that bacteria may be spread by a ventilating system; and Cruickshank and Muir (1940) believed, on bacteriological and epidemiological evidence, that an outbreak of streptococcal infection among a group of men recovering from influenza had been aerially spread. A report of the Medical Research Council's Committee on Tuberculosis in Wartime refers to the effect of reduced ventilation on the spread of tuberculosis, and other Committees of the Council remark that of the general measures which can be taken to reduce the bacterial content of the air of hospital wards, ventilation is probably the most effective.

There may be different opinions as to whether direct droplet infection or airborne infection is the more important mode of transmission of respiratory disease, but there seems to be no real knowledge as to their relative importance under various circumstances. Evidence such as that just referred to indicates that the transmission of disease by airborne organisms does occur. In densely-occupied rooms the risk of droplet infection will clearly be greater than in sparsely populated ones, but unless the ventilation varies in proportion to the density of occupation the concentration of the airborne organisms will also tend to be greater with the denser occupation.

Thirty years after the need for sanitary air control was expressed, one notes that whereas purification of drinking water has almost eliminated the enteric fevers–in this country, infections of the respiratory tract have not been so controlled by what has been termed 'sanitary ventilation'. One reason for this may be that the rate of ventilation in a room may be, say, five changes/hour, which is very small indeed when compared with the rate of ventilation in an equivalent volume of outside air, which may be expressed in changes/sec.

Hence, one writer (Macmillan, 1967) said that surgical operations could be conducted more safely in the open air than in a badly ventilated operating theatre. Yet Williams (1967), when discussing the spread of pathogenic airborne bacteria, concluded that with most of the infections, the reduction in dosage which can be achieved in practice by 'sanitary ventilation' is too small to have a detectable effect on the frequency of occurrence of infection. However, Chamberlain (1967) notes that most of the bacteria and viruses in the atmosphere are carried by particles of dust mostly smaller than 4 μm in diameter which float about like rafts and ventilation is the most effective method of removing the non-viable carriers with their viable passengers.

Dust-borne Bacteria

Micro-organisms in occupied rooms tend to become attached to dust particles; and amongst these pathogenic organisms may be included. Edward (1944) reports that on the average the dusts of two hospital wards contained respectively 300,000 and 250,000 haemolytic streptococci/gram; and the occurrence of a particular type of streptococcus in the dust often resulted from the presence of one or more infected persons in the ward. Viable streptococci can be recovered from the air of rooms in which blankets artificially infected with streptococci have been beaten; and these organisms may be liberated long after infection, for streptococci retain almost their full virulence in blankets for four weeks (van den Ende, Lush and Edward, 1940). Bacteria-laden dust is an obvious risk in hospital wards. A great increase in the number of airborne organisms usually accompanies bed-making, but this increase can be almost eliminated (90 per cent reduction) by treating bedclothes with liquid paraffin dissolved in white spirit, which process does not render the bedclothes objectionably oily (van den Ende, Edward and Lush, 1941; van den Ende and Thomas, 1941).

Sweeping a dry and dusty floor redistributes into the air a large proportion of the bacteria deposited on the dust, but an application of crude liquid paraffin–spindle oil–before sweeping greatly reduces the aerial contamination. Floors, especially in hospital wards, should therefore be sticky rather than highly polished. Data recorded by van den Ende, Lush and Edward (1940) show the high bacterial contamination which may be caused by the beating of blankets, and the results of sweeping the dust deposited on treated and untreated floors.

The experiments were performed in two similar rooms, the floors of both rooms were washed and allowed to dry, and the one floor was treated with spindle oil. A dusty blanket, heavily impregnated with streptococci, was suspended in each room and beaten vigorously. Blood-agar plates were exposed for 10 min after beating to determine the relative numbers of streptococci distributed into the air of the two rooms. Half an hour was allowed for the dust and bacteria to settle to the floor, after which both rooms were thoroughly ventilated for 5 min. A second series of plates was exposed for 10 min after ventilation, to ascertain the number of bacteria still suspended in the air, and then both rooms were swept for 1 min with ordinary household brooms which had been sterilized by autoclaving. More plates were exposed for 10 min during and after sweeping. The bacterial counts obtained are shown in the table:

Effect of floor treatment on the numbers of streptococci raised into the air by sweeping

	Control room (floor untreated)		*Test room (floor treated with spindle oil)*	
Exposure of plates	*Height of plates above floor (ft)*	*Average count*	*Height of plates above floor (ft)*	*Average count*
10 min after beating	2	426	2	435
	5	312	5	557
10 min after ventilating	2	2	2	4
	5	2	5	2
10 min during and after sweeping	2	73	2	8
	5	99	5	0

Sampling Apparatus

The method of exposing plates containing nutrient media for specified periods enables the number of organisms settling on those plates to be ascertained. It gives a measure of the deposition of particles on a given area, but gives no accurate information about the numerical concentration of bacteria in the air.

A great advance in bacterial sampling was made when Wells (1933) introduced his air-centrifuge apparatus. A glass cylinder is rapidly rotated about its vertical axis, and a current of air is thus caused to enter through a central axial tube, and to escape along a thin layer of nutrient medium which coats the walls of the cylinder. Suspended particles in the air thus sampled are deposited on the medium, and colonies develop where the individual bacteria have been precipitated. Since the air flow is measured, the method provides a measure of the concentration of organisms in the air. Many investigations have been carried out by means of the Wells apparatus.

A further notable advance was the introduction by Bourdillon, Lidwell and Thomas (1941) of the slit sampler, an instrument based on the principle of the Owens jet dust counter. In the slit sampler the air is sucked through a narrow slit, 0.25 mm wide, 27.5 mm long, and 3 mm deep, on to the surface of agar or other culture medium contained in a Petri dish just below the slit. The dish is slowly rotated during the sampling so that the colonies are distributed uniformly, and the slit is so arranged relative to the dish that the bacteria are deposited in an annular ring, leaving a clear space of about 18 mm diameter in the centre. The Petri dish can be rotated either by hand or mechanically. The rate of sampling recommended is 1 cu. ft/min. Detailed studies have been made of the influence of various factors, such as particle size, slit-plate distance, rate of air flow, and the shape and size of the slit, on sampling efficiency.

For an aerosol consisting of *Staph. albus* sprayed from distilled water as single cocci, the slit sampler has a collection efficiency of about 96 per cent. Thus the finest bacteria-carrying particles are collected almost as efficiently as coarser ones. When, on the other hand, sampling is done by the exposure of open Petri dishes, it appears that in the air of crowded rooms the larger particles are collected at least 200 times as efficiently as are single washed bacteria. For normal airborne organisms the Wells centrifuge has an efficiency of about 75 per cent. For further information the reader is referred to the review of techniques and instruments available for the investigation of airborne bacteria given by Wolf *et al.* (1959) entitled *Sampling Microbiological Aerosols* published as U.S. Public Health Monograph No. 60 at Washington D.C. by the United States Government Printing Office.

Much work has been done in this field in the past 30 years and some 160 papers are cited by May and Noble in their papers on sampling techniques contained in a symposium 'Airborne Microbes' published for the Society for General Microbiology in 1967 by the Cambridge University Press.

May (1966) has developed a three-stage impinger-type sampler which is designed to simulate the conditions under which deposition of microbes takes place in the human respiratory tract. However, as Noble (1967) points out, the main problem today is that of designing methods for the recovery and identification of virus particles present in air.

Bacterial Content of the Air of Occupied Spaces

The results of a survey of the normal bacterial content of the air of enclosed spaces in New York City are reported by Buchbinder, Solowey and Solotorovsky (1945). The Wells centrifuge was used for sampling, and about 5,000 samples were taken in schools, subway cars, theatres, and in streets and Central Park. The average counts observed in various situations during a period of a year are shown in Fig. 37.

The air of schools yielded the highest number of bacteria; the lowest average for a school was as high as the highest for a theatre. The bacterial counts observed in air-conditioned theatres were only about a quarter as great as those for theatres which were not air-conditioned. In schools, the counts for classrooms and for assembly rooms were twice as high when the rooms were occupied as when they were vacant (Fig. 38).

The two schools which gave the highest counts were old and of inferior construction, and had small windows. Schools 5 and 6, which gave the lowest counts, were of more modern design, with large windows and high ceilings.

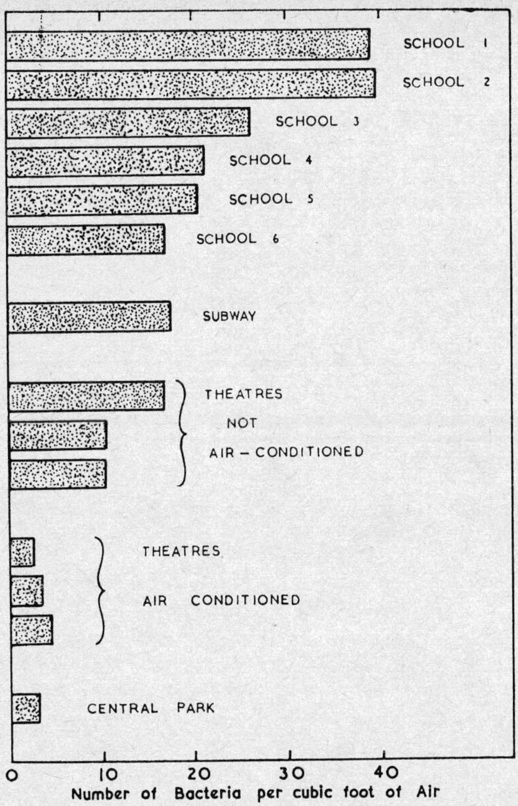

Fig. 37. Averages of total bacterial counts in various situations. (Samples collected on blood agar.) (Buchbinder et al.*)*

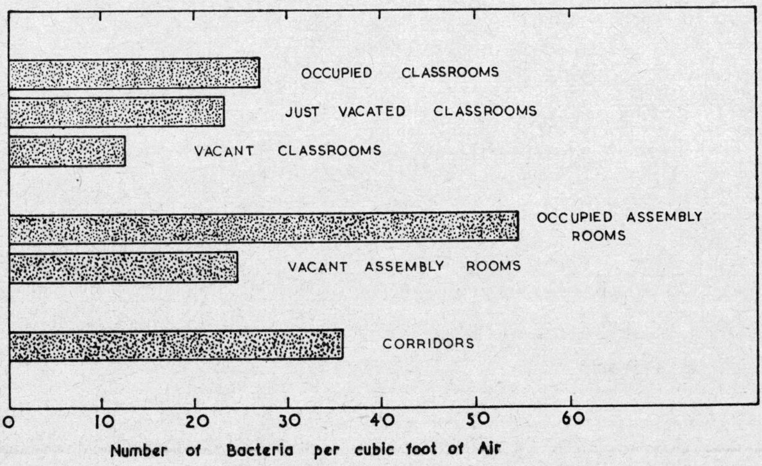

Fig. 38. Bacterial counts under various conditions of occupancy; averages for six schools. (Buchbinder et al.*)*

REMOVAL OF AIRBORNE ORGANISMS

The concentration of pathogenic and other organisms in the air of an enclosed space can be reduced by ventilation, or by the application of some bactericidal agent.

Wells and Fair (1935) demonstrated the high bactericidal power of ultra-violet light. Andrewes and others (1940) found ultra-violet light to have a high germicidal effect in the laboratory, but their practical tests, *e.g.* in hospital wards, did not give such good results. They found ultra-violet rays very effective against bacteria in fine particles, but less so against those on dust. Air filters can be effective against dust, and are thus complementary to ultra-violet lamps. In applying ultra-violet radiation to air sterilization the air should be passed first through filters to remove the dust, and then over the lamps.

Edward, Lush and Bourdillon (1943) achieved rapid sterilization of atmospheres containing atomized particles of influenza virus by using ultra-violet radiation of wavelength 253.7 nm. At least 99 per cent of an aerosol of the virus was killed by exposure for 6 sec at a distance of 2 cm from a Hanovia lamp or a G.E.C. 'Sterilamp'. These findings confirm the earlier work of Wells and Brown (1936). Air disinfection by ultra-violet radiation is discussed by Bourdillon, Lidwell and van den Ende (1948). They survey the chief factors concerned, compare different types of lamps, discuss the value of ultra-violet light barriers, and describe field trials.

Yaglou (1941) concludes that, apart from contagious wards and operating rooms, the value of ultra-violet radiation for sterilizing the air of occupied rooms is still undecided.

The effect of air disinfection with ultra-violet light on illness among school children was investigated in a suburban area near London (Med. Res. Coun., 1954). All the classrooms and assembly halls at each of the three primary schools were equipped with ultra-violet lamps. Three similar schools were observed as non-irradiated schools. The investigation was carried on through three years. During the investigation the general bacterial count in the air was about 16 per cent lower in the irradiated schools than in the control schools. The count of *Str. salivarius,* studied as an index of the degree of mouth pollution of the air, was reduced by as much as 70 per cent, and the count of haemolytic streptococci, made over a period of six months, was reduced by about 80 per cent.

There was no appreciable effect on the total sickness absence recorded. However, examination of individual causes of absence suggested that the irradiation probably reduced the number of absences due to certain diseases by amounts between 15 and 45 per cent. These diseases included mumps and chickenpox; asthma; gastritis, etc.; scarlet fever, acute pharyngitis and tonsillitis, otitis media and earache. Secondary-attack rates following single-case introductions of measles, mumps and chickenpox were apparently lower in the irradiated schools. But absences due to the causes named account for only a small part of the total absence through sickness, and the effect of irradiation on the total absence is therefore small. It was concluded that these results did not appear to justify the wide use of irradiation for the control of infection in primary urban day schools.

The use of germicidal sprays is mentioned by Wells and Wells (1936), who say that in tolerable concentrations commercial germicides raise the death-rate of organisms about three-fold. Much work on bactericidal sprays has been done in Great Britain by Pulvertaft, Lemon and Walker (1939), by Twort and his colleagues (1940), and by Andrewes and others (1940). Andrewes and his colleagues found that many proprietary substances advocated for use as mists are valueless, and that the devices sold for putting up mists do not give fine enough division. Substances proved to be of value as sprays included sodium hypochlorite, chloramine-T, resorcinol, hexyl resorcinol, eugenol carbinol, and catechol. The last four of these are most effective when dissolved in propylene glycol, hexyl resorcinol in propylene glycol being especially potent. Regarding sodium hypochlorite, there is some dispute as to whether this acts as a mist or a gas. Sodium hypochlorite is cheap and in unlimited supply; it is a valuable deodorant and is effective in low concentrations. Care should be taken that it is rapidly evaporated, for coarse drops of sodium hypochlorite may have deleterious effects on fabrics or metal parts.

The results of prolonged research on the use of antiseptic

sprays for air sterilization are given by Pulvertaft (1944). As criteria for a good antiseptic for use as a bactericidal mist he suggests that it should be required to effect a 90 per cent reduction of bacteria, compared with the control, when not more than 5 ml of the antiseptic solution, at a strength not greater than 1 per cent, is nebulized into a room of 1,000 cu. ft capacity, *i.e.* when the concentration in the air is 1 gm in 20,000 cu. ft.

At times of epidemic respiratory disease, Pulvertaft concludes, it may well be expected that antiseptic mists will have a lethal effect on the disease-producing organisms. Mist emitters must be placed near air inlets, but it may be undesirable to place them within air ducts, for the mist particles may be precipitated by impingement against the sides of the ducts. A continuous, or semi-continuous, mist emission is desirable. The value of electrolytic sodium hypochlorite is confirmed. For special cases the more expensive and more efficient glycol solutions of phenolic compounds may be employed.

Later work by Elford and van den Ende (1945) shows the great influence of atmospheric humidity on the bactericidal efficiency of hypochlorite. At ordinary room temperatures killing takes place rapidly at relative humidities of 70 to 90 per cent, but at relative humidities below 50 per cent inactivation of organisms is slight.

The applications of glycol vapours for the control of airborne infections is dealt with by Jennings, Bigg and Olson (1944). They show that bactericidal concentrations of the vapours—0.2 mg propylene glycol, 0.005 mg triethylene glycol, per litre of air—can be maintained in large occupied spaces for long periods of time, and that the vapours can be uniformly distributed in large rooms. One of the troubles in the application of bactericidal mists is that when effective concentrations are achieved there is a risk that some occupants will be exposed to irritating concentrations. Generation of controlled amounts of the vapours was contrived so that occupants of treated rooms experienced no discomfort.

The glycol is vaporized from an aqueous solution. Both water and glycol are given off from the boiling mixture, but the proportions of the two vary with the boiling point, *i.e.* with the concentration of glycol in the mixture. The control system devised for the vaporizer includes a thermostat which controls a solenoid valve in the water feed to the vaporizer, so that the temperature of the boiling point can be controlled. This governs the ratio of the quantities of water vapour and glycol given to the air. Then, by varying the rate of heat input to the vaporizer, the rate at which the liquid is vaporized is controlled. Field tests show that this method of control is effective. The vapours are discharged into the ventilation system and distributed with the ventilating air.

Discussing the control of cross-infection in hospital wards, a Committee of the Medical Research Council (1944) comments that the various methods of air sterilization that are being tried, ultra-violet irradiation, or disinfectant vapours or mists, are ineffective against bacteria carried by gross particles such as dust.

Ozone, in suitable concentrations, is also a bactericide. The use of ozone in ventilation is discussed below.

It is pertinent here to quote a remark by Andrewes and his colleagues: 'No method for the control of infection is as effective as adequate spacing of individuals combined with adequate ventilation. These enormously reduce the risk of infection, whether from droplets, droplet nuclei, or dust.' They say that methods such as ultra-violet light and mists are of value only as second-rate substitutes where proper spacing and ventilation cannot be achieved.

In air-raid shelters and other places, where good ventilation may be impracticable, the use of these bactericidal methods may be justifiable, and, indeed, desirable, but in factories and dwellings adequate ventilation appears to be the preferable way of reducing the risk of airborne infection, as well as of removing odours and providing an invigorating environment.

THE USE OF OZONE IN VENTILATION

Many claims have been made for ozone, especially in commercial brochures. It is said to destroy odours, to sterilize the air, and to have valuable physiological effects. Joseph (1938), indeed, claims that ozone had very beneficial effects on miners working in the Rand mines.

Witheridge and Yaglou (1939) made a careful study of the value of ozone in ventilation, mainly from the standpoint of its effect on body odours, and, consequently, on fresh air requirements for odour control.

In considering the use of ozone in ventilation it must not be forgotten that even in very low concentrations it is a very powerful irritant. Concentrations as low as 0.1 to 1 part/million have a strong irritant action on the respiratory membranes, and exposure to such concentrations shortens the lives of guinea-pigs. The irritation threshold appears to be about 0.05 part/million, and concentrations well below this do not appear to injure health. The olfactory threshold level for those with a keen sense of smell was confirmed by Witheridge and Yaglou as 0.01 part/million parts of air, while a concentration of 0.015 part/million was detectable by most normal persons. These authors found that 0.10 part/million was objectionable to all normal persons and caused irritation to the nose and throat in most; and they recommended that 0.04 part/million should be regarded as the upper permissible limit in occupied rooms.

Ozone and Body Odours

Witheridge and Yaglou found that ozone in concentrations of 0.015 part/million, a concentration that was only just detectable by the occupants of the room, reduced the smell of body odours enough to permit a reduction of at least 50 per cent in the fresh air requirement for odour control. Lower concentrations had no effect in reducing body odours.

Ozone does not destroy body odours. It masks them. This has been said by various earlier workers, and the statement has been hotly contested by commercial interests. But Witheridge and Yaglou adduced convincing evidence that the deodorizing effect is due to masking. After the body odours in an occupied experimental room had been counteracted by ozone, and the room then vacated and the ozonator turned off, the smell of ozone gradually disappeared, while that of body odours again became perceptible (see Fig. 39).

Although ozone in sufficient strength completely obliterated the smell of body odours, it was difficult to balance the two so that neither could be perceived by persons entering the room direct from the fresh air. Ozone had to be present in perceptible concentrations to be of any use.

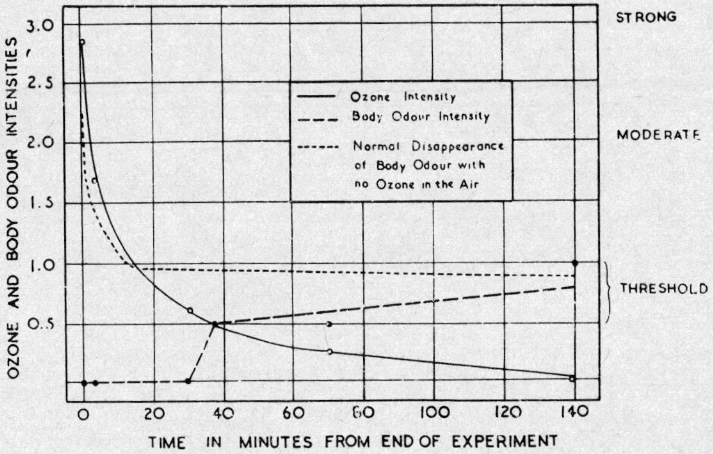

Fig. 39. Reappearance of body odour with disappearing ozone odour after subjects had left room. (Witheridge and Yaglou.)

With a given output of ozone into the air, the concentration may vary within wide limits according to the humidity, so that a constant output of ozone may yield only a trace of ozone in a room on a humid day, while on a dry day it may give an undesirably high concentration. For this reason Witheridge and Yaglou recommend that until adequate means are devised for accurately controlling the concentration of ozone in occupied rooms its use should be discouraged.

Ozone as a Bactericide

Reviewing earlier work, Witheridge and Yaglou remark that it is now generally recognized that ozone is not a satisfactory germicide at concentrations within the range of human tolerance. Wells and Wells (1936) found that in tolerable concentrations it had a lethal effect on airborne organisms. Concentrations of four times the olfactory threshold value raised the death-rate threefold. Now the olfactory threshold for those with a keen sense of smell is 0.01 part/million, so that the concentration referred to by Wells and Wells was presumably at least 0.04 part/million, which concentration is the upper permissible limit for occupied rooms.

Further work on the use of ozone in air sterilization has been done by Elford and van den Ende (1942). They found that ozone in tolerable concentrations (not exceeding 0.04 part/million) was able to inactivate certain bacteria when they were present as unprotected singleton aerosol particles in atmospheres of 60 to 90 per cent relative humidity. When such bacteria were covered with a protective coating of organic matter, however, as in aerosols naturally emitted during a sneeze or cough, then ozone in permissible concentrations was without significant effect. Inactivation of such protected organisms could only be achieved by the use of ozone in decidedly higher concentrations, which must be deemed dangerous for long spells of breathing. Bacteria that have settled on surfaces are generally more resistant to ozone than when in newly formed aerosols.

This evidence leads to the conclusion that ozone, in concentrations that can be breathed over long periods without irritation, cannot be expected to provide any effective protection against airborne bacterial infection, through direct inactivation of the infectious carrier particles.

Ozone, then, is not a practicable germicide for use in ventilation, nor, until its concentration can be accurately controlled, is it suitable for use in combating body odours.

Physiological Effects of Ozone

Weiner (1938), commenting on Joseph's statement that ozonized air had been of benefit in deep mines, where it was implied that it increased metabolism, referred to the results of tests made on the Rand. The experimenters could detect no difference in the working capacity or resistance to fatigue, or in various other reactions of the body, when ozone was added to the air.

VENTILATION AND AIR MOVEMENT

As a broad generalization, increased ventilation means more air movement, but this statement does not always hold true. For instance, in some factories where, for the removal of toxic vapours, ventilation rates of 20 to 30 air changes/hour are demanded, this ventilation is maintained with air velocities of only 30 or 40 ft/min around the occupants. But such velocities are commonly found in other factories during the winter months when the ventilation rate is frequently no more than 3 or 4 air changes/hour. Yet in the same room, air movement generally does increase with increased ventilation. Unpublished observations by Warner, Chrenko and Bedford show that substantial increases in ventilation are necessary to gain appreciable increments in the air movement. Thus, in a closed room with less than one air change/hour the air velocity was 10 or 11 ft/min and to raise the velocity to 20 ft/min by natural ventilation it was necessary to increase the ventilation rate to about 20 air changes/hour. In factories movement of persons and machinery tends to cause increased air movement.

In winter weather, provided there is no overheating, the air in a dwelling or factory building will probably not be very stagnant if the fresh air supply does not fall below the amount required to remove body odours. In summer, on the other hand, much more ventilation than that necessary to remove body odours is very desirable. As the temperature increases the amount of air movement necessary to provide an invigorating atmosphere also increases. Thus, in summer, copious ventilation is desirable on physical grounds as well as for the removal of odours and micro-organisms.

REMOVAL OF NOXIOUS DUSTS, FUMES, VAPOURS AND GASES

A further requirement of ventilation, especially in industry, is the removal of deleterious gases, vapours, and suspended solids from the air. Many industrial operations result in the production or emission of these substances. Every effort should be made to prevent the dispersion of toxic substances into the atmosphere. In some processes it is possible to use non-toxic substances in place of harmful ones; where such substitution is possible it is the best way of controlling atmospheric pollution. Where this cannot be done, the next best method is the total enclosure of the process, so that there is no possibility of the escape of the harmful substance. Failing total enclosure, local exhaust ventilation should be considered. In some cases even local exhaust ventilation is impracticable, and then recourse must be had to increased general ventilation.

For many toxic substances now used in industry standards of maximum permissible concentrations, or 'threshold limit values', have been proposed. Such standards are recommended by various bodies, and there is considerable divergence between the standards accepted in different countries. A valuable list of threshold limit values is published annually by the American Conference of Governmental Industrial Hygienists, and the values given are subjected to regular scrutiny and revision in the light of experience. It is pointed out that the limits given should be used as guides in the control of health hazards, and not as lines sharply dividing safe from dangerous concentrations. They represent conditions to which it is believed almost any worker may be repeatedly exposed, day after day, without ill effect. The values are based on a variety of criteria of toxic effects or of marked discomfort, and they should not be used as a common denominator of toxicity. In the table opposite the limits for some of the commoner toxic substances are set out.

During recent years there has been an increasing risk of pollution of the domestic internal environment. According to Kennedy (1970), the home has been referred to in America as the 'contaminated closed environment'. In 1968, over one and a half billion cans of aerosol spray were sold in the United States of America and each of the products is discharged into the indoor atmosphere. In many cases the active substances and their propellants are non-toxic but there is the risk of thermal and chemical decomposition of these substances with the

Some threshold limit values (American Conference of Governmental Industrial Hygienists, 1961)

Contaminant	*Maximum parts/million parts of air*	*Concentration in milligrams/ cu. m of air*
Ammonia	100	
Amyl acetate	200	
Aniline	5	
Arsine	0.05	
Benzene	25	
Cadmium		0.1
Butyl acetate	200	
Carbon bisulphide	20	
Carbon monoxide	100	
Carbon tetrachloride	100	
Chlorine	1	
Chlorodiphenyls		0.5 to 1
Chromic acid		0.1
o-Dichlorbenzene	50	
Dichlorethyl ether	15	
Ether	400	
Formaldehyde	5	
Hydrochloric acid	5	
Hydrogen cyanide	10	
Hydrogen fluoride	3	
Hydrogen sulphide	20	
Lead		0.2
Mercury		0.1
Methanol	200	
Monochlorbenzene	75	
Nitrobenzene	1	
Nitrogen oxides	5	
Phosgene	1	
Phosphine	0.05	
Sulphur dioxide	5	
Tetrachlorethane	5	
Tetrachlorethylene	100	
Toluene	200	
Trichlorethylene	100	
Turpentine	100	
Xylene, coal tar naphtha	200	
Zinc oxide fume		15

formation of potentially toxic substances. Electric and gas cooking and heating systems may be responsible for such decomposition especially of fluorocarbons.

But some substances are toxic to begin with; for example, sprays containing DDT, chlordane, benzene and petroleum distillates. The following case quoted by Carson (1963) indicates the danger. A housewife went into her basement armed with an aerosol spray containing DDT and a petroleum distillate in order to kill the spiders present. As she abhorred spiders, she made a very thorough job of the spraying. As she finished the work she began to feel quite ill with 'nausea, extreme anxiety and nervousness'. After a few days she felt better and, not suspecting the cause of her malaise, she repeated the programme of spraying a month later, and she experienced a similar malaise. She repeated the procedure yet again but this time it was followed by new symptoms, such as fever, pains in the joints, acute phlebitis in one leg accompanied by a generalized malaise. She was examined by Dr Hargraves of the Mayo Clinic who found that she was suffering from acute leukaemia. She died within the following month. Of course, it does not necessarily follow that this association of the onset of acute leukaemia with exposure to a toxic chemical means that there is a causal relationship.

Carbon tetrachloride is an excellent non-inflammable solvent much used in industry. It is a dangerous narcotic poison which can also damage the liver and kidneys. It is the active agent in many domestic dry-cleaners and serious effects have resulted from such materials when used to clean upholstery.

The risk is increased in recent years because, owing to changes in building methods, the rates of infiltration of air have been markedly reduced. Kennedy (1970) noted that with relatively low wind velocities, *i.e.* not exceeding 5 m.p.h. and with differences between inside and outside temperatures not exceeding 10° F rates of air change varied from 0.3 to 0.1/hour. Therefore any substance discharged into the air inside the home will tend to remain for longer periods than in the past.

One sometimes tends to take ventilation for granted but the following example of the effects of pollution of indoor air should make one realize that in some countries conditions are so primitive that cooking is done over open fires in huts without chimneys. Clifford and Beecher (1964) report cases of post-nasal cancer associated with the inhalation of smoke-laden air in such huts in Kenya. The smoke comes from fires used for cooling. The effects may be seen from the illustrations below.

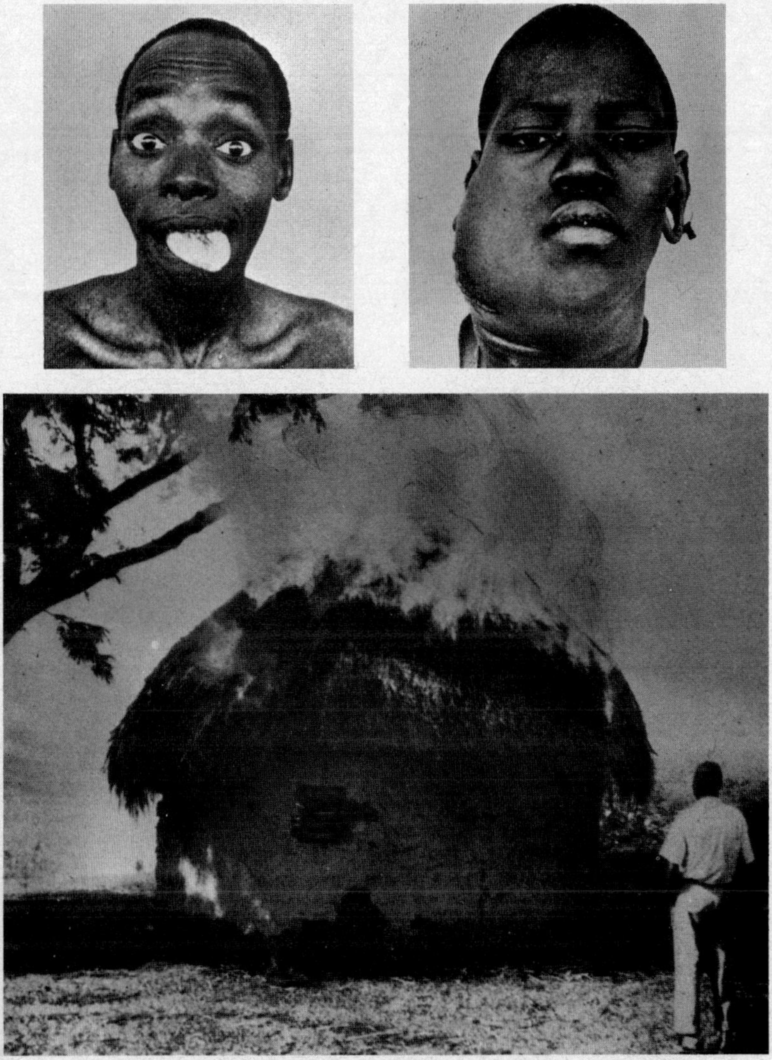

Fig. 40. Effects of inhaling smoke-laden air.

CHAPTER X

THE MEASUREMENT OF VENTILATION

THE ventilation rate can be ascertained directly by measuring the flow of air into or out of a room, or indirectly by measuring the change of concentration of some gas or vapour added to the atmosphere.

DIRECT MEASUREMENT

Direct measurement is usually the method of choice when ventilation is by fans. At fan outlets, or at the air inlets or outlets in individual rooms, the velocity of air flow can generally be measured by means of the vane-type anemometer, and thus the quantity of air passing is rapidly estimated. When measurements of the flow in ducts are required they are generally computed from determinations of pressure differences made with Pitot tubes.

In some buildings with natural ventilation also, direct measurements of ventilation are possible. The velocity of air flow through each individual ventilation opening can be measured with a vane anemometer, and the area of each opening is easily measured. Thus the rate of air flow through each opening is known, and it is usually found that estimates of the total flows through inlets and through outlets correspond remarkably well. Systematic measurements were made in this way by Vernon, Bedford and Warner (1926*a*) in a study of factory ventilation, and during the last war the writer and his colleagues have made many similar determinations of the ventilation rates in factories where shells were filled with explosives.

When, with a mechanical ventilation system, the air is impelled into a room so as to set up a slight positive pressure, a fairly accurate figure for the ventilation rate is probably given by measurements of the amount of air supplied by the fan system, for other openings are likely to serve as outlets unless they are large or numerous. In many factory buildings, however, exhaust fans are used to extract air from the buildings, while windows and doors are also open. These other openings may act as inlets or as outlets, according to their orientation and the direction of the wind, and it may be that the air extracted by the fans represents only a small part of the total air turnover. Thus, Vernon, Bedford and Warner found that in a factory which had a good extraction system, as well as good natural ventilation, there were occasions when the air extracted by the fans represented only about one-third of the total air exchange.

The ventilation rate for a building or an individual room is sometimes expressed in terms of the total amount of air passing in unit time (cu. ft/hour, or per min), or of the amount supplied per occupant/hour or min, and sometimes it is given in terms of the number of air changes, or air turnovers, per hour. As we have seen, from the standpoint of odour control it is the air supply per person that matters, while, if the number of occupants varies from time to time, the air velocity in the room will depend on the total amount of air passing rather than on the air supply per person. The number of air changes per hour is calculated by dividing the total hourly air supply to the room by the cubic capacity of the room. That is, if we say that the ventilation rate in a room is five air changes/hour, we mean that the volume of air supplied each hour is five times the volume of the room.

The heating and ventilation engineer ascertains what is likely to be the number of occupants, and then estimates the number of air changes per hour which will provide a satisfactory fresh air supply. This figure is then used in subsequent calculations of heat requirements and ventilation design.

INDIRECT MEASUREMENT

Ventilation rates in small rooms, such as those in dwellings and in many schoolrooms and offices, have frequently been estimated from changes in the composition of the air, and most commonly from variations in the carbon dioxide content. In the middle of last century, Roscoe (1857) illustrated the diffusion of air from a closed room by liberating carbon dioxide and then tracing the rate at which its concentration diminished.

If the only source of added carbon dioxide is the air exhaled by the occupants, and if occupation has been continuous for a period long enough to ensure that the composition of the air has approximately reached equilibrium, the ventilation rate can be estimated by measuring the concentration of carbon dioxide. For sedentary persons, or those very lightly occupied, the carbon dioxide production amounts to about 0.6 cu. ft/person/hour, and this is the figure most commonly assumed in such estimates. If x is the equilibrium concentration of carbon dioxide in the room, expressed as parts of carbon dioxide/10,000 parts of air, the air supply per person, in cu. ft/hour, a, is found from the relation

$$a = \frac{6000}{x - 4}$$

when the concentration of carbon dioxide in the external air is taken as 4 parts in 10,000, and the production as 0.6 cu. ft/person/hour. The relation between fresh air supply and carbon dioxide content is shown in Fig. 41.

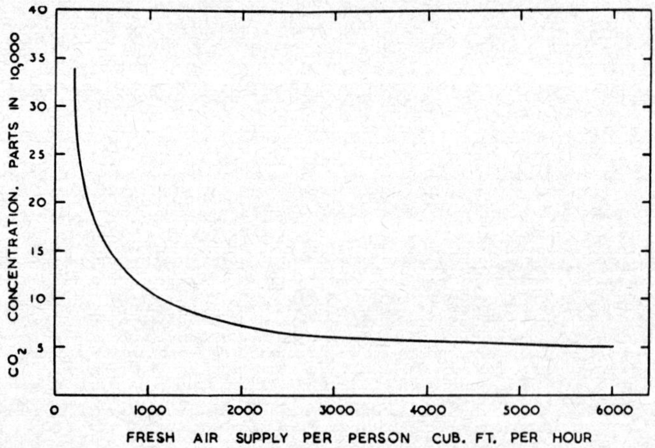

Fig. 41. Relation between fresh air supply and carbon dioxide concentration. (CO_2 production taken as 0.6 cu. ft/person/hour.)

The result obtained by this method is not likely to be very accurate, for the carbon dioxide production of the occupants will not be known precisely, and it may well be significantly different from the assumed figure.

The time taken for the carbon dioxide concentration to reach its equilibrium level depends on the rate of air change as well as on the air supply per person. If it be assumed that all the occupants enter at the same time, and thereafter continue to produce carbon dioxide at the steady rate of 0.6 cu. ft/person/hour, the concentration at any time after the start of the experiment is given by the equation:

$$x = 4 + \frac{6000}{a}(1 - e^{-at/S})$$

where x = the concentration of carbon dioxide in the air of the room, expressed as parts per 10,000 of air;
a = the fresh air supply, cu. ft/person/hour;
S = the number of cu. ft of space per person;
t = the time, in hours, after the beginning of occupation; and
e = the exponential function, 2.718.

In the exponent of e, a/S represents the number of air changes per hour.

The development of the equation for the calculation of the rate of air change from measurements of gas concentration can be illustrated by considering the case when the gas chosen is carbon dioxide, and when its only source, apart from that normally present in the external air, is the respiration of the occupants.

Besides the symbols used above, let

p = the concentration of carbon dioxide in fresh air (parts in 10,000); and
P = the volume of carbon dioxide produced per person/hour (cu. ft).

In a short interval of time (dt) a volume of air, $a \, . \, dt$ cu. ft of air/person enters the room, and a similar quantity leaves it. Carbon dioxide passes from the room with that air to the amount of $ax \, . \, dt/10{,}000$ cu. ft, while there is a corresponding gain from the fresh air of $ap \, . \, dt/10{,}000$ cu. ft. Meanwhile, there is also a gain of $P \, . \, dt$ cu. ft/person from respiration.

The net gain of carbon dioxide/cu. ft of room space in time dt is

$$\frac{1}{S}\left(P + \frac{ap - ax}{10{,}000}\right)dt \text{ cu. ft}$$

and, expressed as parts in 10,000 of air, the increase in the concentration of carbon dioxide, dx, is

$$dx = \frac{1}{S}(10{,}000P + ap - ax)dt$$

Integrating,

$$10{,}000P + ap - ax = ke^{-at/S}$$

where k is a constant.

If x_0 is the concentration when $t = 0$

$$k = 10{,}000P + ap - ax_0$$

whence, denoting the concentration after t hours by x_t, we have

$$10{,}000P + ap - ax_t = e^{-at/S}(10{,}000P + ap - ax_0)$$

When at the beginning of occupation the concentration of carbon dioxide in the air of the room is the same as in the external air, *i.e.* when $x_0 = p$, the equation reduces to

$$x_t = p + \frac{10{,}000P}{a}\left(1 - e^{-at/S}\right)$$

If, instead of carbon dioxide, a gas is used which is not normally a constituent of the air, *e.g.* hydrogen, the terms in p and P vanish, and the equation takes the form

$$x_t = x_0 \, . \, e^{-at/S}$$

The rate at which the carbon dioxide concentration rises after the beginning of occupation, with various rates of air supply and cubic space allowances, is shown in Fig. 42. With low ventilation rates the approach to equilibrium is slow. For

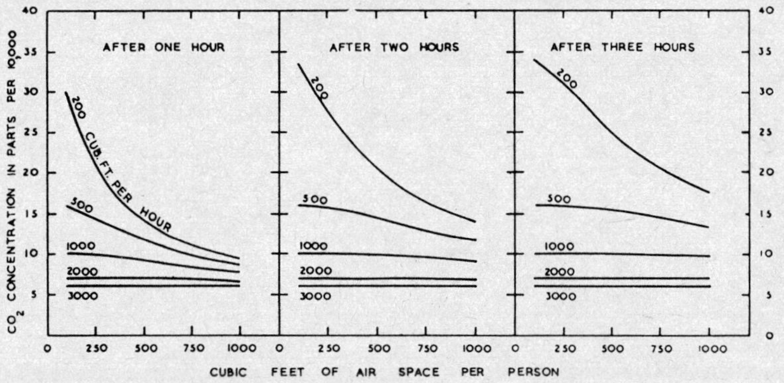

Fig. 42. Increase of carbon dioxide concentration after beginning of occupation.

example, with a very low fresh air supply of 200 cu. ft/person/hour, the carbon dioxide concentration after 3 hours occupation is 23 parts/10,000 when the occupants have 600 cu. ft of air space each, whereas, after the same period and with the same fresh air supply, if the space allocation is only 100 cu. ft/person the carbon dioxide concentration will have practically reached its equilibrium level of 34 parts in 10,000.

Although the average carbon dioxide production of resting persons is taken as 0.6 cu. ft/hour, it may differ from that amount. If the observed carbon dioxide concentration is 6 parts in 10,000 of air, the rate of air change given by the equation for equilibrium conditions, where the production has been assumed to be 0.6 cu. ft, will be 500 cu. ft/person/hour, but if the average production is only 0.5 cu. ft/hour, the rate will in fact be 417 cu. ft. If, on the other hand, the same concentration is observed and the average carbon dioxide production is in reality 0.75 cu. ft/hour, the true ventilation rate is 625 cu. ft/person. When expressed in cu. ft of air supply, these errors seem substantial, but the estimates of ventilation so obtained do, after all, give an indication of the order of magnitude of the ventilation rate which is sufficient for many purposes.

A source of difficulty with high ventilation rates is that a slight difference in carbon dioxide concentration represents a large difference in ventilation. Thus a carbon dioxide concentration of 6 parts in 10,000 indicates a ventilation rate of 3,000 cu. ft/hour/person, a concentration of 7 parts in 10,000 represents a ventilation rate of 2,000 cu. ft/person, and one of only 5 parts a rate of 6,000 cu. ft/person/hour. The measuring burette in the gas analysis apparatus generally used for the estimation of carbon dioxide is graduated to read in parts in 10,000. If useful estimates of ventilation quantities are to be obtained when the carbon dioxide concentration is low, it is necessary to estimate the readings of the burette to the nearest tenth of a graduation. This can be done by a skilled analyst, but much care and practice are required before such accuracy is achieved.

A further source of error is that during normal conditions of occupation, when the ventilation rate is high, a substantial proportion of the air which is admitted to the room may leave it again without passing close to the occupants, and may thus escape without removing a full share of the products of respiration. This is likely to occur unless there is really efficient cross-ventilation. In this respect, Yaglou and Witheridge (1937) computed the relative ventilation efficiency as

$$\text{Ventilation efficiency} = \frac{6000}{x - p} \times \frac{1}{a}$$

where x and p respectively represent the carbon dioxide concentrations (per cent) in the air of the room and in fresh air, and where a is the fresh air supply/person in cu. ft/hour.

They found that in their experimental room the ventilation efficiency was practically 100 per cent when the fresh air supply was below 300 cu. ft/person/hour, and at higher rates of supply it decreased progressively until at 1,800 cu. ft/person/hour it was only about 75 per cent. Hence, it appears that from the standpoint of efficiency large rooms have an advantage over small ones.

Use of a Tracer Substance

A better way of measuring the rate of air change is to discharge into the room a quantity of some gas or vapour and then measure the rate at which its concentration declines owing to the diluting action of the ventilating current. If a gas not normally present in the air is used the relation between rate of air change and gas concentration is

$$x_t = x_0 \, . \, e^{-at/S}$$

or, using logarithms to base 10,

$$\frac{a}{S} = \frac{2.303}{t} \times \log_{10}\left(\frac{x_0}{x_t}\right)$$

Hertz (1896) suggested that the ventilation of a room could be measured by tracing the rate of disappearance of water vapour which had been liberated into the air of the room. Dufton and Marley (1934) investigated the practicability of this method, and found that the great difficulty in the way of its application was that in an ordinary room many hours elapse before the walls, furniture and draperies come into equilibrium with a changed humidity, and if there is no such equilibrium the method is not really applicable.

Later, Marley (1935) liberated hydrogen and, by means of a katharometer, measured the rate at which its concentration declined. In further work (Building Research Board, 1938) he used coal gas as the source of hydrogen. Bedford, Warner and Chrenko (1943) also used coal gas, measuring concentrations with a katharometer.

The katharometer is a convenient instrument for the measurement of hydrogen concentrations. Devised by Shakespear, it has been described by Daynes (1920). Essentially, it is a Wheatstone bridge circuit. The four arms of the bridge are identical platinum spirals, two of them being exposed to dry air in sealed cells, and the other two to the gas under test. Any difference between the thermal conductivity of the standard gas and that under test affects the temperature of the spirals when a current is passed through them, and the resultant difference in electrical resistance unbalances the bridge and deflects the galvanometer needle. Minute changes in gas composition can be accurately measured, and continuous readings of the galvanometer can be obtained without the labour entailed by the usual methods of gas analysis.

With a sensitive galvanometer an initial coal-gas concentration of about 30 parts in 10,000 suffices. This concentration is well below the lower explosive limit for air-hydrogen mixtures, and the low concentration of carbon monoxide which results causes no ill effects in the observer.

The accuracy of this method of measuring air change was tested by Marley in experiments in which katharometer measurements were checked against measurements of air flow made by the orifice plate method; and Warner (1940) reports on a comparison in which the rates of air change in a galvanized tank were determined simultaneously by the katharometer method and by metering the air which passed through the tank. From Fig. 43 it will be seen that the results obtained with the katharometer were virtually identical with those arrived at by the other methods.

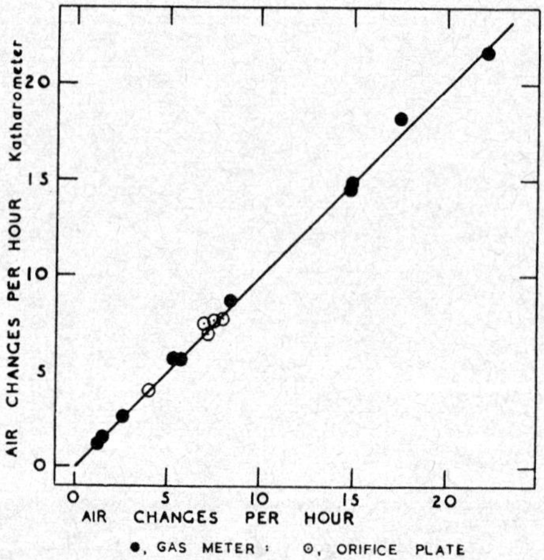

Fig. 43. Rates of change measured by katharometer and by direct measurement of air flow.

Lidwell and Lovelock (1946) describe another method in which the same principle is applied. A highly volatile chemical, *t*-butyl-hypochlorite, is evaporated into the air of the room and allowed to disperse. A drop of another solution (*o*-tolidine and benzoic acid in petroleum ether and chloroform) is placed on a filter paper. The *o*-tolidine reacts with the butyl hypochlorite in the air to give a blue colour on the paper, and the rate of development of this colour is proportional to the concentration of butyl hypochlorite in the air. The times taken to reach three standard intensities of blue–methylene blue stains on test papers–are measured by stopwatch. These times are inversely proportional to the amount of chemical in the air, so that as the hypochlorite concentration is decreased by ventilation the times taken to reach the standard colours increase. From a series of timings the rate of air change can be calculated.

This method has the great advantage of simplicity. The apparatus required consists only of two reagent bottles, filter papers and a stopwatch, and no electrical supply is needed. But the results obtained are not always as accurate as those yielded by some of the other methods. Observers vary in their ability to match colours, and when this matching has to be done with one of the colours increasing in intensity more or less rapidly some observers may find difficulty in making judgments. Further the colour given by the reaction of the hypochlorite with tolidine is not of exactly the same hue as methylene blue, and this difference tends to make the matching more difficult. The method is useful where it is only necessary to ascertain the ventilation rate approximately, but for more exact measurements another method is desirable.

A further method of measuring ventilation is described by Lidwell and Lovelock. Acetyl acetone or ethyl acetoacetate is used as the tracer substance, and its concentration in the atmosphere is estimated by measuring its ultra-violet absorption. This method enables the ventilation rate to be measured with accuracy. It is also used by Lidwell (1946) for measuring the rate at which air circulates within a room. The bactericidal effect of partial irradiation by ultra-violet light depends to a great extent on the vertical circulation rate, and the circulation rate is readily estimated by the use of this method.

In recent years nitrous oxide has been used as a tracer substance with an infra-red gas analyser as a detector which is capable of responding to a concentration of one part of nitrous oxide in a million parts of air. This method has been used for measuring the rate of ventilation of large rooms and other places where the katharometer could not be used owing to the large amounts of tracer required.

A disadvantage of the infra-red gas analyser is its cost and complexity: it is also somewhat bulky and heavy for use in field investigations. Such disadvantages do not apply to methods in which radio-isotopic substances are used as tracers. Since Collins and Smith (1955) first described one such method, several studies of ventilation have been made in which radioactive tracers have been used.

As with any other tracer gas, if C_0 is the concentration immediately after release, and C is the concentration after time t, the ventilation rate R may be obtained from the expression:

$$C = C_0 e^{-Rt}$$

Provided that the half-life of the radio-isotope used is very much longer than the period of measurement, C and C_0 in the equation may be taken as the count-rates measured by a suitable detector. Such a detector will also respond to background radiation when no radioactive gas is present and the above equation becomes

$$C = C_0 e^{-Rt} + C_b$$

where C_b is the background count-rate.

Another correction must be applied because of the delay due to the response time of the apparatus used: the relation between the measured count-rate C' and the true count-rate C is given by

$$C = \frac{C'}{1 - C'T}$$

where T is the so-called 'dead-time' of the equipment used to measure the count-rate.

If the above equation is differentiated with respect to time, and provided that the background count-rate is negligible, the rate of ventilation R is proportional to

$$-\frac{\mathrm{d}}{\mathrm{d}t}(\ln C).$$

Thus the rate of ventilation may be obtained by connecting the detector to a logarithmic count-rate meter, the output of which is connected to a differentiating circuit. The output signal of this circuit is directly proportional to the rate of ventilation and may be indicated by an appropriate meter. This is the basis of an instrument which directly measures the rate of ventilation following the release of a radioactive-isotope tracer gas. It was designed by Evans and Webb (1971) who have recently critically examined the use of radioactive tracers in measurements of the ventilation of simple enclosures. These workers have shown that with such methods the accuracy of measurement of ventilation is better than ±3 per cent. They recommend the use of ^{85}Kr as the radiological hazards in measurements of ventilation with this tracer can be made negligible. The instrument which they have designed is compact and portable and as such it has obvious advantages for use outside the laboratory. It represents, in the opinion of the writer, a major advance in this field.

CHAPTER XI

NATURAL VENTILATION

VENTILATION systems are divisible into two main groups; natural ventilation systems in which reliance is placed on natural forces without the aid of mechanical power, and mechanical systems in which power is used to drive fans. Frequently, mechanical ventilation is employed to augment the ventilation obtained by natural means.

The two agencies that bring about natural ventilation are wind pressure and the force of gravity. The wind drives air in through openings on the windward side of the building so that it passes out on the leeward side. Passing over a chimney top or a roof cowl, the wind has an aspirating action on the air within the flue or cowl. Gravity effects are due to temperature differences. Cool, and relatively dense, air displaces warmer and lighter air, so that convection currents are set up.

An unheated room provided with an open fireplace has a fair measure of ventilation due to the passage of the wind over the chimney, but when a fire is lit in the grate, and the air in the flue is heated, the convection effect is added to the aspirating influence of the wind, and the ventilation rate is substantially increased. As another example, consider a factory building with roof cowls and with window openings in the walls. When, as is usual, the temperature within the factory is above that outside, the warm air inside is displaced by the cool air entering at the windows, and it passes out through the roof cowls. This gravitational effect is supplemented by the influence of the wind blowing over the cowls.

LAWS OF AIR FLOW

Shaw (1892, 1907) enunciated four laws of the flow of air in circuits. Such a circuit begins and ends in the external air, and

Shaw regarded it as an inlet channel and an outlet channel between the outside air and the ventilated space, with some motive power, natural or artificial, to maintain the flow. An air circuit may be simple, with a single inlet and outlet and a single motive power, or it may be complex with several inlets or several outlets, any of these being provided with an independent source of motive power.

The process of continuous ventilation consists in the admission of a quantity of air to a room or other enclosed space through an inlet or inlets, and the simultaneous removal of a similar quantity of air from the ventilated space. The inlets may be open doors or windows, or even the crevices which always exist around doors and windows. *This process of the continuous entry and egress of air is referred to by Shaw as the Law of Continuity of Flow. As he remarks, it is the first fundamental principle in the theory of ventilation, and needs only to be stated for its importance to be allowed. Yet it is one of the principles most often disregarded, for houses and other buildings are frequently erected with no proper provision for the entry of air to replace that which is removed by chimneys and other outlets.* Thus objectionable draughts are often caused. On the other hand, the inlets may be satisfactory and the outlets insufficient for adequate ventilation as the following example shows.

During a visit to one of the newly-created universities during winter, the present writer noticed a series of open doors along a corridor in one of the residential buildings. The occupants of the rooms explained that the system of heating produced uncomfortably hot conditions. The opening of windows was insufficient to reduce the temperature to a comfortable level so the doors were opened as well. There were no adjustable 'fan lights' over the doors.

If air is to be moved in a ventilating system some motive power is necessary. Shaw's second fundamental principle is the law of relation between the 'head', or aeromotive force, and the flow of air through an orifice. Considering an orifice in a thin plate, the law is stated: The head is equal to the product of the resistance of the orifice and the square of the flow. This may be written

$$H = R \,.\, V^2$$

when H is the head, R is the resistance, and V is the flow. If the head is expressed in ft-lb/lb of air, or feet of air, and the flow in cu. ft/sec, R is equal to $1/27a^2$, when a is the area of the orifice in sq. ft.

When air flows along a duct there are losses of head due to friction against the sides of the channel, to sudden bends or changes in diameter, besides the losses due to friction on entering the duct, and possibly to the contraction of the duct where the air emerges. Each of these losses is proportional to the square of the flow, so that the final relation between head and flow is of the form:

$$H = (a + b + c + d + \ldots)V^2$$

where a, b, c, ... refer to the losses of head due to the several causes, and if the sum of these losses is called R the equation is still

$$H = R \,.\, V^2 .$$

This total resistance is conveniently expressed in terms of the equivalent orifice in a thin plate, the equivalent orifice, a, being connected with the resistance, R, by the relation, $R = 1/27a^2$.

The third law states that when two or more ducts are in parallel, and themselves produce no head, their effect is the same as that of a single duct whose equivalent orifice is equal to the sum of the equivalent orifices of the separate ducts.

Shaw's fourth law relates to instances where air flows through a number of ducts in series, and is stated as: The resistance of any number of ducts in series is the sum of the resistances of the separate ducts, provision being taken for the non-interference of their rapids.

The natural forces which produce head are: (*a*) variation of barometric pressure, or of temperature; (*b*) direct impact of wind on an opening; (*c*) wind blowing across an orifice; and (*d*) hot air or smoke in flues. Head may also be produced by (*e*) the injector effect of steam or compressed air jets; (*f*) ventilating fans; and (*g*) blowing engines. The following table showing the wind velocities or temperature differences required to produce various heads is adapted from a table given by Shaw (1907).

Showing 'head' produced by various natural agencies

Head, expressed in			*Head produced by*		
			Wind velocity (ft/sec)		*Internal temp. of a 40 ft. shaft, °F with external temp. of 41°F*
Ft of air	*Ins. of water*	*Lb. wt./ sq. ft*	*Direct impact*	*Passing over mouth of duct*	
1	.015	.08	9.0	7.9	54
2	.03	.16	12.6	11.3	63
3	.05	.23	15.5	13.8	78
4	.06	.31	17.0	16.4	91
5	.08	.39	21.6	17.9	103
6	.09	.47	23.3	19.8	115
7	.11	.55	25.9	21.1	128
8	.12	.62	27.4	22.4	140
9	.14	.70	39.0	23.7	153
10	.15	.78	54.9	25.0	165
20	.30	1.56	61.1	35.8	291
30	.45	2.34	94.9	43.7	–

MEASUREMENTS OF NATURAL VENTILATION IN DWELLINGS

Infiltration

Even when a room is closed, *i.e.* when the door and windows are closed and any flue or other fixed ventilator is sealed, there is always some ventilation due to infiltration through cracks and joints, and, in less degree, through the fabric of the building. In the absence of exact information, the amount of ventilation occurring in closed rooms has tended to be overestimated. Bedford, Warner and Chrenko (1943) made many determinations of the ventilation rates in closed rooms, and their data are tabulated below. The table shows, for various groups of

Frequency of observations of various ventilation rates in closed rooms

Description	Size of rooms (cu. ft)	No. of observations	Average no. of air changes/hour	Percentage of observations in which no. of air changes/hour was			
				Under 0.4	0.4 to 0.99	1.00 to 1.99	2.0 and over
London flats (1939)	940	84	0.44	37	63	–	–
London flats (1936)	535 to 938	33	0.69	15	63	22	–
London flats (1939)	1070	45	0.69	5	84	11	
London flats (1888)	900 to 1435	24	0.71	–	88	12	–
London cottages (1860)	620 to 1280	55	1.51	2	23	55	20
Other dwellings (1898-1928)	410 to 1810	117	1.04	1	59	32	8
All above		358	0.80	11	60	23	6

dwellings, the average ventilation rates and the proportions of observations in which various ventilation rates were found.

In the flats built in 1936, for instance, where all the rooms examined were of less than 1,000 cu. ft capacity, 78 per cent of the observations showed ventilation rates of less than one air change/hour, and in 15 per cent of the observations rates below 0.4 air change/hour were found. In one of the blocks of flats the average ventilation rate in the closed rooms was only 0.44 air change/hour, and no observation gave a figure exceeding one air change/hour.

At the other end of the scale, in the cottages built about 1860, much higher ventilation rates were found. The average for closed rooms was 0.5 air changes/hour, and in 20 per cent of the observations more than two air changes/hour were recorded. The bulk of the observations in all these dwellings were made in rooms little if any larger than 1,000 cu. ft, but even when the old cottages are included the average ventilation rate was only 0.8 air change/hour.

In the London flats built in 1888 the ventilation rate averaged only 0.71 air change/hour, and in only 12 per cent of the observations did it exceed 1 air change/hour. With this exception there is a broad relationship between the age of the houses and the amount of ventilation obtained; in the older dwellings the greater ventilation rates were found. This result is probably to a large extent due to the shrinkage of woodwork. A further factor is the lower permeability of the fabrics in modern buildings, due to the use of hard plaster finishes.

In some of the old London cottages mentioned in the preceding table, estimates were made of the leakage areas around doors and windows. All the rooms considered were of about 650 cu. ft capacity. In three groups of rooms the average leakage areas were respectively 17, 29 and 79 sq. in, and the average ventilation rates for these rooms were 1.3, 1.6 and 2.1 air changes/hour.

Dick (1949) reported on studies carried out during the winter months in four experimental houses in which provision was made for the entry and exit of air. The replacement air for a room without convection ducts or ducts from outside was obtained through a ventilator above the door to the hall, which in turn had a similar ventilator beside the front door. In a flueless room the exit for air was a ceiling ventilator and duct to the ventilated roof space. The houses were situated on an exposed site.

In six bedrooms of 920 or 1,025 cu. ft capacity the average ventilation rates observed ranged from 1.1 to 2.8 air changes/hour. The standard error of the mean of 1.1 air changes was only 0.08, while that of the mean of 2.8 air changes was 0.31 air changes. In three smaller bedrooms of 520 cu. ft capacity the averages were 1.2, 1.2 and 3.1 air changes/hour with standard errors respectively of 0.08, 0.36 and 0.15 air changes. In four living rooms of 2,035 cu. ft capacity the average rates and their standard errors were 1.4 ± 0.06, 2.6 ± 0.26, 2.7 ± 0.07 and 2.9 ± 0.13 air changes/hour.

The ventilation rates could generally be expressed as a linear function of wind speed, although in some instances the direction of the wind was also a significant factor.

A substantial amount of leakage may occur around windows which are apparently close-fitting. In a room in a university building, with well-fitting steel frame windows, the rate of air change with the door and windows closed was 0.36/hour, but when the cracks around the windows and door were sealed with glued paper the ventilation rate was reduced to 0.08 air change/hour. In this instance there was very little leakage through the fabric.

When air gratings, flues, or other ventilators are open, the wind has an important influence on ventilation, but the effect of wind speed on the ventilation of closed rooms is generally very small, and is sometimes undetectable. In such rooms little or no provision is made for continuity of flow. Where such provision is made the leakage through crevices may be considerable. Emswiler and Randall (1928) built a steel-framed window into a wind tunnel and subjected it to a series of pressure differences corresponding to wind velocities up to 25 m.p.h. The leakage/hour/ft run of orifice was 12 cu. ft when the opening was 1/64 in. Cracks of 1/32 and 1/16 in respectively gave leakages of 20 and 40 cu. ft/hour/linear ft. The tightening of a locking device on windows may reduce infiltration by 17 per cent (Rusk, Cherry and Boelter, 1932). In Canada and the United States sash windows are weather-stripped to reduce infiltration. Leakage rates for various types of window at different wind speeds are tabulated by Harding and Willard (1929).

Roscoe (1857) showed that gases can diffuse through brick, and Larson and others (1929) observed the rates of air infiltration through 13-in brick walls of various types. These latter observers noted that the leakage through hard brick walls in cement-lime mortar was only one-quarter of that through porous bricks set in lime.

Internal wall finishes reduce the amount of infiltration. Houghten and Ingels (1927) noted that brick walls (9 in or 13 in) allowed some infiltration, but when the walls were finished with hard gypsum plaster they became practically air-tight. Papered walls admit less air than distempered walls. According to Wolpert (1905), the ventilation rate in a standard room was one-sixth greater when the walls were distempered than when they were papered. Data obtained by Wellner (1932) and Warner (1940) show the hindrance to ventilation caused by treating walls with oil paint.

When infiltration through the fabric of a building is a significant factor it is to be expected that with rooms of similar shape the ventilation rate, expressed in air changes/hour, will decrease with increasing size of room, for the surface area will vary not as the volume but as the two-thirds power of the volume. Haldane (1899) made experiments in rooms of various sizes, and found a close relationship between the volume of ventilation and the two-thirds power of the room volume, and various other observers have noted a tendency to a similar relationship. Warner noted such a tendency, but in later unpublished observations by Bedford, Warner and Chrenko no such relationship could be detected in some series of observations, while it could in others. In a great many buildings infiltration through crevices is of far greater import than diffusion through the fabric.

WALL VENTILATORS

The ventilation rate in a closed, flueless room tends to be small, and if a reasonable fresh air supply is to be ensured when the windows are closed some permanent ventilator is necessary. Both inlets and outlets should be provided, but often the only specific provision for ventilation is some form of opening in the wall—a louvre, air grating, or other fitting.

Quite often air gratings which provide only a small opening for the passage of air are used. Such gratings do not, as a rule, exert any great effect on the ventilation rate. Bedford, Warner and Chrenko (1943) mention observations made in two sets of rooms. The *A* rooms were of 700 cu. ft capacity and the *B* rooms 1165 cu. ft. The former had no flues, and were each provided with a small louvred ventilator of which the effective opening area was 10.5 sq. in; and each of the latter had a flue together with a lattice grating of only 6 sq. in available opening. Observations were made in the *B* rooms with the flues sealed, and in all the observations in both sets of rooms the doors and windows were kept closed. The average ventilation rates are shown in the table.

Effect of small ventilators on ventilation of unheated rooms

Rooms	*Room volume (cu. ft)*	*Effective opening in wall ventilator (sq. in)*	*Average ventilation rate (air changes/hour) with air-grating*		*Percentage increase due to open air-grating*
			Open	*Closed*	
A	700	10.5	0.83	0.55	51
B	1165	6.0	1.04	0.71	46

With the the air-gratings open the ventilation was increased by about 50 per cent, but even that increased rate of air change can scarcely be described as good ventilation.

Effects of Wind Speed and Direction and of Size of Ventilator

The former London County Council By-Laws required that any room used as an office or for habitation, if not ventilated by a ventilation system installed in the building, shall have a flue or other aperture of at least 50 sq. in area communicating with the outside air, or if the opening communicates with a lobby or corridor it must be 100 sq. in in area.

Observations were made by Bedford, Warner and Chrenko in flueless rooms in a block of flats where wall ventilators giving an opening area of 50 sq. in had been fixed. The rooms were of 940 cu. ft capacity. Each had a louvre in the external surface of the wall with an effective opening area of 76.5 sq. in. The aperture in the wall was 9 in x 9 in, and into this, on the inner aspect, was fixed a sheet metal hopper which deflected the entering air upwards and provided the opening of 50 sq. in. Measurements of ventilation were made with doors and windows shut. On each day of observation the ventilation rate was ascertained when the ventilator was completely sealed, and with the hopper fully open to 50 sq in, and further measurements were made when the hopper was partially closed. Further trials were made when the hopper was removed and the opening in the outer wall varied. For a given size of orifice the ventilation rate was the same whether the restriction of the opening was on the inner or the outer face of the wall.

During each set of experiments the direction and strength of the wind were observed. These determinations were made at a height of 4 ft from the ground, in the middle of the road between blocks of flats. The wind speeds thus measured were usually substantially lower than those recorded by the Meteorological Office in exposed positions at the same time. In a densely built-up area buildings shield each other and reduce the force of the wind at street level. A comparison of wind speeds observed near street level in London with those measured by the Meteorological Office at Kew shows that, on the whole, when the wind speed at Kew was 5 m.p.h. that in London streets was only 1.8 m.p.h., and when the Kew value was 20 m.p.h. the speed near the street level in London averaged 4.1 m.p.h. Probably because of the influence of variations in wind direction on the shielding effect of buildings, the correlation between the two sets of data is not a very close one.

For the consideration of the effects of wind on the ventilation afforded by wall ventilators the air speed recorded on the site was used.

Figure 44 shows, for a wind speed of 2.5 m.p.h., the relation between the ventilation rate and the size of ventilator, (*a*) when the wind was blowing directly on the ventilator, *i.e.* at an angle of 90° with the wall; (*b*) when the wind direction made an angle of 45° with the wall; and (*c*) when the wind direction was parallel with the wall in which the ventilator was fixed. When the grating was on the leeward side of the building the ventilation rate was the same as when the wind blew parallel to the grating. Up to an area of 50 sq. in an increase in the size of the ventilator was associated with a proportionate increase in the ventilation rate. When the hopper was removed and the full louvre opening of 76.5 sq. in was utilized, the same proportionate increase in the ventilation rate was observed. With the wind speed of 2.5 m.p.h. and the wind blowing directly on to

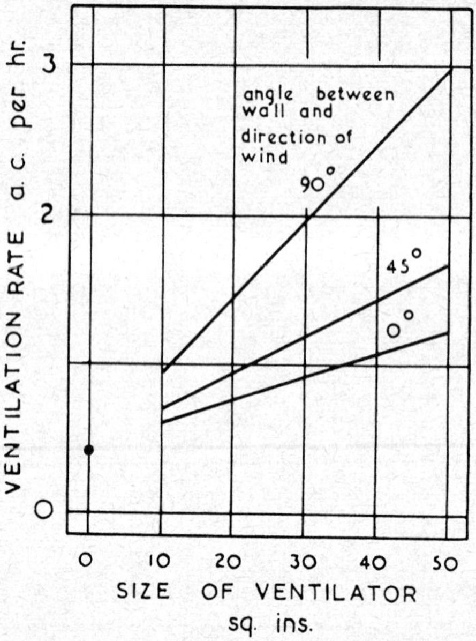

Fig. 44. Effect of size of wall ventilator and of wind direction on ventilation rate (wind speed of 2.5 m.p.h.).

the grating the ventilation rate was only 0.95 air change/hour with a 10 sq. in opening, and when the opening was increased to 50 sq. in the rate increased to 3 air changes/hour. With the wind at an angle of 45° with the wall the ventilation rates for openings of 10 and 50 sq. in were respectively 0.7 and 1.7 air changes/hour; while with the wind parallel with the wall the corresponding rates were 0.6 and 1.2 air changes/hour. When the ventilator was closed the average ventilation rate was rather more than 0.4 air change/hour.

In Fig. 45 the effect of wind speed on ventilation is shown for a wind at 45° with the wall. With the 50 sq. in opening, the ventilation rate was 3 air changes/hour with a wind speed of 5 m.p.h., and barely 1 air change/hour when the wind was only 1 m.p.h. For the same wind speeds the corresponding rates with the 10 in opening were about 1.0 and 0.5 air changes/hour. Wind had little effect on the closed room: an increase in velocity of from 1 to 5 m.p.h. only increased the ventilation rate from 0.4 to 0.5 air change/hour.

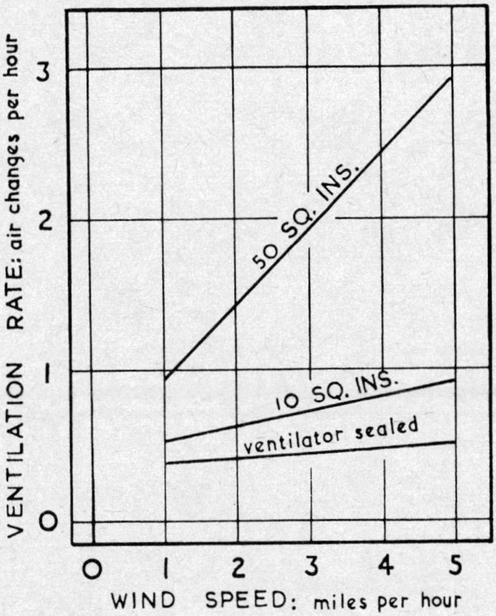

Fig. 45. Effect of size of wall ventilator and of wind speed on ventilation rate, with wind at angle of 45° with wall.

The ventilation rates with an opening of only 10 sq. in are in close agreement with the rates observed in other rooms with small air gratings to which reference has already been made.

It has been said that the 50 sq. in ventilator required by the by-laws in flueless rooms may give rise to objectionable draughts, but observations give no support to this contention. During every determination of the ventilation rate measurements were also made of the air velocity at head level in the middle of the room. The velocities were always low, and the highest velocity found with a 50 sq. in opening was only 12 ft/min. In no single observation was the rate of air movement such as would be expected to provoke complaints of draught even from sensitive persons.

EFFECTS OF UNHEATED FLUES

When a room is unwarmed and the flue unheated the temperature of the room is not greatly different from that of the outside air, and temperature then exerts relatively little effect on the ventilation afforded by the flue. The main ventilating agent in such circumstances is the wind.

Ventilation with the Standard Flue

Even when there is no fire at its base the ordinary flue is a valuable ventilator. Measurements were made by Bedford, Warner and Chrenko in a room provided with a flue of the standard 9 in × 9 in cross-section, and with a chimney throat of 53 sq. in. The average ventilation rate when the flues were open was nearly three times as great as when they were closed. With the flues closed the number of air changes/hour averaged 0.72, and when they were open 2.06. The corresponding volumes of air passing through each room were 820 and 2300 cu. ft/hour. These measurements were made as paired comparisons, *i.e.* during each observation period determinations were made with the flue open and with it closed, so that the differences between the ventilation rates cannot be ascribed to differences in wind strength.

The effect of wind speed on ventilation through an unheated 9 in × 9 in flue is shown by the data obtained in another series of rooms. The throat apertures here were 51 sq. in. The Meteorological Office figures for wind speed are used, as these give a better measure of the velocity at chimney level. The rooms were in 5-storey buildings on an exposed site.

The results of observations made with doors and windows closed are summarized in the table. On the whole, an increase of 10 m.p.h. in the wind speed was accompanied by an increase of about 1.3 air change/hour in the ventilation rate. Even at the lowest wind speeds encountered in these observations the ventilation with the flue open was much more than when it was sealed, and with a wind of 20 m.p.h. it was about 5 times the closed room value.

When a flue is heated, additional height of flue causes additional ventilation, for the ventilating head varies with the height of the column of hot air in the flue. In an unheated flue, however, when the head is due to the aspirating action of the wind, increase in flue height imposes additional resistance to the

Showing effect of wind speed on ventilation through unheated 9 in × 9 in flue

Wind speed, m.p.h. (M.O. value)			*Average ventilation rate/hour*	
Range	*Mean*	*No. of observations*	*No. of air changes*	*Volume of air (cu. ft)*
7 to 10	8.6	10	1.82	1,947
11 to 14	12.8	19	2.01	2,151
15 to 18	15.4	10	2.92	3,124
19 to 24	21.7	6	3.53	3,777
	Mean with flue sealed		0.69	738

flow of air, and thus tends to reduce ventilation. This is illustrated by observations made in a 5-storey block of flats. On each day observations were made on the ground, second, third, and fourth floors, and on some occasions also on the first floor. The average wind speeds for the observations on the several floors were substantially the same.

The average ventilation rates are shown in Fig. 46. It is clear that increasing height of flue was associated with reduced ventilation. On the whole, an increase of 10 ft in the flue height was accompanied by a reduction of 0.26 air change/hour in the ventilation rate. The rooms were of 1070 cu. ft capacity, so that for every 10 ft increase in flue height there was a reduction of 278 cu. ft/hour in the volume of fresh air entering the room.

Ventilation with Restricted Flues

If, with a 9 in × 9 in flue, the chimney throat is restricted, the amount of ventilation is reduced. The extent of this reduction was measured in two rooms (Fig. 47).

Observations were made with the throat of 50 sq. in fully open, with reduced throat apertures of 30 sq. in and 20 sq. in and with the flue sealed. Doors and windows were shut. The volumes of the rooms were 1725 and 1420 cu. ft. The differences between the ventilation rates for the two rooms were due to differences in wind speed. Reduction of the flue throat from

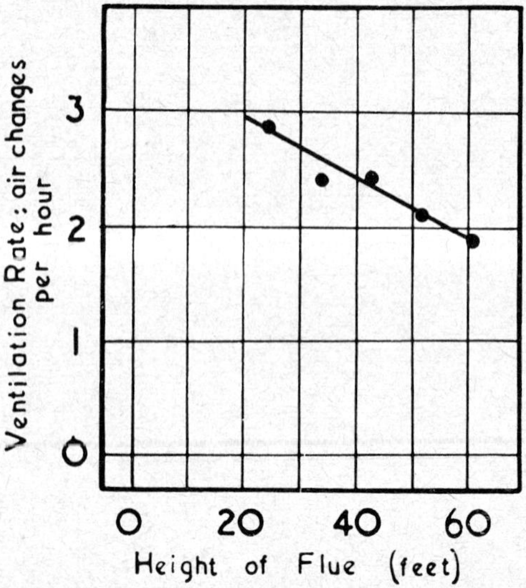

Fig. 46. Ventilation in relation to height of unheated flue.

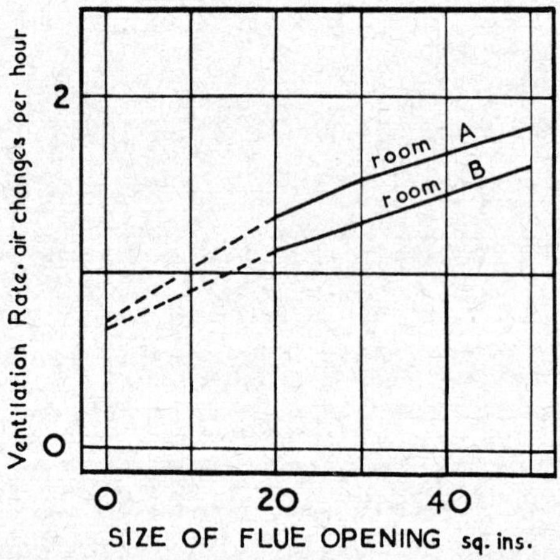

Fig. 47. The effect of constriction of the flue throat on ventilation.

50 sq. in to 20 sq. in reduced the ventilation rate by about 30 per cent.

Warner (1940) made observations in rooms of 900 cu. ft capacity, which had 4-in diameter flue pipes leading from gas fires. In the closed rooms, with the flues sealed, the ventilation rate averaged 0.84 air change/hour, and when the flue was open the average was increased only 1.17 air change/hour. This increased rate was only one-half the average rate observed in nearby buildings in which the flues were of standard size, when the average wind speeds were practically the same. The flue of 4 in diameter only increased the ventilation by about 40 per cent above the closed room value, whereas the standard flue nearly trebled it.

Ventilation with Wall Opening and Flue

The observations on unheated flues so far discussed were made with doors and windows, and any fixed ventilators, closed. The available inlet area was small, and the flow of air up the chimney was thereby restricted. Additional inlet area would have allowed of the removal of greater volumes of air. Warner made observations in three groups of rooms which had wall ventilators of 6 sq. in or 10.5 sq. in opening area as well as flues. In each instance the ventilation rate with the wall ventilator and flue both open was greater than with either one open and the other closed. The excess of the rate with both ventilators open over the closed room value was practically the same as the sum of the separate excesses of the two means of ventilation taken separately. It appears that in these observations the effects of the wall ventilators and of the flues were additive.

VENTILATION WITH HEATED FLUES

Standard Flues

At the Building Research Station (1939*b*) observations in a room heated by an open fire gave a ventilation rate of 4.5 air changes/hour, while the rate when the flue was open but unheated was 1.7 air changes. Bedford, Warner and Chrenko made similar measurements in a room of 1420 cu. ft capacity, with similar results. When the door and windows were closed and a fire was burning in the grate the average ventilation rate was 4.5 air changes/hour; when there was no fire and the flue was cold the ventilation rate was 1.6 air changes/hour, and when the flue was sealed 0.6 air change. These rates represent air flows of 6,390, 2,272 and 852 cu. ft/hour respectively.

Restricted Flues

The effects of restriction of the flue on the ventilation of an unheated room have already been shown. Any such restriction also hinders ventilation when a fire is burning at the base of the flue. Dufton (1942) records that when similar temperatures were maintained in a room heated (*a*) by an open fire, and (*b*) by an experimental fire employing an adaptation of the *cheminée de Nancy,* the number of air changes/hour was 3.9 with the open fire and 1.5 with the experimental one. The fuel consumption in the first case was 3 lb of coal/hour and in the latter, 2 lb. The substantial reduction in the rate of air flow up the chimney was undoubtedly due mainly to the restriction of the flue caused by the fitting of the tapering metal trumpet over the experimental fire, for the flue opening was probably restricted to about 15 sq. in.

Other examples of the effect of restriction of a flue are given in a report from the Building Research Station (1939*b*). When an anthracite stove was fitted into the fireplace, and alight, the ventilation rate was only 0.7 air change/hour, whereas it was 1.7 air changes without heating when the flue was unrestricted. With a gas fire in the fireplace the ventilation rate with the fire alight was 3.1 air changes/hour.

VENTILATION WITH OPEN WINDOWS

If liberal window openings are provided in suitable positions any reasonable amount of ventilation can generally be obtained. The amount of ventilation provided by a given area of window opening depends on the strength and direction of the wind, as well as on the disposition of the windows. If windows on opposite sides of a room are open, through currents of air are set up and the room is thoroughly scavenged. In experimental rooms in a university building, of a size commonly found in dwellings, Chrenko measured ventilation rates as high as 30 air changes/hour due to cross-ventilation.

In the ordinary living room the ventilation effected by an open window is much modified by the presence or absence of an open flue. Thus, Warner (1940) found that in one room when a sash window was open to the extent of 2 in the ventilation rate was 1.8 air changes/hour when the flue was sealed. The flue was then opened, the window remaining untouched, and the ventilation rate increased to 5.4 air changes/hour. In another room with similar arrangements the corresponding ventilation rates were found to be 1.9 and 4.3 air changes/hour.

Formerly, sash windows were in general use in houses in Great Britain, but in recent years casement windows have come into considerable favour. The casement window has the advantage that its whole area can be opened, whereas only one-half of the area of a sash window is available for ventilation.

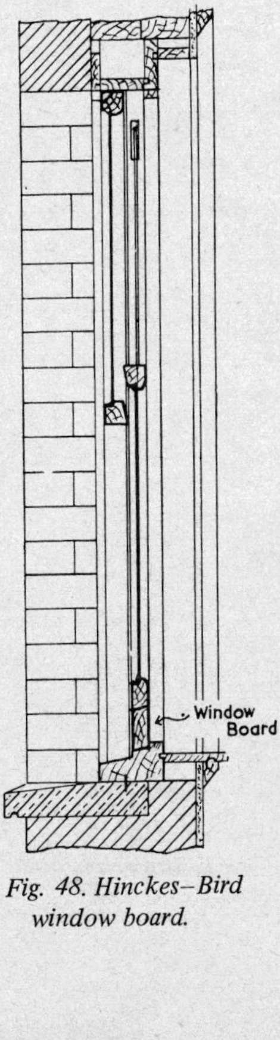

Fig. 48. Hinckes–Bird window board.

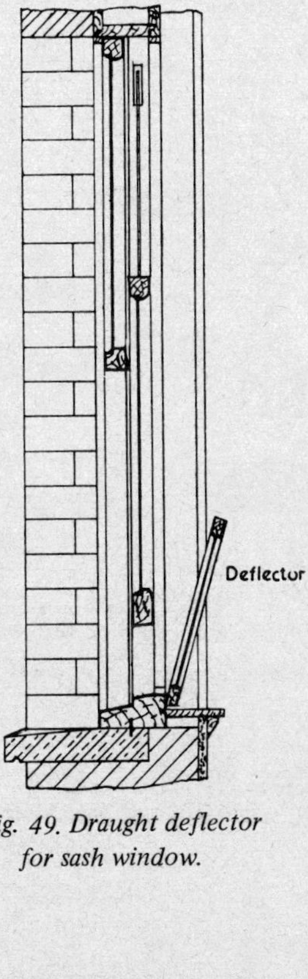

Fig. 49. Draught deflector for sash window.

Nevertheless, the sash window has considerable merit in that it can be arranged to allow of the entry of air without causing objectionable draughts. A simple contrivance which has long been used is the Hinckes-Bird board (Fig. 48). This is a board, 3 in or so in width, which is made to fit in the window frame below the bottom sash. When it is inserted an opening is left between the two sashes, and the air which enters through the opening is deflected upwards so that it passes over the heads of the occupants.

Figure 49 illustrates a deflector which can be fixed to the bottom portion of the window frame. The window can be opened to varying extents so that the amount of ventilation can be varied. Deflectors are not easily fitted to casement windows.

VENTILATION OF POST-WAR HOUSES

The Committee of the Building Research Board which has advised the Ministry of Works (1945) on heating and ventilation in post-war buildings, recommends that the minimum fresh air supply in rooms in dwellings should be 600 cu. ft/person/hour. This figure is lower than the 1,000 cu. ft suggested as the desirable minimum in Chapter IX. It is not a lavish standard, and the recommendation was probably influenced by the need to conserve fuel and keep down fuel costs. The recommended standard does, however, represent an improvement on the ventilation now achieved in a great many houses during cold weather. For instance, many bedrooms which are occupied by only one person have a cubic capacity of less than 700 cu. ft. Warner (1940) gives data relating to bedrooms in modern flats which had a capacity of 670 cu. ft. The average ventilation rate in these rooms was 0.83 air change, or 580 cu. ft/hour, when the air gratings were opened. This average figure was below the 600 cu. ft now recommended, but in 55 per cent of the observations the ventilation rate was less than 500 cu. ft/hour, and in 5 per cent it was below 335 cu. ft. When the air gratings were closed 89 per cent of the observations gave ventilation rates of less than 500 cu. ft/hour, and 18 per cent below

270 cu. ft. Thus, if the 600 cu. ft minimum standard can be achieved a distinct advance will have been made.

The former Ministry of Works standard requires 600 cu. ft of fresh air/hour in a bedroom with one occupant, 1,200 cu. ft in a double bedroom, and 2,400 cu. ft/hour in a living room where four occupants can be expected. If these ventilation rates are to be attained in cool weather some form of permanent ventilator will be necessary. It is useless to rely on windows for, unless they are fitted with draught deflectors, draughts will be experienced. Moreover, to obtain sufficient ventilation in cold weather, but not too much, requires frequent adjustment of the sizes of the window openings to compensate for wind variations, and most people will prefer to keep their windows closed. Earlier in this chapter the effects of varying wind conditions on the ventilation given by a 50 sq. in wall ventilator have been shown. From these data it is evident that if the desired minimum rate of ventilation is achieved by such openings when the ventilators are not exposed to the direct impact of the wind, when they are so exposed the ventilation rate will be far above that minimum. Then, even if the air velocity in the room is too low to cause draughts, the heat loss from the room may be excessive. Similarly, if reliance is placed on unheated flues and smaller wall ventilators the amount of ventilation will be much influenced by vagaries of the wind.

Some arrangement is necessary whereby natural ventilation can be automatically controlled, and compensation made for changes in wind. The Colt 'constant flow' ventilator appears to achieve the desired end (Fig. 50). This type of ventilator is available in a variety of sizes and capacities, but for domestic use the standard model allows of an air flow of 600 cu. ft/hour. Ordinarily, it will act either as an inlet or as an outlet, but it can be arranged that air is allowed to flow in only one direction. The ventilator is so contrived that an increase in the wind pressure on it raises the resistance to flow, and thus a constant flow is secured. A design requirement was that when the ventilator acted as an inlet there should be no perceptible draught. The maximum air velocity 6 in away from the inlet is said to be only 100 ft/min, and at 4 ft distance it is only

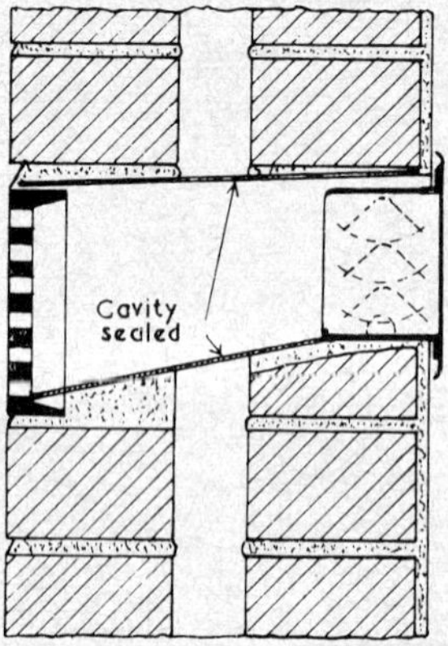

Fig. 50. Sectional view of Colt constant flow ventilator.

30 ft/min—an air velocity commonly found in satisfactorily ventilated factories.

Figure 50 shows one of these ventilators fixed in a cavity wall. On the external face of the wall the ventilation opening is guarded by an air grating. It is important that this grating should have an opening area as large as possible, for any restriction of the opening will reduce the flow of air.

NATURAL VENTILATION OF FACTORIES

Window Ventilation

In narrow factory buildings of the width common in the multi-storey type, good cross-ventilation can usually be obtained by the provision of large and suitably placed windows, or by a combination of windows and wall ventilators. Figure 51 shows a good arrangement of windows in such a factory.

The factory is 280 ft long and 46 ft wide. The sides are almost wholly of glass, and in each glazed section there is a centre-hung hopper at high level. These top hoppers are 3.5 ft × 3.75 ft in size, so that each gives an opening of 13 sq. ft. Below these are bottom-hung hoppers, each 3.5 ft wide and capable of being opened to a width of 10 in, giving a maximum opening of nearly 3 sq. ft. A valuable feature of the

Fig. 51. Good arrangement of hopper windows. Tobin tube also shown. (I.F.R.B. Rept. No. 35, H.M.S.O.)

lower hoppers is that they are fitted with gusset pieces at the sides. The air entering at these hoppers is deflected upwards, and the gussets prevent draughts at the sides.

It is a useful working rule that the opening area available for natural ventilation should be not less than 5 sq. ft for every 100 sq. ft of floor area. The openings should, of course, be suitably placed. In the factory depicted in the illustration the area of the window openings is more than 7 sq. ft/100 sq. ft of floor. Between the windows to the left of the photograph will be noticed a Tobin tube ventilator. Such air inlets are provided at intervals of 20 ft along the sides of the building. Inside each tube there is concealed a hot-water radiator, whereby the entering air is warmed in winter, and the inlets are controlled by dampers so that they can be closed if necessary. By means of these ventilators tempered air can be admitted in winter without objectionable draughts being felt.

Measurements made in this factory during summer weather showed that, with only 40 per cent of the available window area open, the ventilation rate was 20 air changes/hour, which figure represented a fresh air supply of 12,000 cu. ft/hour/operative (Vernon, Bedford and Warner, 1926*a*). Judged by the suggested minimum fresh air standard of 1,000 cu. ft/person/hour this is lavish ventilation, but the windows were as the operatives had opened them from choice, and the conditions were distinctly pleasant. The temperature was 68° F and the average air velocity 61 ft/min.

There are occasions in winter when it is difficult to achieve satisfactory cross-ventilation by means of open windows. Draughts are likely to be experienced, especially in windy weather, unless considerable care is taken that windows in the most suitable positions are open. Often, in order to avoid draughts, the workers close all windows, and the ventilation is then likely to be inadequate. With the object of securing satisfactory ventilation even in winter, in the factory under discussion provision was made for either natural or mechanical exhaust ventilation. Along the middle line of the factory there are four stacks, each stack having a separate flue from each of the four floors (Fig. 52).

The exhaust stacks terminate in louvred outlets erected on

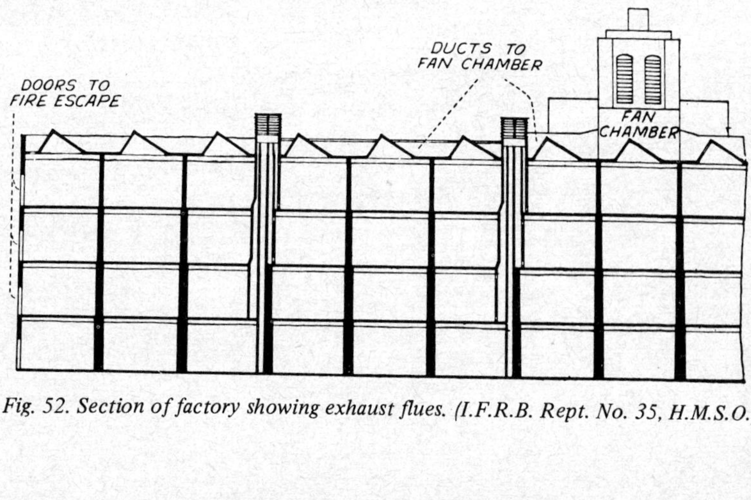

Fig. 52. Section of factory showing exhaust flues. (I.F.R.B. Rept. No. 35, H.M.S.O.)

the flat roof, so that by the combined influence of gravity and the aspirating power of the wind air is extracted upwards from the workrooms. By means of regulators placed in each workroom flaps are controlled by which any single flue can easily be shut off from the roof louvre and, instead, connected with a horizontal duct which leads to a large centrifugal fan situated beneath an outlet tower. Thus, on occasions when the natural ventilation by Tobin tubes and gravity exhaust is inadequate, and when it is not desired to throw the windows open, satisfactory ventilation can be achieved by mechanical extraction. In winter, when nearly all the windows were shut and the fan was running, the air was changed every 7 to 15 min.

Vernon, Bedford and Warner (1926*a*) made a large number of observations in factories ventilated in a variety of ways, and they noted the extent to which window ventilation was utilized. The table shows, for six workrooms with natural ventilation, the percentage of the available opening area of windows which was open at various outside temperatures.

Window area kept open in various workrooms

Workroom	*No. of workers*	*Floor area (sq. ft)*	*Sq. ft vent. area per 100 sq. ft of floor*	*Percentage of window area open when outside temperature (°F) was about*				
				35	*45*	*55*	*65*	*75*
A1	274	12,880	9.9	1	5	15	33	65
A2	188	12,880	7.3	1	7	25	54	85
A3	120	12,880	7.3	0	14	20	39	–
E	236	6,480	4.5	14	18	52	81	–
F1	280	5,840	6.0	–	12	63	75	–
F2	230	21,530	3.7	–	3	35	43	–

The area of windows open was greatly increased in mild and warm weather, and, on the whole, more advantage was taken of the available ventilation facilities in the rooms which were the more densely occupied. When the external temperature was below 40° F very few windows were open. In a fully occupied workroom some people must generally work near windows, and in cold weather they are liable to be exposed to unpleasant draughts. Even when the windows are closed, the warm air of the workroom coming into contact with the cold window surfaces is cooled and falls rapidly towards the floor. This descending current of air often provokes complaints of draughts, and workers quite naturally object to further avoidable draughts from open windows. The provision of sources of heat below large window areas can do much to prevent draughts due to convection, and the use of draught deflectors can sometimes mitigate the troubles which arise from open windows. Much benefit would often be reaped from the installation of draught-proof wall ventilators.

Roof Ventilators

In wide buildings, especially those of the single-storey type, window ventilation alone is likely to be quite inadequate. In such factories, although the localities near open windows have good ventilation, there are often considerable 'dead' areas where the air is relatively stagnant. Frequently conditions can be much improved by providing openings of liberal size in the roof, so that advantage can be taken of gravity effects. Such roof outlets are often in the form of louvres (Fig. 53).

Double-sided ridge louvres can be made to provide a large outlet area. Roof louvres suffer from the disadvantage that, except when the wind blows along the ridge parallel to the

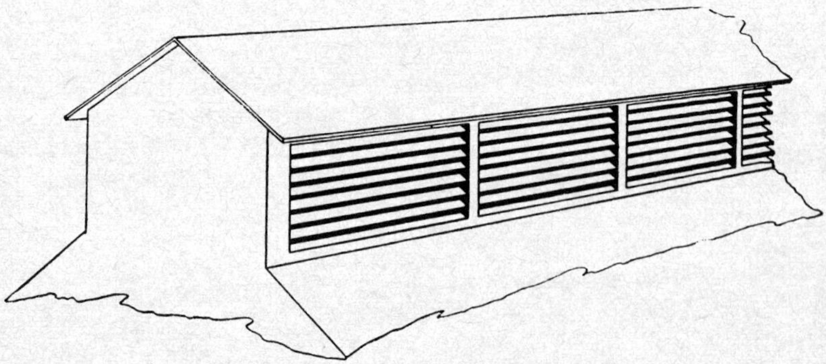

Fig. 53. Louvred roof ventilator. (Factory Dept. Welfare Pamphlet No. 5, H.M.S.O.)

louvres, there is direct wind pressure against the louvres of one side, so that the exit of air is hindered. The louvres may, in fact, become inlets, and the downdraughts thus caused may give rise to strong complaints in cold weather. Shutters can be provided so that the louvres facing the wind can be closed, but unless there is careful supervision trouble is still likely to arise.

A better arrangement for natural extraction from the roof is to provide modern roof cowls with a guaranteed performance. Exaggerated claims were formerly made by the manufacturers of roof ventilators. Forty years ago, a Committee of the Sanitary Institute in Great Britain made tests of the performance of a variety of cowls, and later tests were made by the Bureau of Standards of the U.S. Department of Commerce. The performance of a roof ventilator depends on the difference between the temperatures of the external and of the internal air, on the height of the ventilator above the air intake, on the velocity of the wind, and on the cross-sectional area of the ventilator. Harding and Willard (1929) cite the following equation for the quantity of air removed by a cowl:

$$Q = A\left(\frac{36\sqrt{H(t_i - t_0)}}{6 + V} + 20V\right)$$

where Q is the number of cu. ft of air exhausted/hour from a ventilator with a free area at the throat of A sq. in, mounted on a roof at a height H ft above the centre of the air inlet, and with a wind velocity of V m.p.h., when the temperatures inside and outside are respectively t_i and t_0. The value of Q thus obtained is a conservative one, and ventilators of the highest class may discharge 25 per cent more air than is indicated by the equation.

An example of a roof cowl is the Robertson ventilator (Figs. 54(*a*) and (*b*)).

The data book issued by the makers shows the performance of cowls of various sizes in relation to temperature difference, height of ventilator and wind velocity. Some of the data given for a 24-in diameter cowl are shown in the table.

The substantial effects of differences in temperature and in wind velocity will be noted. With the same temperature

Fig. 54(a). The Robertson ventilator.

Fig. 54(b). The Robertson ventilator.

Exhaust capacity of 24-in Robertson ventilators

Temperature difference (°F)	*Height of ventilator above intake (ft)*	*Ventilation volume (cu. ft/min) wind velocity (m.p.h.) of* 2	4	6	8	10
0	All heights	193	387	580	775	968
5	10	435	629	822	1017	1210
	20	533	727	920	1115	1308
	40	674	868	1061	1256	1449
20	10	674	868	1061	1256	1449
	20	875	1069	1262	1457	1650
	40	1155	1349	1542	1737	1930

difference an increase in the wind velocity from 2 to 10 m.p.h. roughly doubles the ventilation. The ventilation volumes indicated in the table are those which may be expected when there is no considerable obstruction to the entry of air into the building. If the air is to be extracted it must first be admitted—the law of continuity of flow must be observed. Too often the provision of proper air inlets is neglected, and then it is wondered why extraction systems fail or cause draughts in undesirable places. If roof cowls are relied upon for the extraction of air, inlets with a total area of at least twice the outlet area should be placed in suitable positions. It is often found desirable to place the inlets 4 to 5 ft above the floor, and to construct them as variable louvres so that the entering air can be directed upwards or downwards as desired.

The values in the Robertson data book are closely fitted by the equation:

$$Q = A(10.82\sqrt{H(t_i - t_0)} + 30.83V)$$

where Q is expressed in cu. ft/min, A is the cross-sectional area of the ventilator in sq. ft, H is the height of the ventilator above the inlets in feet, t_i and t_0 are the internal and external temperatures in °F, and V is the wind speed in m.p.h.

These ventilators can be supplied with baffles whereby they can be closed as required, and thus by manual control the ventilation can be regulated to compensate for changes in the wind or in external temperature. Such control is clearly necessary for the successful application of this method of ventilation.

In some factories considerable heat is generated, or fumes may be liberated, intermittently. Then it is of great advantage if the ventilation can be increased above the usual level. This can be achieved with Robertson cowl outlets if propeller fans are fitted in the throats of the cowls. In this way large increases of ventilation can be obtained.

Another appliance which is employed as a roof outlet is the Colt ventilator. The constant-flow type can be used to ensure that a steady volume of air is extracted despite considerable change in the wind speed, and additional controls, operated thermostatically or by hand, can also be fitted. While changes in wind speed have little effect, a simple thermostat can cause the outlet orifice to be increased when the temperature in the roof space exceeds a certain chosen figure, and to be reduced to a minimum when that figure is not reached. Thus, it appears, if desired the roof ventilators can give a measure of automatic control over the temperature within the factory, providing the heat input is sufficient.

It often happens that in large single-storey factories, even when the air turnover is sufficient to maintain an atmosphere which is satisfactory from the chemical standpoint, the air is relatively stagnant and gives rise to feelings of oppression. The air admitted at the inlets is often drawn upwards towards the roof outlets, so that in positions remote from windows such air movement as there is takes place above the heads of the occupants. The Colt Ventilation Company has introduced a roof inlet ventilator, or inducer, so designed that when the wind impinges against it air is driven downwards towards working level. The entering air can be diffused so that strong draughts are not caused.

Although the use of mechanical ventilation and especially air-conditioning is becoming more widespread it might be considered that ventilation of buildings by natural means would tend to become obsolete. Such is not the case, especially in factories where considerable heat is generated, because the very

Fig. 55. A large Robertson Streamline Ventilator installed at Stewarts and Lloyds Ltd. at Corby.

Fig. 56. A large Triple Ventilator installed at the Rockware Glass Company Ltd. at Wheatley in Doncaster.

heat produced will result in a large difference in temperature and therefore a considerable difference in density between the air inside and the air outside the factory. The essential motive force for natural ventilation is thus already provided by the industrial process and ventilation, which will not depend upon the vagaries and variability of external wind speed, is ensured by the provision of suitable inlets and outlets. In fact, experience indicates that, provided that there is at least a 15° F difference between the temperatures of the inside and outside air, such a system of natural ventilation is a practical proposition.

But even if there is little heat generated inside the building by some industrial process, considerable differences between inside and outside temperatures may occur in tropical countries and these may be augmented by the heat gained by solar radiation. One British company (Robertson's) were able without recourse to fans to reduce the difference between inside and outside temperatures in a commercial building in Calcutta during the summer to less than 5° F by means of ventilators installed in the roof of the building with suitable intakes for fresh air provided in the side walls.

Modern roof ventilators are shown in Figs. 55 and 56.

CHAPTER XII

MECHANICAL VENTILATION

IN large buildings ventilation by means of windows is not always practicable, and it is necessary to use mechanical methods. A system of ventilation in which the air is impelled into a room is known as a plenum or propulsion system, and when air is withdrawn from the room the installation is known as an extraction or exhaust system. Either system may be used alone, or a ventilation scheme may incorporate both impulsion and extraction.

FANS

During the eighteenth century several mechanical ventilation appliances were devised. The outstanding one of these, the centrifugal fan (Fig. 57), was invented by Dr. Desaguliers and used by him about the year 1736 for ventilating the House of Commons.

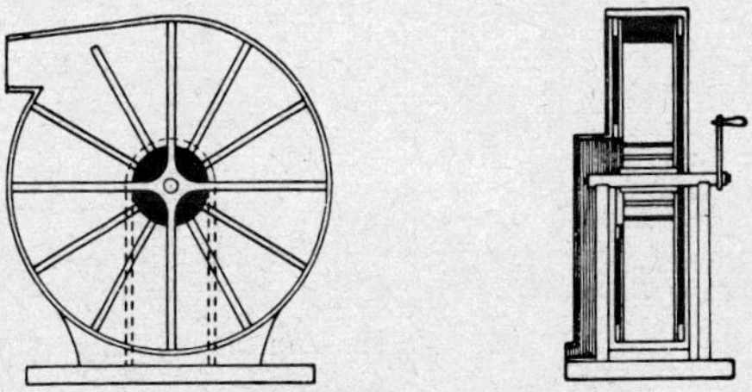

Fig. 57. Desaguliers' centrifugal fan.

The fan could be used either for extraction or for propulsion, and was rotated by hand. At about that time an outbreak of sickness in the Royal Navy directed attention to the want of ventilation in warships, and Desaguliers was invited to demonstrate his centrifugal wheel and air pipes in a ship which was lying off Woolwich. Not one of the Lords of the Admiralty deigned to witness the experiment, which, though a success, was adversely reported upon by underlings, and the Surveyor of the Navy condemned the fan. Today the Royal Navy uses large numbers of centrifugal fans which are its direct descendants.

Fig. 58. Types of centrifugal fans.

Centrifugal Fans

Centrifugal, or radial flow, fans are used in situations where relatively large quantities of air have to be moved against appreciable resistances. They are often known as pressure fans. They vary considerably in type, but basically they consist of a fan wheel which rotates in the fan casing. On the circumference of the wheel are a number of blades. There may be only five or ten of these blades or, in the multiblade type, there may be thirty or more of them. The blades may be straight, or may

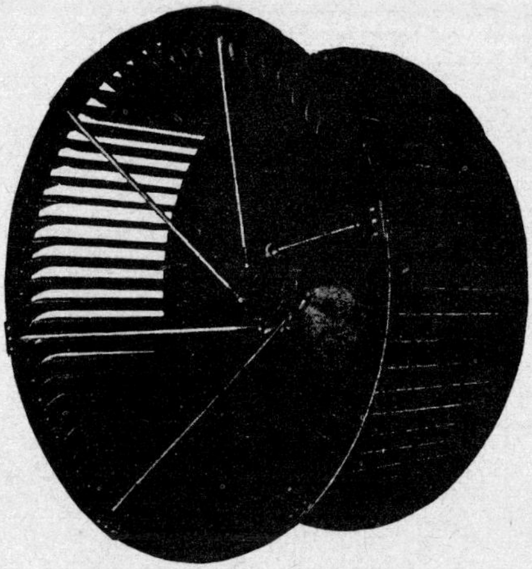

Fig. 59. Centrifugal fan wheel (multiblade).

curve either backwards or forwards. The direction of curvature influences the characteristics of the fan. From the diagram of Desaguliers' fan it will be noticed that his fan casing was concentric with the fan wheel. In modern fans the casing is eccentric, and thereby the efficiency of the fan is greatly increased. The air is drawn into the fan near the centre of the fan wheel, and under the influence of centrifugal force is thrown towards the periphery of the wheel and discharged tangentially through the fan outlet.

In deciding upon the type of fan required for a particular installation, it is necessary to estimate the resistances which the fan must overcome. With the fan running at a given speed increase in the resistance results in the delivery of a decreased volume of air. The variation of the output of a fan in relation to changes in pressure is shown by the characteristic curve of the fan, and is determined under test at constant speed. If resistances are kept as low as possible the power required to move a given volume of air is relatively small, and a fan of suitable overall dimensions working at 2 in water gauge may discharge over 100,000 cu. ft of air/hour/b.h.p. (Min. of Lab., 1940). With a given resistance, the volume discharged, the static pressure and the b.h.p. vary as the speed, the square of the speed, and the cube of the speed respectively. It is not economical to supply the air for a ventilating scheme by means of small high-speed fans connected to small ducts. Fans should be operated as nearly as possible under the conditions of maximum mechanical efficiency, which conditions are specified in makers' catalogues.

Propeller or Axial Flow Fans

In situations where the resistance to the flow of air is negligible, propeller or axial flow fans are capable of moving large volumes of air with a low expenditure of power. Air is drawn from all directions on the intake side and propelled forward at high velocity on the discharge side as a jet or beam. Propeller fans which have been in use for many years include the open-blade and box-blade types. Some of the open-blade types are referred to as disc fans.

The volume discharged by fans of similar design, when run at suitable speeds and under free air conditions (*i.e.* when there is no resistance to flow) is proportional to the speed and to the cube of the diameter. Thus, the relation of the volume of air discharged/min, V, to the diameter, D, and the speed of the fan, N rev/min, is

$$V = kD^3N$$

where k is a constant. For many types of fan k is about 0.6, but for highly efficient fans it may be 0.7 or more (Min. of Lab., 1940), while according to Joselin (1934) it may be as high as 1.0. The Ministry of Labour pamphlet on the ventilation of factories gives suitable speed ranges for sizes of propeller fans in common use, and the corresponding approximate outputs based on a value for k of 0.6. These are set out below.

Output of propeller fans

Diameter in	*Speed range (revs./min)*		*Volumes discharged (cu. ft/min at)*	
	Minimum	*Maximum*	*Minimum speed*	*Maximum speed*
12	1,300	2,100	780	1,260
15	1,050	1,685	1,230	1,970
18	850	1,400	1,720	2,835
20	760	1,265	2,110	3,510
24	630	1,050	3,020	5,040
25	610	1,015	3,310	5,510
30	510	840	4,780	7,875
35	440	715	6,550	10,640
36	425	700	6,890	11,340
40	380	625	8,440	13,890
42	360	590	9,260	15,180
45	340	550	10,760	17,400
48	320	525	12,290	20,160
50	305	510	13,240	22,140
54	295	490	16,130	26,790

Makers' catalogues show the free air discharge volumes for fans of various sizes running at different speeds, and these indicate the decline in output when fans run against specified resistances. It is undesirable that propeller fans should be used when the resistance exceeds about $\frac{1}{4}$ in water-gauge. If these fans are used to extract air from rooms the air supply must be adequate, and the flow of air to and from the fans must be unimpeded. They should not be used to drive air through ducts unless these are very short, and then the ducts should be smooth internally and free from sharp bends.

As an outcome of the aerodynamical research which has been carried out in recent years, a new type of axial flow fan has been devised. Sometimes it is referred to as the 'airscrew' or 'airfoil' fan, and sometimes merely as the axial flow fan. This type of fan resembles an aeroplane propeller in the shape of the cross-section of its blades and in the twist given to the blade as the centre of the rotor is approached. The central hub may have a diameter 50 to 70 per cent of the total diameter of the fan. Figure 60 illustrates one type of airscrew fan.

Fig. 60. Airscrew fan.

These fans differ from the older propeller or disc type in that they will operate efficiently against considerable pressures. Indeed, Heath and Criqui (1944), discussing the uses of the axial flow fan in ventilation, remark that a single stage aerofoil fan is capable of meeting any pressure demand imposed by any ordinary ventilation requirement. MacFarlane (1938) discusses the aerodynamics of airscrew fans, and the advantages and construction of various types. He comments that the airscrew fan should be regarded as the complement rather than the competitor of the centrifugal fan, for there are many applications to which the centrifugal fan is the better suited. The greatest advantage of the screw fan is that of economy in power consumption. This type of fan is particularly suited for use in the ventilation of mines.

EXHAUST VENTILATION

One example of mechanical exhaust ventilation which could be used to assist natural ventilation, or operated in the absence of it, has been described in the preceding chapter. This method of ventilation can be used to ensure effective cross-ventilation in a narrow room by placing propeller fans in exhaust openings at suitable intervals in one side wall and providing adequate air inlets in the opposite wall. The air inlets should be of large size,

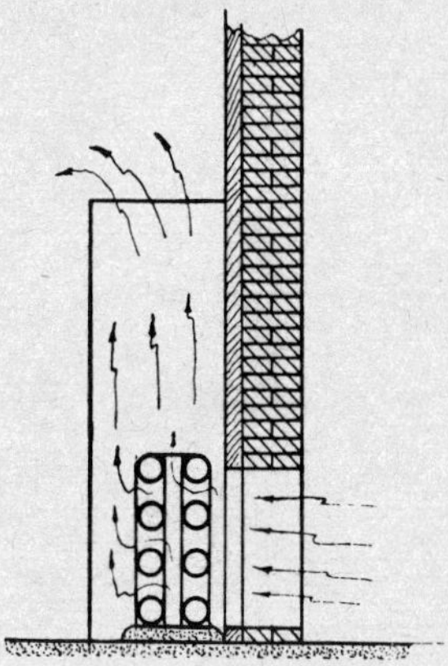

Fig. 61. Screened air inlet with heater.

so that inlet velocities are kept low and draughts thereby avoided. Provision should be made for warming the incoming air. When, as is very often the case, exhaust fans are mounted in openings in external walls, they should be protected by wind shields so that wind pressure will not substantially reduce their efficiency. More effective protection from winds is obtained if the fans discharge into large ducts which are carried vertically some distance above the eaves.

Windshields for fans should allow of free air discharge, as is indicated in Fig. 62.

Extraction fans of the propeller type are often used to

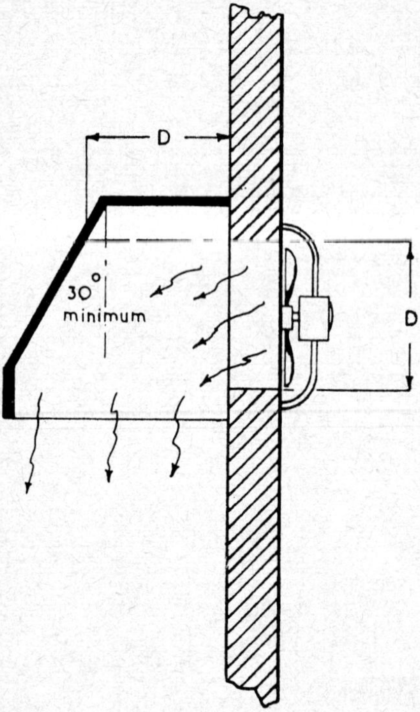

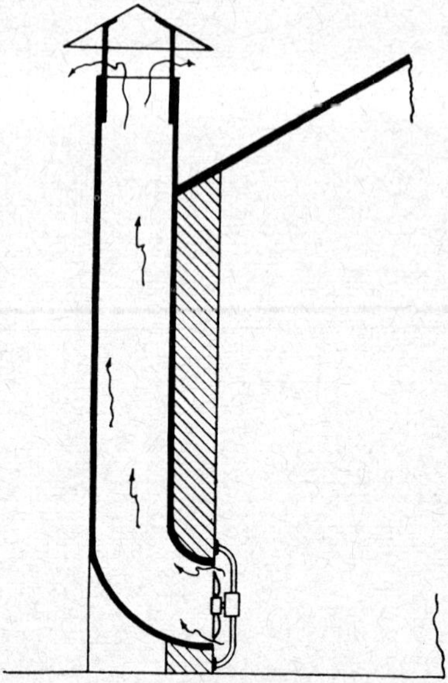

Fig. 62. Wind shield for propeller fan.

Fig. 63. External discharge duct for propeller fan.

(Factory Dept. Welfare Pamphlet No. 5, H.M.S.O.)

ventilate single-storey buildings of small or medium size. Sometimes no suitable provision is made for the entry of air. Then, either the efficiency of extraction is diminished, or the incoming air enters by any available opening or crevices, thus creating unpleasant draughts. Proper inlets should be provided and so placed that a thorough scavenging of the air in the building is ensured. The total inlet area should be at least three times the total disc area of the fan.

Windows and other apertures near the fans should be kept closed, for if they are open the fans will draw their air supply from them, and this short-circuiting will prevent efficient ventilation.

As long ago as 1907 the Home Office Departmental Committee on the Ventilation of Factories and Workshops drew

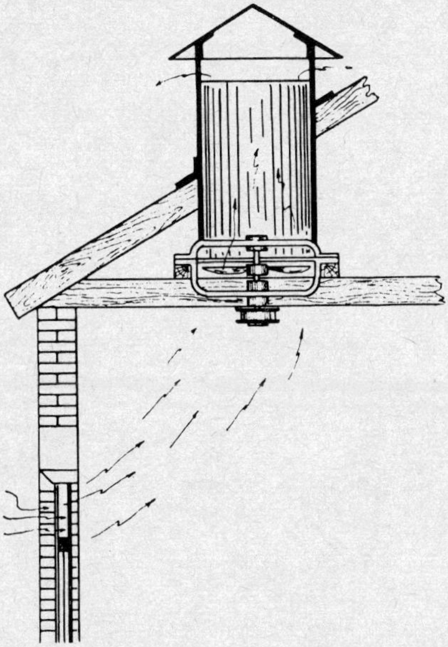

Fig. 64. Short-circuiting of exhaust fan. (Factory Dept. Welfare Pamphlet No. 5, H.M.S.O.)

attention to the failure of ventilation due to short-circuiting (Fig. 64).

Yet such short-circuiting often occurs today. It is not uncommon to find that in a single-storey factory exhaust fans are placed at intervals in the gables, while intermediately there are louvred openings in other gables. The fans take their air supply from the nearby openings, and the workers below reap little or no benefit from them.

In a paper discussing common cases of failure of ventilating equipment in factories in Detroit, Witheridge (1944) refers to the short-circuiting of ventilating fans, which he describes as 'ventilating the outdoors', and he remarks that the frequency with which this fault occurs is amazing.

PLENUM VENTILATION

In the plenum system of ventilation air is driven into the building, generally by means of a centrifugal fan, and is distributed through ducts. The system has various advantages. The slight positive pressure set up within the building helps to prevent the leakage of air inwards, and thus cold draughts near doors and windows are to a large extent avoided. The incoming air can be readily warmed by the incorporation of a heating battery in the duct system, and if desired, the humidity can be controlled and the entering air can be filtered to remove soot and other suspended matter. The air can be distributed throughout the building as required. In most of the plenum installations found in industry the air is untreated except that it is warmed in cold weather.

Duct Design

In designing air ducts for a plenum system the aim should be to convey the air as directly as possible at suitable velocities, with due regard to economy of power and of material. Ducts should be smooth, *e.g.* of sheet metal, and as straight as possible. Sharp bends and sudden enlargements or contractions should be avoided. When the direction or dimensions of a duct must be changed the angle between the sides and the axis of the ducts should never exceed 30°, and preferably should not be greater than 15°. To keep down frictional losses, the ducts should have a minimal surface area, if they are rectangular in shape they should preferably be of square cross-section.

Duct velocities recommended by the American Society of Heating and Ventilating Engineers are shown below.

Recommended duct velocities

Type of building	*Recommended velocities, ft/min in*		
	Main ducts	*Branch ducts*	*Branch risers*
Residences	700-900	600	500
Schools, theatres, public buildings	1000-1300	600-900	600-700
Industrial buildings	1200-1800	800-1000	800

The velocities depend on the nature of the installation and the degree of quietness required. Velocities about 10 per cent higher than those recommended are given as the maximum values, but the velocities indicated in the table should not be exceeded unless noise is not a factor of importance.

Some systems of air distribution, however, depend on high inlet velocities, and then duct velocities higher than those mentioned in the table are necessary. The duct velocities should be not less than the throat velocities at the inlets. When high duct velocities are to be used special care is necessary in the design of the ducts, to keep resistances as low as possible and to limit noise.

Air Inlets

In plenum ventilation systems the air inlets should be so designed, and the velocity of discharge so arranged, that suitable eddying currents are set up. The air velocities at working level should not be strong enough to cause complaints of draughts, but there should be currents of such strength and variability as will provide a stimulating atmosphere. When plenum air is discharged overhead, inlet velocities are often of the order of 700 to 1,500 ft/min, and may sometimes be well over 2,000 ft/min. If high velocity streams are discharged towards nearby walls, the rebounding air may cause serious discomfort.

When cool air is discharged horizontally it falls to a lower level and thus mixes with the general atmosphere, but when hot air is so discharged, there is a strong tendency for steep temperature gradients to be set up. This aspect of hot air distribution will be discussed later. When, on the other hand, the plenum air is discharged downwards towards the floor, more or less satisfactory mixing of the air is achieved when warm air is delivered, and the relatively high air velocities set up at floor level are largely offset by the raised air temperature.

In many non-industrial buildings—schools, offices and public buildings—the air is discharged through grilles or registers. These are so situated that the entering air does not cause unpleasant draughts, but their size should be such that the inlet velocity is sufficient to create adequate air movement throughout the room. One sometimes encounters elaborate installations which supply lavish quantities of fresh air, but which still cause complaints of 'dead' air. In such instances it has been observed that the inlet velocities were low, with the result that the entering air simply cascaded from the grille towards the floor and did nothing to create turbulence. In my own experience desk fans have had to be employed to create air movement which should have been set up by the air entering from the plenum registers.

Better results are often obtained by the use of diffusers or jets. Sometimes a circular plate is fixed in front of the air inlet, and a few inches from it, so that the entering jet of air impinges against the disc and is spread. This is a primitive form of diffuser, and there are now on the market various types which are more efficient. These are generally in the form of a series of coaxial cones or concentric rings of diminishing diameter. When these types are used the air is diffused so that although the air velocity at the throat of the inlet may be quite high, the velocity in the room 2 or 3 ft from the inlet is greatly reduced. The entering air is spread evenly in all directions around the inlet so that good distribution is thus ensured, the risk of draughts is small, and yet stagnation and stuffiness can be avoided.

In ship ventilation punkah louvres are much used. These inlets are of varying sizes, but frequently they are of 3 in or 4 in diameter, and they can be rotated about a joint so that the direction of discharge can be altered at will. Inlet velocities are mainly of the order of 1,000 to 2,000 ft/min. In hot weather the air stream can be directed on to the occupants, who then derive much benefit from the cooling effect of moving air, while at other times the air can be admitted without causing undue draughts.

This principle of jet ventilation has useful applications in industry. Sometimes ejector nozzles from which air is delivered at high velocity are used. A longer throw is obtained than would be got with lower inlet velocities, and more disturbance of the air is thereby caused. If these inlets discharge horizontally a little distance above the heads of the occupants undue draughts are avoided, and the turbulence caused by the jets promotes freshness. Jets of this type were used with notable success in some basement air-raid shelters during the war.

INDUSTRIAL PLENUM INSTALLATIONS

In many factories where plenum systems are used no air outlets are specially provided. The air makes its escape through such doors or other openings as are available, and it generally happens that these openings are sufficient to allow the installations to operate satisfactorily. Figure 65 gives a plan and section of a single-storey factory showing the arrangement of a plenum ventilation system.

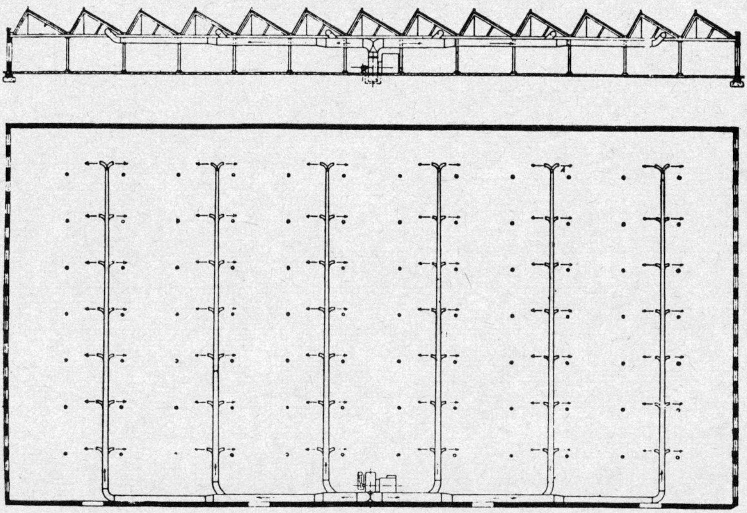

Fig. 65. Overhead plenum system. (Factory Dept. Welfare Pamphlet No. 5, H.M.S.O.)

The casing of the centrifugal fan stands on the floor near one wall. From the fan the main air trunk rises vertically to the level of the roof members and then branches into two large ducts running along the wall almost the whole length of the factory, and at intervals subsidiary ducts branch off to run across the building. Air is discharged from outlets in these branches. All the ducts are overhead, and the air is discharged in a horizontal direction. With this arrangement there is a tendency in cold weather for the warm air to accumulate in the roof spaces, and thus set up a steep temperature gradient.

In some installations the air is delivered through ducts which are brought down to within 3 ft or so of the floor. The air is discharged vertically downwards or at an angle of about 45° with the floor, and thus the warm air is brought down to the working zone and the temperature gradients are less than when the air is discharged overhead. Such an arrangement is portrayed in Fig. 66.

Fig. 66. Plenum ducts discharging downwards.

Angus (1923, 1936, 1938) has used jet ventilation for creating air movement in factories where the atmosphere had previously been too stagnant. His jet tube ventilator (Fig. 67) consists of a small centrifugal fan which drives air through a short length of duct the shape of which is gradually and smoothly changed so that the discharge orifice is a long and narrow slot. The air is discharged overhead at high velocity.

Originally this ventilator was intended merely to recirculate and agitate the air of the factory, but it is a simple matter to arrange that the fan takes its supply from the external air, and if required a heater can be incorporated.

At the beginning of this century plenum systems of ventilation became popular. In some of these installations air outlets were planned to ensure that the air impelled into the building would be properly circulated and, to avoid short-circuiting,

Fig. 67. Jet tube ventilator.

windows were kept shut as far as possible. Observations were made in a boot and shoe factory with such a rigid plenum system (Vernon, Bedford and Warner, 1926*a*). There were three floors, each 252 ft x 52 ft in area. In the middle line of each floor were ten plenum inlets each 7 ft x 1.75 ft, and at an average height of 6.3 ft from the floor. These inlets were connected by vertical ducts to a large chamber in the basement along which air was driven by a propeller fan 10 ft in diameter. The air was drawn through a battery of steam pipes, and there was an auxiliary steam battery at the foot of each shaft. The air was discharged from the inlets horizontally towards the sides of the factory and returned to the outlets which were of similar size to the inlets, and likewise situated in the middle line of the room but at floor level. From the outlets the air passed up vertical ducts which ended in louvres on the roof. The inlet velocity was about 200 ft/min and the ventilation provided 8 air changes/hour. There was no provision for recirculation.

The ventilation was completely shut off during the night and only started about 45 min before work was begun in the morning. The air often smelt stale and the factory appeared stuffy, especially in the early morning in summer. Since the windows could not be opened there was no possibility of flushing the building with fresh air, so that in summer the cool night air could not be used to full advantage for cooling the building, and the response to changes in the outside temperature was very slow.

Air Heating

When air is supplied by plenum systems provision is almost invariably made for warming it in cold weather. In many installations, and especially those found in industrial buildings, the circulation of hot air is the only method of warming. Then, it is not sufficient to warm the entering air to the temperature which it is desired to maintain within the building: it must be heated to a distinctly higher temperature to allow for the heat losses which take place through the structure.

It is the general practice to recirculate considerable volumes of the air from the building and to mix this with the fresh air taken in from outside. In buildings which are not heated through the night, it is often the custom to bring the temperature up to the required level as rapidly as possible first thing in the morning by supplying only warmed and recirculated air, and then, when the required temperature has been reached, the desired proportion of fresh air is taken in and mixed with the recirculated air.

Recirculation has the advantage that the volume of air delivered by the fan is much greater than could be afforded on grounds of economy if only cold outside air were drawn into the fan system, and the circulation of a greater volume of air helps to create more air movement. Further, if only the fresh air taken from outside were passed over the heating batteries it would be heated to an undesirably high temperature, proper distribution of this hot air would be difficult or impossible, and discomfort would often be caused.

TEMPERATURE GRADIENTS

One of the most serious faults to be found with the older types of plenum installation is that when they supply warm air there is often a considerable difference between the temperatures at head level and at foot level. We have seen earlier that such temperature gradients are a source of discomfort.

In one factory in which the work-rooms were 50 ft wide, the plenum ducts were 9 ft above the floor and the air was discharged through strips of perforated zinc, 2 in wide, running along the sides of the ducts. The air was discharged from the ducts with a velocity of about 700 ft/min, and in winter its temperature was 120° to 140° F. Figure 68 shows the average

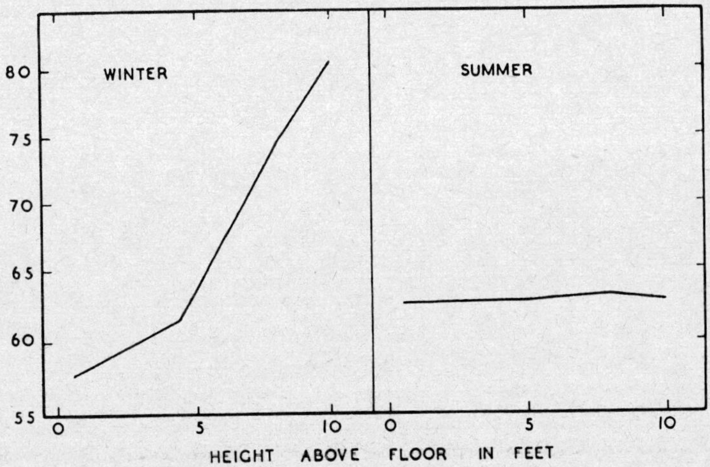

Fig. 68. Temperature gradients caused by overhead plenum system.

temperature gradients observed in one of the rooms in this factory (Vernon, Bedford and Warner, 1926*a*).

In winter there was a difference of 20° F between the temperatures at the floor and at a height of 10 ft, whereas in summer, when cool air was circulated, there was no noticeable gradient.

In another workroom the plenum duct ran 9.5 ft above the floor, and the air was discharged horizontally at that level from a number of inlets in the sides of the duct. Here, also, steep temperature gradients were caused when the hot air was introduced. To improve the temperature distribution, the air supplied by the duct was led down to near the floor and discharged through $\frac{3}{4}$-in perforated strips in the sides of ducts which were run along the floor, while the overhead inlets were shut. When the air was admitted at the overhead inlets the temperature gradient between the floor and the 9.5 ft level was 11.6° F. When the inlets were 6 in from the floor the tempera-

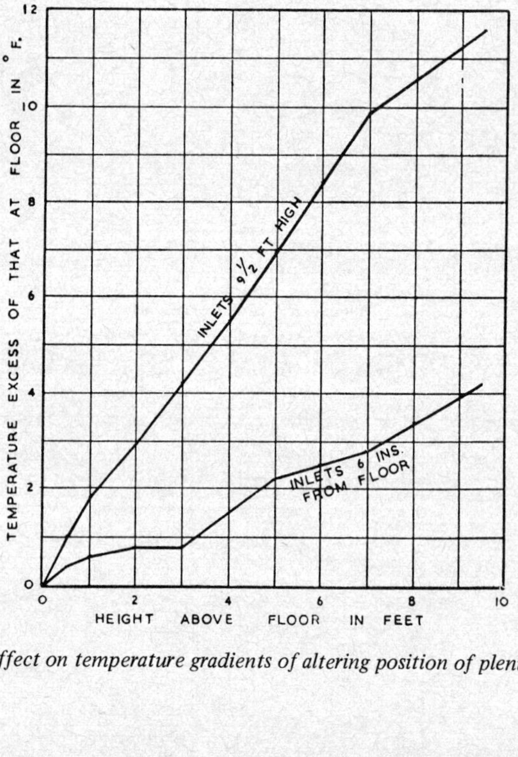

Fig. 69. Effect on temperature gradients of altering position of plenum inlets.

ture gradients from the floor to the 9.5 ft and the 5 ft levels were respectively about 4° and 2° F.

In some other factories where the warm plenum air was discharged downwards to the floor, or where the discharge was partly towards the floor and partly above head level, the temperature gradients were much less striking. The average differences between the temperatures of the air at 6 in from the floor and at a height of 4.5 ft were, in three factories, approximately 1°, 2° and 3° F.

AIR VELOCITIES

When air is discharged at high velocities from plenum inlets, currents are set up which have a noticeable influence on the air velocity in the general atmosphere of the apartment. If the air is discharged in suitable directions the increased air velocity may be beneficial to the occupants, whereas with badly arranged inlets unpleasant draughts may arise. Vernon, Bedford and Warner noted that in two factories the average air velocity at head level when the fan was shut off was 25 or 26 ft/min and when the fans were running at their usual speeds, delivering air above head level, the average velocities rose to 30 ft/min in one factory and 43 ft/min in the other. In the first of these factories one fan supplied the air to four floors. When the ducts leading to three of the floors were closed, and the whole output of the fan delivered to one floor, the air velocity in that room was increased from 30 to 39 ft/min. These increased velocities were not high enough to evoke complaints of draughts, and they imparted a sense of freshness to the atmosphere.

In some single-storey factories, where the workrooms were much wider than in the multi-storey buildings just referred to, the air issuing from the overhead plenum ducts had a smaller effect on the average air velocity; the mean velocity was increased by the fan only from 32 to 36 ft/min. In these factories the plenum air supply had but a slight effect on ventilation during warm weather. Large loading doors and skylight windows were open in summer and these provided ventilation rates of 12 to 20 air changes/hour. At such times it was observed that the air from the plenum system appeared to rise and pass directly from the trunks to the open skylights. Yet the ventilation was far more adequate, and pleasanter conditions were maintained, than would have been found if doors and windows had been closed and the plenum system only had been available.

When plenum inlets discharge downwards towards the floor high air velocities are caused locally, and these floor draughts may cause discomfort to persons who work near the air inlets. Observations made in one engineering shop with a downward plenum system, with ducts 9 in in diameter, showed that on the floor 1 ft away from the line of the inlet, the air velocity was 700 ft/min, while 4.5 ft away the velocity was 200 ft/min. The air velocity 6 in above the floor was at a maximum about 5 ft away from the duct, where it was about 200 ft/min, but 10 ft away it was sometimes as high as 100 ft/min. At head level there was no appreciable influence of the fan, and the air velocities were about 20 to 30 ft/min. Ducts were spaced 30 ft apart, and Fig. 70 shows the velocities noted at a height of 6 in from the floor along the line between two ducts.

Draughts at floor level provoked many complaints from the workmen, who often used boxes or boards to deflect the air

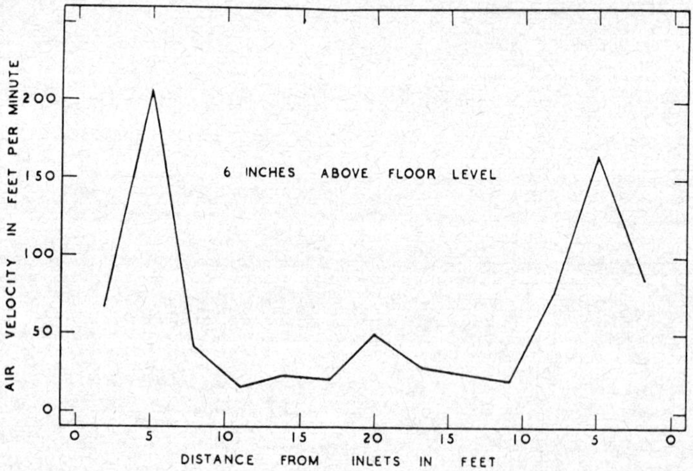

Fig. 70. Air velocities between plenum ducts (air discharging towards floor).

currents. Sometimes men working as far as 10 ft from ducts protected themselves in this way, and on some occasions the inlets were tied up with sacks.

The high velocities along the floor were due not only to the high inlet velocity, but also to the size of the inlets. The volume of air discharged from each inlet was large. With inlets of smaller diameter discharging towards the floor, the air velocities a short distance from the duct are lower. This was noted in another factory where air was blown downwards from 4-in and 6-in ducts with an inlet velocity of 1,100 ft/min. Whereas in the preceding example the air velocity along the floor at a distance of 2 ft from the duct was about 480 ft/min, with the 6-in ducts the velocity 2 ft from the duct was 320 ft/min, and with the 4-in ducts only 130 ft/min, in spite of the substantially higher velocities at the smaller inlets.

When air is driven down to the floor at high velocity it is liable to redistribute dust which has fallen to the floor. In one factory 11-in ducts discharged air at an angle of 45° with the floor. The inlet velocity was 1,100 ft/min and the air velocities observed at a height of 6 in above the floor at distances of 3, 4, 5, and 6 ft from the inlet, and in line with it, were 310, 471, 330, and 166 ft/min. When dust was placed on the floor, and the fan then switched on, the dust was blown away immediately. When the discharge velocity was reduced to 470 ft/min the current was still enough to raise dust. At another factory also, it was noted that the downward blast from the air inlets blew dust into the air. We have seen that airborne bacteria are often attached to dust particles, and that any stirring up of dust greatly increases the bacterial count. It seems evident that plenum systems which discharge air at high velocity on to the floor are undesirable.

Tuve and Priester (1944) have studied the behaviour of air streams projected from orifices of various shapes and sizes, and they have been able to derive an equation expressing the relationship between the velocity at the outlet, the size of the outlet, and the maximum residual velocity at any cross-section of the air stream in a large room. The maximum velocity at a given cross-section of the air stream was about three times the average velocity at that cross-section. Beyond a distance from the outlet equal to 25 times the diameter of the outlet, the maximum residual velocity in the stream was inversely proportional to the distance from the outlet, until the maximum velocity fell below 500 ft/min, and at greater distances from the outlet the reduction in velocity took place more rapidly. The proportion of room air entrained in the air stream is greater when the stream is discharged from a long and narrow slot than when the orifice is proportionately wider. Tuve and Priester measured the proportion of air entrained by the air streams from slots 24 in long, and of widths varying from $\frac{1}{4}$ in to 3 in, the air velocity at the outlet being in each case 800 ft/min. The table shows the entrainment ratios measured 9 ft from the inlets.

Outlet size in inches	*Entrainment ratio*
24 x 3	4.62
24 x 2	6.37
24 x 1	8.47
24 x $\frac{1}{2}$	11.85
24 x $\frac{1}{4}$	17.75

By doubling the ratio length/width of the slot the entrainment ratio was increased by about 50 per cent. For a given air supply more turbulence is set up when the air is delivered through narrow slots than when it is admitted from wider orifices. It is pointed out that when multiple slots are placed only a few inches apart they produce an air stream which performs almost exactly like the stream from a single outlet of the same total area. Hence, if full benefit is to be obtained from the turbulence set up by narrow slots these must be well spaced. Quarter-inch slots should be more than 6.5 in apart.

These laboratory findings support the use of jet ventilation as advocated by Angus (1923–38).

TEMPERATURE OF PLENUM AIR

The temperature of the air discharged from plenum ducts is often as high as 140° F. When air at such a high temperature is admitted into a room where the temperature is only in the region of 65° F, the hot air tends to rise very quickly, and it is extremely difficult to ensure that the entering air becomes properly mixed with the air already in the room. More often than not undesirable temperature gradients are set up with resulting discomfort, while the pooling of hot air near the ceiling, or in the roof space, involves much wasteful loss of heat. As we have seen, even when the air is discharging into the room at high velocity this stratification is liable to occur. When the hot air is discharged in a downward direction the occupants are often exposed to the unpleasant hot blast from the inlets.

In one authoritative work (Allen and Walker, 1931) it is remarked that the hot-blast system is often the best one for warming factories, and it is certainly a very convenient method, for the air trunks can be suspended from the roof members and kept out of the way of machinery. Nevertheless, it is remarked, an ever-present problem is the tendency for the hot air to accumulate at the higher levels, and, it is said, it is therefore preferable to circulate a larger quantity of air at a moderate temperature rather than a small quantity at a high temperature; and it is further said that the maximum desirable air temperature, which should be allowed only in the coldest weather, is 130° F. My own experience leads me to endorse and reinforce this statement. In any air-heating system high inlet temperatures should be avoided. As far as possible the temperature of the inlet air should not greatly exceed that of the room into which it is to be admitted. Unless large volumes of air are circulated by the fans this result cannot be achieved when the plenum system must supply all the heat required for warming the building. A more satisfactory arrangement, which is now much in favour, is to warm the plenum air to the temperature which it is required to maintain in the building, and to provide other sources of heat to compensate for the loss through the structure.

COMBINED PLENUM AND EXTRACTION SYSTEMS

Better control of a ventilation system is often attained by combining plenum ventilation with extraction. Many years ago Bedford made observations in a long and fairly narrow factory in which the fresh air was supplied by an overhead plenum trunk, the air being discharged horizontally above head level. At either end of the factory there were propeller-type extraction fans. Fresh and comfortable conditions were maintained in this factory both in warm and in cold weather.

More elaborate arrangements are now often seen in which there are duplicate duct systems, one set of ducts supplying air through diffuser inlets and the other set extracting through similar openings. In extensive buildings, especially, this arrangement works very well. The system can be so regulated that large volumes of air are supplied to, and collected from the more densely occupied areas, while in parts where there are few occupants the supplies can be restricted. Distribution is under complete control.

DIRECTION OF VENTILATION

When reliance is placed on natural ventilation the air currents in a room generally travel either across from one side to the other, as when the air passes from windows on one side to other openings in the opposite wall, or from a window to a flue, or they travel across and upwards, as when air enters at windows and passes out through roof openings. This upward ventilation takes place because the air inside the building is more often than not warmer than that outside. Convection effects cannot be ignored if a ventilation scheme is to be planned successfully.

In an occupied room, apart from the effects of any warm parts of the structure, such as a chimney breast, the occupants themselves act as convectors. The surface temperature of the clothed human body at normal room temperatures is several degrees warmer than the environment–in a room of 65° F the average temperature of the clothed human body is 75° F or rather more–and close to the body there is a vigorous upward current. We have estimated the speed of these upward currents at 35 to 40 ft/min.

Shaw (1907) wrote: 'The dominant physical law in the ventilated space is the law of convection. It is at once the condition of success and the cause of most failures. Without convection ventilation would be impossible; in consequence of convection nearly all schemes of ventilation fail. . . . The convection currents are interfered with by momentum currents; but the momentum currents only interfere seriously with the horizontal distribution in the middle regions of a room. The vertical distribution of air currents in a ventilated space, and their horizontal distribution at floor and ceiling are to be ascribed generally to convection. Ventilation would be much easier if warmed air or cooled air could be carried along at any height required; but the law of convection is inexorable, warmed air naturally finds the ceiling, cooled air the floor.'

In the past there has been much discussion and controversy concerning the relative merits of upward and downward ventilation. Shaw remarked that when he was asked by those who were distracted by opposing schemes of ventilation, whether air inlets should be set at high or low levels, he invariably asked whether the air was to be supplied warm or cold. If warm air were to be supplied it would come to the top wherever the inlet were placed, and if cold, it would find its way to the floor. The answer depended entirely upn the extent to which convection could be managed. Shaw calculated that air at a temperature of 98° F saturated with moisture and containing 4 per cent carbon dioxide has the same density as surrounding air at a temperature of 81° F. Hence, it follows that exhaled air rises if the room temperature is less than 81° F and falls if the room temperature is greater than 81° F.

Vernon (1934) discusses upward and downward ventilation and describes many installations where one or other type of ventilation is used. He mentions that in the ventilation of theatres and assembly rooms, where large numbers of persons are congregated, a considerable degree of cross-ventilation can be established, but that generally the ventilation scheme is so arranged as to induce the air currents to flow chiefly in an upward or a downward direction, so to combine one of these directions with cross-ventilation. He remarks that the problem of deciding which is the best system is an extremely difficult one, and that it is impossible to give a definite conclusion which will embrace all types of building under all sorts of conditions. Each case must be decided on its merits.

Upward ventilation has long been employed in the Houses of Parliament. The old House of Lords was not sufficiently warmed by the flues built under the floor, and its ventilation was unsatisfactory. Sir Humphry Davy conveyed fresh air through brick flues placed alongside the heating flues, and the warmed air was admitted into the Chamber through small perforations in the floor. The air outlets were in the ceiling, and the extraction of the air was aided by heating the exhaust flues. Bernan (1845) noticed that Davy's exhaust ducts did remove some air from the Chamber, for on one occasion he approached within a few feet of the outlet cowl, and; 'The whiff we had of the gaseous evacuation caused the only indelible impression ever made on our olfactory organs . . . after a lapse of thirty years, the recollection of it is as pungent as ever.' It does not appear that Sir Humphry Davy was rewarded for his efforts, for Bernan quotes a couplet:

'For boring twenty thousand holes,
The Lords gave nothing–damn their souls.'

Interesting information concerning the ventilation of the House of Commons is recorded by Baines (1931). In the House of Commons Chamber, which was destroyed by German bombs, outside air was drawn into the basement where it was warmed. It then passed into an equalizing chamber beneath the Debating Chamber and thence through the perforated floor into that chamber. The air outlets were in the ceiling, and ducts from these led to the Clock and Victoria Towers which acted as exhaust shafts. Furnaces in the upcast shafts supplied the aeromotive force. In 1892, the system was converted from the exhaust to the plenum principle, for in that year a fan was installed to propel air into the heating chamber, while the upcast furnaces were still used. In 1904 or 1905, a fan was added for extraction purposes so that the system then combined plenum with mechanical extraction.

Much of the air supplied to the Chamber entered through slots near the feet of the members, and the draughts thus caused were a source of discomfort. A Select Committee, appointed in 1913, took evidence from Sir Leonard Hill and others and recommended that the slots in the floor under the seats should be closed, and that air inlets should be placed in the face of the gallery so that more air movement would be created at head level. Owing to the outbreak of war in 1914, these recommendations were not carried out. Later, experiments were performed by the staff of the National Physical Laboratory, some of the experiments being done on scale models and some in the Chamber itself. As the result of this investigation it was recommended that new air inlets should be constructed under the side galleries to admit air at a velocity of 420 to 600 ft/min, that most of the floor inlets should be closed, that the fresh air supply to the Chamber should be admitted at a temperature a few degrees lower than previously, and that additional heat should be supplied by means of hot-water pipes under the seats. Thus it was proposed that the upward system should be modified by admitting much of the fresh air above head level in such a way that considerable turbulence would be caused (Cmd. 3871, 1931).

Faber (1944) outlined the system that was later installed in the rebuilt House of Commons, and a fuller description was given by Faber and Kell (1951). The air inlets for the floor of the House consist of openings from ducts placed immediately under the side galleries and discharging horizontally. The inlets for the upper portion of the chamber are apertures at the sides near the intersections of the sloping ceiling with the side walls. Separate plants control the condition of the air entering through these two sets of inlets. The bulk of the air extracted from the Chamber is taken out through the ceiling, but a minor proportion is withdrawn near the floor, where three sets of exhaust openings are placed. To ensure that the feet of the Members can be kept warm there is provision for warming the floor by means of hot-water pipes. From the ducts below the galleries some 6,000 to 8,000 cu. ft of air/min can be introduced, while from the upper ceiling inlets 18,000 to 20,000 cu. ft/min can be supplied. The sum of these two volumes, which can be reduced at will, represents a ventilation rate of 9 air changes/hour.

The temperature of the air entering below the galleries is

controlled steadily at a desired level. Adjustment of the temperature of the air entering at the roof openings is relied upon for maintaining the required temperature within the Chamber. However, through this arrangement quite cold air was introduced: it fell rapidly, and in fact cascaded down to the floor of the Chamber creating a degree of turbulence which might have been acceptable out of doors but not indoors. Papers began to blow about. The result was a very stimulating environment for senior Members of Parliament especially those sitting on the front benches who obtained the full benefit of this shower of air. The back-benchers, sheltered by the galleries, were in a zone of completely calm air.

While a case could be made out for putting senior Members in a stimulating environment and *vice versa,* when the stimulus becomes excessive it becomes distracting and there were many complaints. This was in the early days of the installation. Nowadays the defect of temperature at the roof inlets below the temperature near the Speaker's chair is not allowed to exceed a fixed value, and there appear to be no more complaints.

Calver (1930) discussed the ventilation of the U.S. Halls of the Senate and Congress at Washington. Initially it was intended that the direction of ventilation should be downwards, but the design was reversed, and an upward system was installed. This was replaced by a modern air-conditioning installation, when the arrangements were again reversed and the downward system restored. With the old system, dust lodged in the floor inlets and was then blown into the faces of members. The inlets served in the dual capacity of ventilators and cuspidors, and some of them were found to have a coating of 1.5 in of dry saliva and tobacco juice.

While, in America, mechanical ventilation, or a combination of plenum ventilation with gravity exhaust, is often used for relatively small rooms such as those in schools and commercial buildings, in Great Britain natural ventilation will generally be found effective, for in our milder climate more constant use can be made of open windows. Even in Great Britain, however, the problem of ventilating small rooms by mechanical means arises, especially in large commercial buildings which are subdivided into large numbers of small offices. In such buildings where natural ventilation is difficult or impossible to contrive, the problem is to admit air in such a way that a fresh atmosphere will be maintained without objectionable draughts. If warm air is supplied at low velocity through high level inlets undesirable temperature gradients are likely to be caused, whereas if cool air is admitted near floor level at even moderate velocities there will be complaints of draughts. In the ventilation of such rooms diffuser inlets offer promise of good results, provided full care is taken to eliminate noise or to reduce it to a very low level.

CHAPTER XIII

THERMAL PROPERTIES OF BUILDINGS

THERMAL TRANSMITTANCE

WHEN a building is kept at a temperature above that prevailing outside, heat passes from the interior and is lost from the external surfaces of the walls, windows and roof, and through the floor.

When heat passes outwards through a simple wall it is first transmitted from the air of the building to the internal surface of the wall, it is then conducted through the fabric of the wall to the outer surface, and then transferred from the surface by radiation and convection and, when the surface is wet, by evaporation. The rate of heat transfer/unit area of wall, for a given difference between the internal and external air temperatures, depends on the thermal conductivity and thickness of the wall fabric, and on the film or surface conductances at the inner and outer surfaces.

The *thermal conductivity* (k) is the number of Btu transmitted/hour through 1 sq. ft of a 1-in thickness of material for a difference of 1° F between the temperatures of the two surfaces. The *thermal conductance* is the amount of heat transmitted/hour through 1 sq. ft of material, of whatever thickness is under consideration, for a 1° F temperature difference. The *surface conductance* is the amount of heat transmitted from surface to air, or from air to surface, per hour/sq. ft/1° F. difference between the temperatures of the surface and of the air. The reciprocals of these conductivities and conductances are respectively known as *thermal resistivities* and *resistances.*

For practical purposes the heating engineer wants to know the amount of heat which will be transmitted from the internal air to the external air, and he is primarily interested in the overall air-to-air transmittances of the various parts of the structure. The total air-to-air resistance of a structure is the sum of the resistances of the component layers *plus* the resistances of the internal and external surfaces. Thus, when

R is the air-to-air resistance of the structure;
R_{si} is the resistance at the internal surface;
R_{se} is the resistance at the external surface;
L_1, L_2 etc., are the thicknesses in inches of the individual layers of material;
k_1, k_2 etc., are the thermal conductivities of those materials; and
R_a is the resistance of any air space which is incorporated;

the air-to-air resistance is given by

$$R = R_{si} + R_{se} + \frac{L_1}{k_1} + \frac{L_2}{k_2} + \cdots + R_a$$

The air-to-air transmittance (U) is the reciprocal of the total resistance: $U = 1/R$.

Comprehensive tables of the thermal conductivities and resistivities of materials used in building construction are given in the booklet on 'The Computation of Heat Requirements for Buildings' issued by the Institution of Heating and Ventilating Engineers, and in the *Annual Guide* of the American Society of Heating and Ventilating Engineers. Some values taken from the former publication, many of them having been determined by Griffiths (1941), are given in the table following.

The surface resistances for plane internal surfaces are taken as:

Walls	0.70
Floors or ceilings:	
Heat flow upwards	0.60
Heat flow downwards	0.85
Roofs	0.60

The external surface resistances vary according to orientation and exposure, but for ordinary conditions the following values may be assumed:

Walls	0.30
Roofs	0.25

For corrugated surfaces the resistances are about 20 per cent less than the above values for plane surfaces.

For air spaces of a minimum width of $\frac{3}{4}$ in the resistance may be taken as 1.0, and for an air space of similar width formed between corrugated material and a lining, 0.9. When a lining is in contact with a corrugated surface the resistance of the resulting air space is about 0.5.

Thermal conductivity and resistivity of building materials

Material	*Conductivity (k) Btu/sq. ft/hour/°F difference/in thickness*	*Resistivity (1/k) °F/in thickness/ Btu/sq. ft/hour*
Asbestos cement sheeting	1.9	0.527
Asphalt	8.7	0.115
Bitumen	1.1	0.91
Bricks, common	8.0	0.125
Bricks, diatomaceous	0.99	1.01
Concrete, ballast	7.0	0.14
Concrete, light clinker	2.3	0.44
Concrete, foamed slag	1.7	0.59
Cork slab, 8-9½ lb/cu. ft	0.29	3.45
Cork slab, 12½-13½ lb/cu. ft	0.37	2.70
Cork slab, 19-23 lb/cu. ft	0.43	2.33
Glass	7.3	0.137
Linoleum, battleship	1.26	0.79
Plastering	3.2 to 4.5	0.31 to 0.22
Roofing felt	4	0.25
Sandstone	9	0.11
Slag wool	0.3	3.33
Tiles	5.8	0.17
Timber, deal	0.87	1.15
Timber, oak	1.11	0.90
Wallboard, fibre board	0.38	2.63
Wallboard, plaster board	1.1	0.91

Consider, for example, a 9-in brick wall with $\frac{3}{4}$-in plaster on one surface. The thermal resistance is

$$R = R_{si} + R_{se} + \frac{L_1}{k_1} + \frac{L_2}{k_2}$$

$$= 0.7 + 0.3 + \frac{9}{8} + \frac{\frac{3}{4}}{4}$$

$$= 2.31$$

and the transmittance (U) is the reciprocal of this, *viz.* 0.43.

If the internal and external air temperatures are respectively 65° and 30° F the heat transmission through the wall will be at the rate of 0.43 x 35 = 15.1 Btu/sq. ft/hour.

For a wall $4\frac{1}{2}$ in thick and similarly plastered, U is 0.57, and the heat transmission is 33 per cent greater than through the 9-in wall.

Now consider the effect of adding to the plastered 9-in brick wall a lining of fibre board ($k = 0.38$) $\frac{1}{2}$ in thick, with an air space of $\frac{3}{4}$ in separating it from the plastered surface. The air-to-air resistance becomes:

$$R = R_{si} + R_{se} + \frac{L_1}{k_1} + \frac{L_2}{k_2} + R_a + \frac{L_3}{k_3}$$

$$= 0.7 + 0.3 + \frac{9}{8} + \frac{\frac{3}{4}}{4} + 1.0 + \frac{\frac{1}{2}}{0.38}$$

$$= 4.63.$$

In this example the addition of the air-spaced lining doubles the resistance, and halves the heat transmission through the structure.

Some typical heat transmittance values for walls, roofs and floors, extracted from the Institution of Heating and Ventilating Engineers' handbook, are given overleaf.

CALCULATION OF HEAT LOSSES

Before the amount of heating required for a given room can be calculated it is necessary to decide on the temperature that is to be maintained when the outside temperature is at a certain datum level—usually 30° F. Then the heat losses through the structure are computed by taking into account the areas of the various parts together with their thermal transmittances. The amount of heat required to warm the air entering the room is calculated, and any necessary additional allowances for height of room or for intermittent heating are made.

Air infiltration

No ordinary building is airtight. Even in buildings of good modern construction air infiltrates through crevices around

Heat transmittance coefficients

	Thermal transmittance	
Walls:	(*U*)	
Brick, solid, 4½ in unplastered	0.64	
Brick, solid, 9 in unplastered	0.47	
Brick, solid, 13½ in unplastered	0.37	
Brick, solid, 4½ in plastered	0.57	
Brick, solid, 9 in plastered	0.43	
Brick, solid, 13½ in plastered	0.35	
Brick, cavity, unventilated, 11 in plastered	0.30	
Brick, cavity, unventilated, 15½ in plastered	0.26	
Concrete, 4 in	0.64	
Concrete, 6 in	0.54	
Glass, single windows	1.00	
Glass, double windows	0.50	
Flat roofs:		
6 in concrete, asphalted	0.57	
Pitched roofs:		
Corrugated asbestos	1.40	
Corrugated asbestos, lined ½ in boards	0.50	
Corrugated iron	1.50	
Corrugated iron, lined 1 in boards and felt	0.35	
Tiles on battens	1.50	
Tiles on battens, felted	0.70	
Tiles on boards and felt	0.35	
Floors, on ground*:		
Ventilated wood floor on joists, air bricks on more than one side	0.40	
In contact with earth or hardcore, wood block on concrete	0.15	
In contact with earth or hardcore, concrete and tiles	0.20	
	Heat flow downwards	*Heat flow upwards*
Floor, intermediate:		
Wood floor on joists, plastered ceiling	0.22	0.29
Concrete, 6 in, with 2 in screed	0.35	0.43
Concrete, 6 in, with wood flooring	0.26	0.30

* With these coefficients the difference between the temperatures of the internal and the external air should be used.

windows and doors, and the amount of this air leakage may vary with the force of the wind. Further, in some kinds of construction there may be infiltration through the fabric. In some buildings the infiltration may be enough to satisfy the fresh air requirements of the occupants, but in others it is necessary to admit air purposely for ventilation. Desirable ventilation rates have been discussed, and measurements of ventilation described, in earlier chapters. The ventilation rates advocated for current practice are set out in the Institution of Heating and Ventilating Engineers' booklet on the computation of heat requirements.

Allowance for Height

With most kinds of heating installations the air at the higher levels in a room tends to be warmer than that near the floor. The steepness of these temperature gradients varies; in some instances the temperature-rise with increasing height is small, while in others it is considerable. In a lofty room the temperature of the air at the higher levels is likely to be several degrees higher than that at the level of the occupants, and therefore, when the heat required to maintain a specified temperature in the zone of occupation is being calculated, some allowance is necessary for the extra heat loss due to the higher temperature at the upper levels. It would be desirable to take the type of heating into account in making such allowances, but sufficient data for the formulation of a detailed scheme of allowances are not available. The Institution of Heating and Ventilating Engineers' handbook gives a table of corrections for height which are especially applicable to low-pressure hot-water systems with radiators located at floor level. These allowances are expressed as percentage additions to be made to the total heat loss calculated for the losses through the structure *plus* the air change losses. The additions range from 2 per cent, for heights of 13 to 14 ft, to 40 per cent when the height is 66 to 70 ft. A graph of these corrections is shown in Fig. 71.

Allowance for Intermittent Heating

The heat requirements of a building calculated as described above are based on equilibrium conditions; they represent the

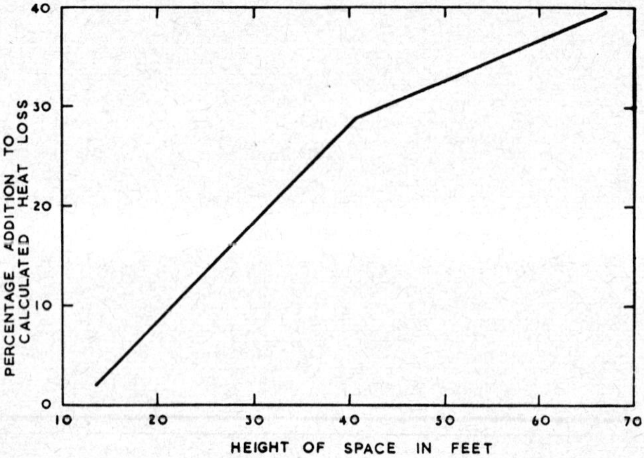

Fig. 71. Addition to calculated heat loss to allow for height of heated space.

amount of heat needed to maintain the building at the required temperature with the specified external temperature, once the required temperature has been attained. If the building is not continuously warmed, however, a substantial amount of heat is needed to warm the structure after the heating appliances are brought into operation, and in calculating the heat requirements, or the heating costs, of a building which is to be intermittently heated, some allowance is necessary for this extra heat consumption. The magnitude of these allowances will clearly depend on the type of heating installation in use, on the nature of the structure, and on the intervals of time which elapse between periods of occupation. Heavy structures of large thermal capacity will need more heat than buildings of lower thermal capacity. Recommendations are made by the Institution of Heating and Ventilating Engineers as to the percentage allowances for intermittent heating which should be added to the heating loads calculated on the basis of continuous heating.

Heat Requirements for a Living Room

As an example of heat-loss calculation consider the amount of heat required to maintain a temperature of 65° F in a living room in a typical British house when the external temperature is 30° F. Assume that the room is 18 ft long, 12 ft wide, and 10 ft high; that two adjoining walls are external, one having a window 6 ft x 6 ft; that the other two walls are internal ones, and that the adjoining rooms and the room above are at a temperature of 45° F. The external walls are of 9 in brick, plastered internally, while the internal walls are of 4½ in brick plastered on both sides. The ceiling is plastered, with joists and a wooden floor above; while the floor is of wooden boards on joists with a ventilated space beneath. Assume that the ventilation rate is 2 air changes/hour, and take the specific heat of air as 0.02 Btu/cu. ft/° F. The heat losses can be calculated as follows.

Part of structure	*Area sq. ft*	*Thermal transmittance, 'U'*	*Temperature difference, °F*	*Heat transmitted, Btu/hour*
External walls	264	0.47	35	4,343
Window	36	1.00	35	1,260
Internal walls	300	0.43	20	2,580
Ceiling	216	0.29	20	1,253
Floor	216	0.40	35	3,024
				12,460
Warming of incoming air: 2,160 x 2 x 35 x 0.02				3,024
Total requirement				15,484

In this example the heat required to warm the ventilating air is about one-fifth of the total heat requirement of the room. If the room were warmed by an open fire the ventilation rate would probably be nearer 4 air changes/hour than 2, and in that event the warming of the incoming air would account for about one-third of the total requirements.

THERMAL INSULATION

Building materials vary greatly in thermal conductivity. The conductivity (k) for common brickwork is 8.0, while that for lightweight cork or for slagwool is 0.3, and that for fibre board 0.38. Materials with low conductivities are used as insulators. It is probably due to the fact that fuel was formerly cheap and plentiful in Great Britain that in this country insufficient attention has been paid to the thermal insulation of buildings. The proper use of insulating material results in substantial savings in heating costs. During the war of 1939-45 the acute shortage of fuel caused attention to be focused on means of fuel economy, and, for the guidance of industrial consumers, the Ministry of Fuel and Power (1943) issued a Bulletin on the thermal insulation of buildings. Valuable papers have also been contributed by Pallot (1943) and Newcombe (1944).

Economy of Insulation

We have noted earlier in this Chapter that if to a plastered 9-in brick wall there is added an air-spaced lining of fibre board, the heat transmission through the structure is halved. The overall heat requirements of a room so treated would not be reduced to quite the same extent, for the insulation would not affect the heat required to warm the ventilating air, and there would still be substantial losses through the window glass. Yet the economy in the heat required to maintain a given air temperature would be well worth while.

Pallot (1943) refers to the saving that can be effected by roof insulation. The thermal transmittance of a corrugated asbestos roof, uninsulated, is 1.50, while a similar roof insulated with fibre board ½ in thick, and with an air space, has a thermal transmittance of 0.31. The instance chosen is a factory having a roof area of 10,000 sq. ft, and it is assumed that an internal temperature of 60° F is to be maintained by a central heating plant throughout the winter. Pallot compares the costs when the roof is (*a*) uninsulated, and (*b*) insulated.

The cost of insulating the roof would be about £500. In this instance, insulation of the roof is a good business proposition. The capital cost of the insulation is practically offset by the saving in the cost of the heating installation, while the annual saving in fuel costs approaches £200.

The Bulletin issued by the Ministry of Fuel and Power (1943) gives detailed calculations of the savings effected by insulating a factory building. It is assumed that the roof is 100,000 sq. ft in

	Uninsulated	*Insulated*
Heating surface to provide for heat transmission through roof with 30° F outside	2,250 sq. ft	470 sq. ft
Approximate cost of heating surface	£750	£155
Approximate fuel consumption per annum with mean external temperature during heating season of 43° F	80 tons	17 tons
Annual fuel cost (at £3/ton)*	£240	£51

* In about 30 years, thc cost of solid fuel has increased approximately by a factor of 10.

area and is of corrugated asbestos sheeting, and that the walls are of 6-in concrete, with a surface area of 20,000 sq. ft. Continuous heating throughout the winter, so that the average difference between inside and outside temperatures is 18° F, is assumed, and the insulation taken into account consists of ½-in thickness of fibre board above the purlins of the roof and on the walls, an air space being provided. The total cost of insulation is estimated at £6,000, and this results in a saving in the initial cost of the central heating plant amounting to £7,130. The annual saving in fuel amounts to more than 600 tons.

INSULATION AND COMFORT

Thermal comfort depends on the radiation from the surroundings as well as on the air temperature. An increase in the temperature of the boundary surfaces of a room tends to make one feel warmer. Figure 72 shows the combinations of air temperature and mean radiant temperature which, according to Bedford's data (1936), yield various equivalent temperatures when the air velocity is 16 ft/min.

When walls are well insulated the temperatures of their inner surfaces are raised to a level much nearer to the internal air temperature than when they have a high thermal transmittance: thus, for a given internal air temperature, the better the insulation the higher the equivalent temperature.

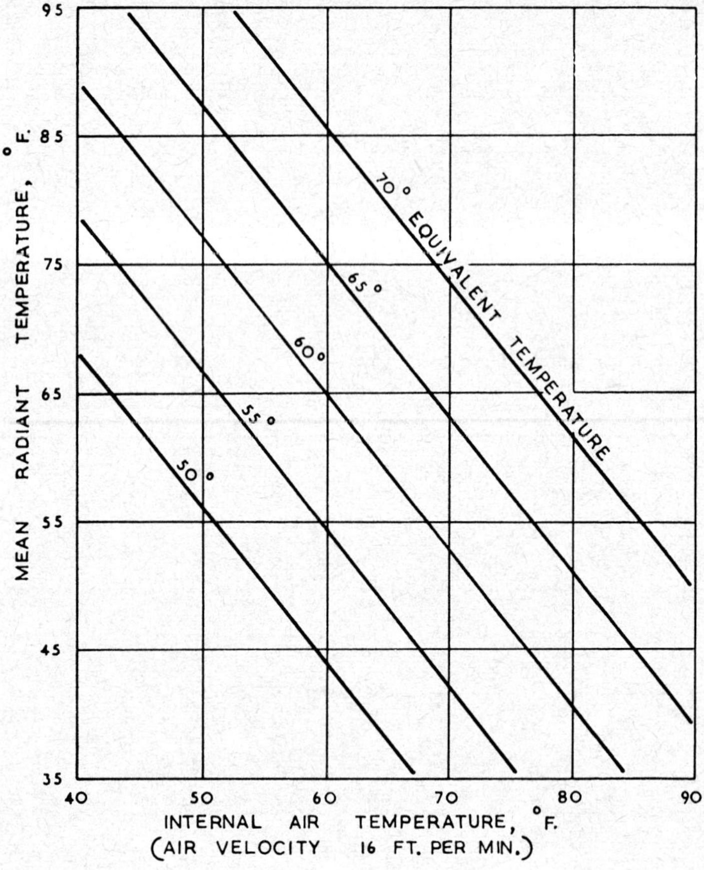

Fig. 72. Relation between equivalent temperature, air temperature and mean radiant temperature, when the air velocity is 16 ft/min.

Consider the example of the 9-in brick wall previously cited, in which we noted that the application of a fibre-board lining with an air space halved the heat transmission. Figure 73 shows the equilibrium wall surface temperatures, and the temperature gradients through the wall, (*a*) when it is unlined, and (*b*) when it is lined.

With the assumed air temperatures of 65° F inside and 35° F outside, the temperature of the internal wall surface is 56° F without lining, and 60.5° F with lining.

If the mean radiant temperature were raised by this 4.5° F,

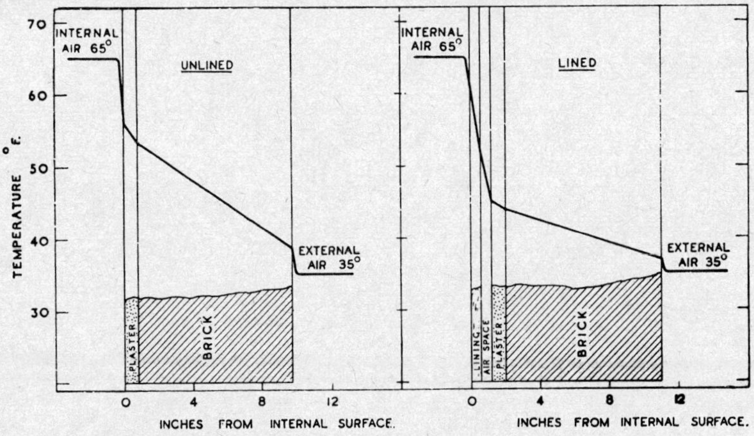

Fig. 73. Surface temperatures and temperature gradients through plastered 9-in brick wall, (a) *unlined, and* (b) *with air-spaced lining of fibre board.*

the equivalent temperature would be increased by about 2° F. The actual rise would not be so great, for the radiation from windows and floor would be unaltered, but it would probably be worth while, especially during the period required for the walls to become warm.

Figure 74 is an elaboration of a graph given by Pallot (1943). It shows, for various internal air temperatures, the inner surface temperatures of walls in relation to their thermal resistances when the external air temperature is 30° F. These also are equilibrium values.

Heating of Walls

Another benefit arising from suitable and adequate insulation when heating is intermittent is that wall-surface temperatures rise rapidly after heating is begun. Dufton (1927) calculated the rate at which the surface temperature of a homogeneous wall rises when heat is applied to it at a constant rate. The rate of heating assumed was twice that necessary to maintain the desired temperature difference between the faces of the wall. The time taken for a 9-in brick wall to reach the desired temperature was calculated to be 6 hours, while for a 1-in wall

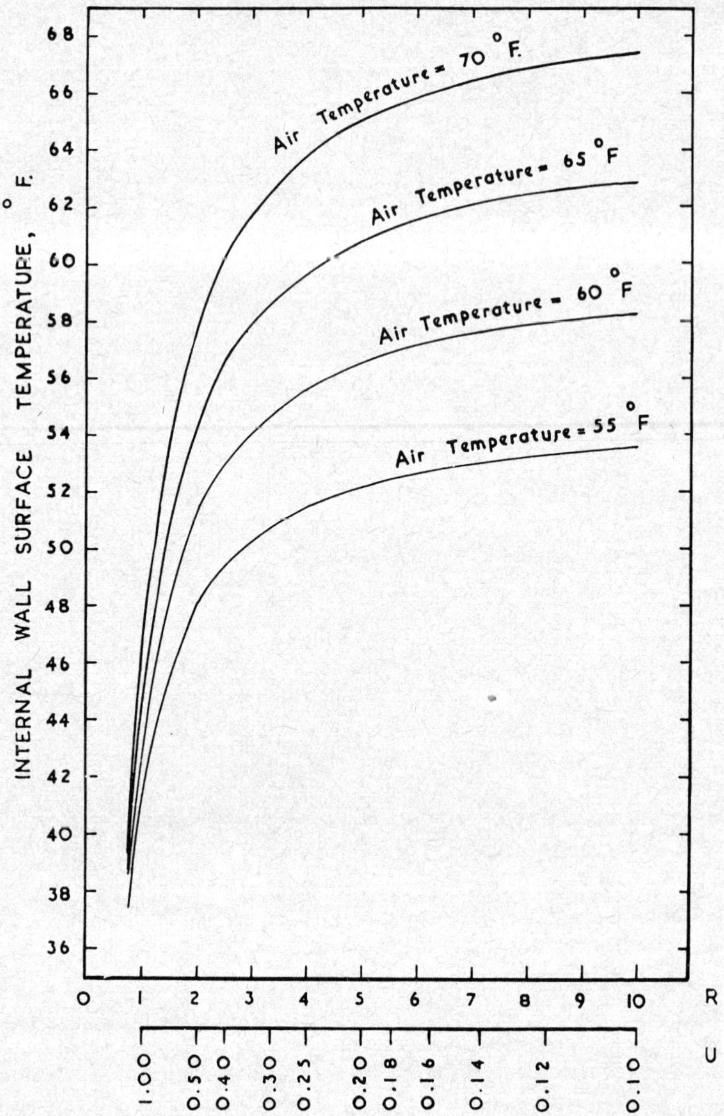

Fig. 74. Inner surface temperatures of walls in relation to thermal resistance and internal air temperature. External air temperature 30° F.

of wood, which confers the same insulation, the time was only 10 min. The rise in temperature of a laminated wall depends largely on the material at the heated surface, and a room with masonry walls is much more readily heated if lined with wood or other insulating material of low thermal capacity.

Dufton (1931*a*) followed up this theoretical exploration with a practical experiment in the dining room of his home. The room was heated by a gas fire burning 0.2 therm/hour. Records of equivalent temperature were taken by means of a eupatheoscope placed 8 ft from the fire. Then the room was lined with oak panelling. It was observed that the panelled room was warmer after half an hour of heating than the unpanelled room after an hour and a quarter, and records of equivalent temperature, taken with the eupatheoscope screened from the fire, confirmed this. When the room was required for only half an hour (*e.g.* for a meal) it was not necessary to light the fire so early after panelling, and for this short occupation the required temperature was obtained with half the fuel consumption of the unpanelled room. It should be noted that the gas fire did not attain its full intensity during the first quarter of an hour, and that only half the area of the boundary surface of the room was changed by panelling.

Newcombe (1944) gives a valuable table of specific heats, densities, heat capacities, diffusivities and conductivities for a large number of building materials, including many of those used as insulators. To secure low thermal transmittance values, he urgues that in house construction there should be used just enough of the outer material for the required strength and stability, with a factor of safety of 4 or 5, and that high-grade insulating material should be placed inside to attain the desired low transmittance values.

The Committee of the Royal Institute of British Architects, which was convened to advise the Ministry of Works (1944*b*) quotes the transmittançe values recommended by the Building Research Station. Two standards are advocated: Standard A when heating appliances are uncontrolled, *e.g.* open fires; and Standard B for controlled heating appliances such as slow combustion stoves, electric heaters, or appliances of a central heating system. The recommended transmittance values are:

Standard A		*Standard B*	
External walls, general	0.30	External walls, general	0.20
External walls, living rooms	0.20	External walls, living room	0.15

Lower standards are suggested for uncontrolled heating because the full benefits of thermal insulation cannot be realized unless the heating appliances are fully controlled. The Committee emphasizes the value of materials of low conductivity and low thermal capacity permitting the inner surface of a room to be heated before the whole fabric attains the desired temperature.

Alternatives are put forward for achieving the transmittance of 0.20 for external walls of living rooms. One is the use of different materials or methods of construction from those normally used, without special linings; and the other is the use of internal linings with or without air spaces. Examples of the different methods or materials suggested are:

5 in pumice concrete, rendered externally, plastered internally;

9 in foamed slag concrete, rendered externally, plastered internally;

10 in clinker concrete, rendered externally, plastered internally;

12 in no-fines concrete, rendered externally, plastered internally;

or combinations incorporating an air space, such as:

4½ in brickwork, 2 in air-space, and 4 in pumice concrete plastered internally;

Similar combinations using 7 in foamed slag, 8 in clinker, 6 in hollow clay brick, or 3 in diatomaceous earth block, in place of the 4 in layer of pumice concrete.

For the use of linings the 11 in sealed cavity brick wall plastered internally is considered as a standard backing. The required transmittance value can be obtained with a lining of ½ in of fibre board, or of ⅜ in of cork, without an air space; or with a lining of ½ in of softwood panelling or plasterboard with a 1-in air space. The use of these linings would permit of a more rapid heating of the inner surface than would occur with the methods which would yield the same transmittance values without lining. Of the linings which have been tried fibre board allows of the most rapid rise in temperature; after that comes ply-board, and then plasterboard and hardboard. A drawback to the use of air spaces is that they provide harbourage for vermin.

For internal walls there is no special call for insulation, except when they separate heated rooms from cold ones. They should, however, be so constructed that their surfaces heat up rapidly.

Cooling of Structures

After a hot spell, buildings of high thermal capacity cool more slowly than those of low thermal capacity. Thus, when a building of the former type is not artificially cooled, the inside air temperature falls only slowly after a sudden change to cooler weather.

An example of this slow response is cited by Vernon, Bedford and Warner (1926*a*). Observations were made in a multi-storey factory, a brick building which was ventilated by a plenum system. The windows were not made to open, and the ventilating fan was shut off during the night. In the middle of one July the outside temperature fell from 74.4° to 59.1°F between 3 p.m. one day and 4 p.m the next day. Meanwhile the indoor temperature fell only from 75.6° to 70.5°, and even after another 19 hours of the cool weather it had fallen only to 66.4°. Since the windows could not be opened, variable natural ventilation could not be held accountable for the results observed. Figure 75 illustrates the slow response of the indoor temperature.

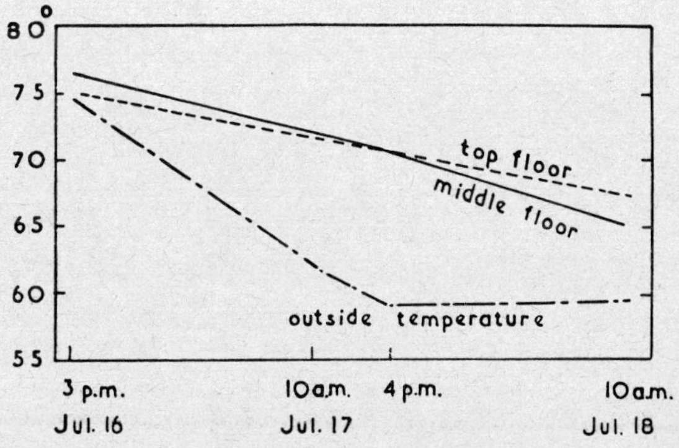

Fig. 75. Slow response of indoor temperature to fall in outside temperature.

Condensation and its Prevention

Unless the air of an occupied room is dehumidified it contains more water vapour than is present in the external air; for to the water vapour already present in the air entering from outside is added vapour exhaled by the occupants. Further sources of added moisture may be the steam arising from cooking or from flueless heating appliances. The effect on the atmospheric humidity of a given rate of vapour production from these sources clearly depends on the ventilation. The more rapid the ventilation the more will the water vapour concentration be reduced, and the lower will be the humidity of the air.

If the dew-point temperature of the air rises above the temperature of the internal boundary surfaces, condensation will occur. The two obvious methods of preventing condensation are: (i) to provide adequate thermal insulation, so that the internal surfaces are only slightly cooler than the air; and (ii) to secure sufficient ventilation to ensure that the dew-point temperature of the air does not rise as high as the internal wall temperature.

As stated here these methods of prevention of condensation seem to be obvious. Yet there have been many complaints of damp in modern houses which, on investigation, have been caused not by faulty construction but inadequate insulation, ventilation or both, resulting in condensation.

EXCLUSION OF SOLAR HEAT

If there were no intervening atmosphere, the intensity of radiation incident upon the earth's surface normal to the sun's rays would be about 430 Btu/sq. ft/hour. Owing to absorption by the atmosphere, the intensity of radiation reaching the earth's surface is only about one-half of that value. At Kew, values as high as 298 Btu/sq. ft/hour have been observed several times, but during a 10-year period the average of the nine highest values each year was 269 Btu (Fishenden and Dufton, 1929).

In winter the heating effects of solar radiation lessen the load on a heating installation, but in summer they throw an additional burden on air-conditioning systems, or, where there is no air-cooling, they may cause the internal temperatures of buildings to rise to unpleasantly high levels.

Usually, when solar radiation impinges on a building most of it is absorbed at the surface of the structure, and the amount of heat passing inwards depends on the degree of insulation afforded by the fabric. Thus, insulation, especially of a roof, is a valuable protection against overheating in summer. Further protection may be gained by treating the external surface to enable it to reflect a higher proportion of the incident radiation.

Owing to the thermal capacity of a structure, there is a time-lag in the transmission of heat through it. With a roof constructed of corrugated iron only this lag is very small, while with some other types of structure it may be considerable. Thus, for a plastered 13 in brick wall the lag is 12 hours; for a 6 in horizontal concrete roof with waterproofing and slag finish, 5 hours; and for a 1 in yellow pine roof, waterproofed with smooth black finish, 1 hour (A.S.H.V.E. Guide, 1945).

Numerous investigations made at the Building Research Station have shown that when a roof surface is whitened it absorbs far less solar heat than when it is dark in colour. On a July day the temperature of an unshielded asphalt roof was 110° F, while on the same day the temperature of a similar roof covered with a 1 in mortar screed whitewashed two coats was 36° F lower (Building Res.. Bd., 1935).

Many factories are built with thin roofs, and unless these are specially treated much solar heat is transmitted through them. Crowden (1934*a, b*) has advocated the use of reinforced aluminium foil for thermal insulation. The metallic foil is adhered to both sides of a reinforcing medium, such as asbestos paper or bitumen felt, and the reinforced foil is used as a diaphragm placed medially in an air cavity. Thus the air space is divided into two separate cells, and this in itself increases the thermal resistance of the gap. The reflective properties of the polished aluminium surfaces assist further, for these surfaces absorb little radiation and emit little. Crowden advises the use of this system for the inner surfaces of walls and roofs. A panel consisting of two sheets of asbestos cement 1 in apart, with the air space divided by one sheet of reinforced aluminium foil has a thermal resistance approximately equal to that of 1 in of cork.

At the Building Research Station (1934) the transmission coefficient for a corrugated asbestos cement roof was found to be 1.41 Btu/sq. ft/hour/° F. When the roof was lined with plywood faced on both sides with aluminium sheet the transmission coefficient was reduced to 0.31. Further tests were made when the corrugated asbestos sheet was lined with one sheet of reinforced aluminium foil, and then when the lining consisted of two layers of the foil $\frac{3}{4}$ in apart. With one layer of foil the transmission coefficient was 0.39/sq. ft/hour, and with two sheets it was 0.19. From rough calculations it was concluded that a closed air space not less than $\frac{3}{4}$ in wide, and bounded on one or both sides by a plane bright metallic surface, affords insulation equivalent to $\frac{1}{2}$ in of cork.

The tendency nowadays is to bring solar heat into buildings rather than to exclude it. Even if the fabric of the building is well insulated, this may be offset by increasing the amount of glass. The Building Research Station found that the average ratio of the area of glass to the area of floor was 1 : 5 in pre-war and 1 : 3 in post-war buildings. It was also found that there was a 9 per cent incidence of complaints of overheating in summer in the older buildings compared with 40 per cent in the post-war buildings. It has been suggested that the very large areas of glass are not based upon the need for adequate daylighting but on some undefined psychological need.

A more likely explanation is that buildings with large areas of glass are in many ways cheap to construct.

Langdon and Loudon (1970) discuss the nature of discomfort in schools from overheating in summer and point out that increased solar gains are more than off set by the cooling effects of cross-ventilation. But, even if windows are openable, one may be reluctant to do this if traffic noise as well as air enters.

Chrenko (1962) noted that the mean radiant temperature at head level in a south-facing office in London during March exceeded 100° F and this was due to solar radiation entering the room via the windows. On lowering the Venetian blinds, the

MRT was reduced to 90° F, with the slats horizontal, and 80° F with the slats vertical. The temperature of the air was about 72° F. Shading of windows is thus important for the occupant and it is also important for the building itself. The heat gain for north-facing glass is only about 10 per cent of that for east-facing or west-facing glass or 15 per cent of south-facing glass. External blinds may reflect 75 per cent of the solar heat load, and specially designed glass, which may be coloured, may reduce the solar load by up to 65 per cent, but specialist firms like Pilkingtons should be consulted in order to achieve the best results. However, reliance should not be placed upon ordinary double glazing, which, although it may reduce the incoming thermal load by 10 to 20 per cent, will not prevent discomfort. One of the worst cases of discomfort from solar radiation experienced by the present writer was in a hospital in which double glazing was specified by the architect to offset solar radiation.

Considerable attention must be paid by environmental engineers to the rôle of solar heat in their estimation of the thermal load on buildings when air-conditioning systems are designed.

REFLECTING WALLS

Dufton (1931*b*) suggested as a possibility that if, in a room with two internal and two external walls, the external walls were covered with metal foil, they would act as reflectors, and that the occupants of the room should feel as warm as if the room were bounded not by the external walls but by the images of the internal walls reflected in them. The loss of heat from the room would be reduced, while greater warmth would be experienced. He later (1934) investigated this point by covering the inner surfaces of the external walls with aluminium foil. The whole house was heated to an equivalent temperature of 65° F, day and night, and the windows were kept closed. The heat input to each room during each 24-hour period was measured, and it was found that the use of this metallic wallpaper resulted in a saving of heat amounting to 8 per cent.

Lorenzi, Herrington and Winslow (1946) have measured the influence of a coating of aluminium paper on heat economy. The ceiling and the brick walls of a room were lined with aluminium foil. The amount of heat required to maintain a temperature of 70° F was 20 per cent less than before the surfaces were lined. With the aluminium lining the response to heating and cooling was much more rapid than before. In these experiments the heat was supplied by infra-red lamps.

Crowden (1934) quotes accounts of Admiral Byrd's expedition to the Antarctic, in which it is reported that the walls of his hut had a bright metallic surface so that radiation losses from the occupants would be largely compensated for by the reflection of heat back to their bodies. He refers to Dufton's experiment, and avers that the effect of reflecting back the heat radiated by the occupants would probably permit of a still greater reduction in the amount of heating required than the 8 per cent saving measured. In Dufton's experiments, the eupatheoscope by means of which the radiation was measured took into account the effects of any reflection back of its own heat, but only two of the four walls were lined with metallic foil. The saving in heat reported by Lorenzi and his colleagues (*vide supra*) was ascribed to the reduction in the inside film coefficients due to the reduced radiation transfer. The mean air temperature was 0.6° F lower, and the mean globe thermometer temperature 1.2° F higher, in the lined room than in the unlined room.

Mills and Ogle (1938) made observations in small experimental rooms, some of which were completely lined with aluminium foil while others were not. They reported that their subjects felt much warmer in the lined chambers than in the unlined ones, when the air temperature was low. It is unfortunate, however, that they apparently made no measurements of the mean radiant temperature or of the equivalent temperature. Further careful investigations are needed before one can decide whether these reflective surfaces have any profound influence on thermal comfort.

Many people would find a room with walls and ceiling completely lined with bright aluminium foil unpleasant to live in; and any ordinary decorative treatment which covers the foil is likely seriously to reduce the reflective powers of the foil for long wavelength radiation. Over 140 years ago, Leslie showed the reduction in reflectivity when polished metal was covered with a very thin coating of transparent non-metallic material. It is said, however, that certain dyes are now available which can be applied to metallic surfaces without much influencing their infra-red reflective properties, and some paints have been manufactured which reflect much of the incident infra-red radiation.

CHAPTER XIV

THE HEATING OF BUILDINGS

HISTORICAL

IN the early days of Greece the usual means of heating the homes of the wealthy was the charcoal brazier. The Greeks introduced the hypocaust system, by which the smoke and hot gases from a furnace were led through flues beneath the floor. Thus, the whole floor was heated, and this in turn heated the air of the apartment. Later, flues of hollow wall-tiles were also used, so that the walls as well as the floor could be warmed. A still further development was to pass heated air into the room by providing openings from the wall-ducts. These were kept closed until the fire had ceased flow, so as to avoid smoke, and then they were opened to allow the entry of heated air. From Greece the hypocaust passed to Persia and Rome, and the later developments of the system were due to the Romans. As Markham (1942) says, the Romans quickly became the world's supreme plumbers and heating engineers. The houses of the

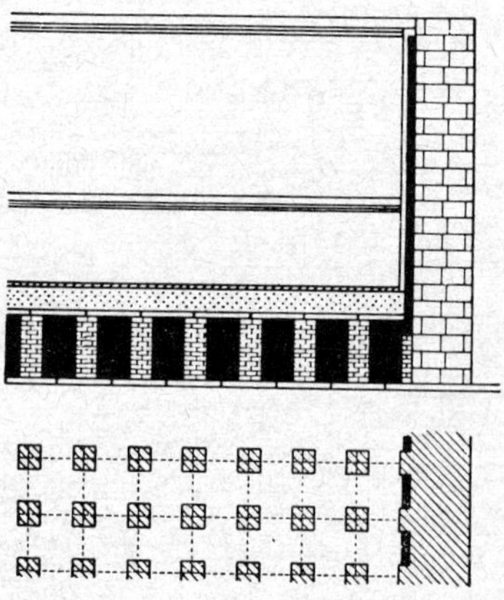

Fig. 76. Diagram of hypocaust in Roman Bath, showing wall duct.

wealthy had central heating, windows were glazed, water was generally laid on, and drains were good.

But although this elaborate method of heating was used in the houses of the rich, and in the public baths, the homes of the poor were heated by fires which generally lacked flues. Commonly the smoke was left to escape through a hole in the roof, or through windows and doors. Yet there is evidence that in Rome and North Italy chimneys were used in dwelling houses (Markham).

After the fall of Rome chimneys seem to have been forgotten, and it was long before they were invented afresh. No chimneys are seen in representations of Saxon houses. The fire was made against a hob of clay on the earthen floor, in the centre of the hall or kitchen, and the smoke flowed through a hole in the roof, or through the window if there was one.

In the strong castles built near the time of the Norman Conquest hearths appeared, recessed into the walls. Such a hearth was provided with an inclined flue that led outwards through the thickness of the wall and discharged the smoke at a height of 10 or 15 ft above the hearth. Bernan (1845) describes such fireplaces in the castles at Conisburgh, Rochester, Norwich, Colchester and Hedingham.

Vertical chimneys were in use in Italy by about the middle of the fourteenth century. They were used in England a hundred years or so later, but in some parts of the country chimneys were lacking in the homes of the poor until the seventeenth century.

In 1189 the monks of Newbattle Abbey were given a grant of the coal between Whiteside and Pinkie; in 1239 Henry III granted a charter to the inhabitants of Newcastle to dig for coal; and fifty years later permission was given for the digging of coal in Dunfermline. Towards the end of the thirteenth century coal was brought to London for use in breweries and other industrial establishments of that time.

Yet, even in the middle of the seventeenth century wood and charcoal were the fuels of choice. These were burnt on large hearths which were ill-suited for the burning of coal, and as coal came into domestic use efforts to cure smoky chimneys brought about the development of the open coal grate. Sir Hugh Platt, in 1594, advised that the width of the hearth recess should be diminished, and the throat of the flue much contracted. It is interesting to note that in the effort to get a more pleasing fire from coal, Platt mixed powdered coal with loam or sawdust, or even cow-dung, and formed the wet mixture in fuel balls. Coal and cow-dung 'maketh a sweete and pleasing ficr'. Dried cow-dung is widely used as a fuel in India.

Early in the seventeenth century Louis Savot described a fireplace used at the Louvre (Fig. 77).

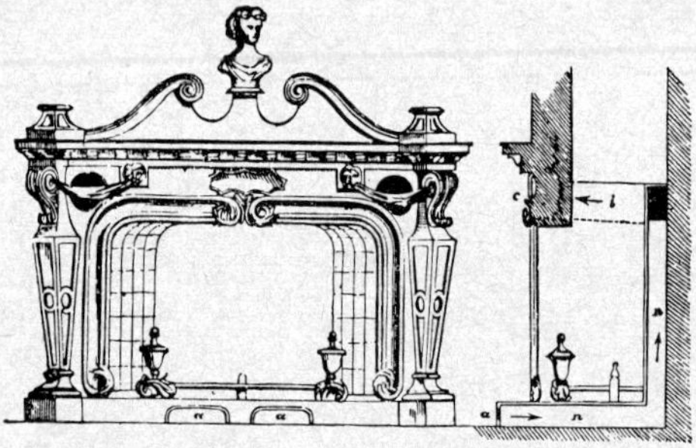

Fig. 77. Savot's fireplace in the Louvre.

Air drawn from near floor level passed under the hearth and behind the fireplace, where it was heated and then re-entered the room through a grille in the mantelpiece. Thus in addition to being warmed by the radiant heat from the fire, the occupants of the room were surrounded by tempered air.

Savot's fireplace heated and recirculated the air in the room. Nearly a century later Gauger (1714) introduced a new form of fireplace by which air either from outside or from the room itself could be heated and passed into the room If desired, external air could be admitted directly without being heated. Gauger also introduced other improvements, and he claimed that his fireplace was a perfect specific against the annoyance of smoke in rooms, which smutched everything in the apartments, and, he said, 'particularly the lace, linen, skin and eyes of the ladies'.

A notable type of fireplace which was in use in France during the early part of the eighteenth century was the cheminée de

Nancy. A canopy is fixed over the hearth, of the same width as the grate and projecting forward at its lower part. The hood diminishes as it rises and at its upper end is contracted into a tube 7 in × 4 in in cross-section. This tube is carried up 2 or 3 in above the level of the mantel soffit and is terminated with a valve. No air can enter the flue from the room except that which flows under the hood and over the fire. The amount of air passing up the chimney is much less than with the ordinary type of fireplace, so that less heat is removed from the room. Moreover, the trumpet-shaped canopy heats the air which comes in contact with it, while the fire radiates into the room in the usual way. Further reference is made to this fire in the next chapter.

With many fireplaces in use two hundred years ago, smoke was a great nuisance. One remedy was to increase the draught up the chimney so as to make certain that no smoke could escape being taken away. Such brisk ventilation was obtained with the Bath fireplace. The flue was of ample dimensions and the gap above the upper grate bar was only 9 in. Bernan (1845) wrote of this fire: 'No great nicety of dimension is necessary, and inch more or less need not be regarded; spare not your coals, and if the chimney shaft is of a fair height, you will have a roaring fire that will melt gold, and a room in which no corner will be without its cutting current; so that whoever . . . impelled by the merciless severity of the frost comes near the grate, he will find his front fired and his rear frozen. . . .'

Great improvements were made in fireplaces by those notable men Dr. Benjamin Franklin and Count Rumford. Rumford enunciated the principles of fire-grate design. He pointed out that the distance between the top of the fire and the throat of the flue was a factor of great importance; if the distance were too great much of the air passing up the chimney did not come into contact with the burning fuel, while if it were too small combustion was accelerated unduly. He also designed grates of improved radiation efficiency. In more recent times Teale (1883) and Harcourt (1915) have pronounced rules for fire-grate construction.

The development of open fires from the earliest times up to a century ago was described by Bernan (1845), and more recent improvements in design are dealt with by Fishenden (1925).

These improvements made for increased comfort in living rooms and other small apartments, but open fires were not adequate for maintaining comfortable warmth in the larger rooms met with in factories. Such rooms were commonly heated by the agency of closed stoves or cockles.

The next step was to place the stove in a separate chamber adjoining or underneath the room to be warmed. The external air admitted to this chamber was heated by contact with the stove, and was then passed through a pipe or channel into the building.

In 1792 Strutt introduced an improved form of cockle stove, so arranged that the air was brought into more intimate contact with the heating surface and was thus efficiently heated. He heated large cotton factories with these stoves, and a few years later improved forms were used for heating hospitals. Hot-air furnaces are still used. Churches, for instance, are sometimes heated by this method, and the hot-air furnace is in common use for house-heating in North America.

James Watt appears to have been the first to use steam heating on a practical basis. In 1784 he used a steam apparatus for heating the room in which he worked. In 1791 a Mr. Hoyle of Halifax obtained a patent for heating by means of steam pipes. Steam heating came rapidly into vogue, and in the early years of the nineteenth century it was widely used in factories. In his book published in 1824, Tredgold gave drawings of a steam-heating system which he installed in a silk mill at Watford. The factory had previously been heated by means of a number of closed stoves. The main steam pipe led from the boiler to the ceiling of the uppermost storey of the mill, and from this pipe there was on each floor a branch pipe of cast iron running the whole length of the mill and suspended from the ceiling. These branch pipes joined a common return pipe which brought the condensed water back to the boiler.

Early in the eighteenth century a scheme for heating a greenhouse by a hot-water pipe was described, and before the French Revolution a hot-water circuit was used in France for heating an incubator in which chickens were hatched for the Paris market. Heating by hot-water pipes was introduced into Great Britain in 1816. Since that time there have been continual improvements in the design of steam and hot-water heating installations. Pipe circuits have been improved, radiators have been introduced, and in recent years the method of panel-heating has come into use.

MODERN METHODS OF HEATING

The heat used is derived from the combustion of fuel–solid, liquid or gaseous–within the building, from electrical energy, or from steam conveyed to the building from some central generating station. The heat may be delivered where it is required wholly as convected heat, or mainly as radiation, or by both radiant and convective methods. Warming is purely convective when heated air is driven into a room and it is mainly convective when the heat is supplied by appliances such as steam or hot-water pipes, or so-called radiators, placed in the room which is to be heated. The open fire, whether it burns solid fuel or gas or is heated electrically, heats a room mainly by radiation, and the closed stove while supplying some radiant heat transmits much of its heat by convection also. When a room is heated by an open fire, intense radiation is emitted from a small but very hot source; but there is another method of heating which is becoming increasingly popular in recent years–panel heating–in which reliance is placed on the radiation from more extensive non-incandescent surfaces.

Open Fires

Although in other countries in northern latitudes stoves and warm-air furnaces are widely used for heating houses, in Great Britain the open grate, which our friends from overseas regard as the most primitive form of heating apparatus, maintains its popularity. Most of the useful heat from the open fire burning solid fuel is delivered by direct radiation to the occupants and the contents of the room. The air passing up the chimney is heated and this in turn warms the chimney breast, and from this there is some heat transference to the air of the room by convection. The fires in common use have no special provision for air heating, but we have noted that, long ago, fireplaces were devised which gave a measure of air heating, and there are now on the market modern ventilating grates. When the ordinary fire grate is used the temperature of the air of the room is below that of the objects which have received radiation from the fire, and the mean radiant temperature is higher than the air temperature.

The appearance of the open fire is another point in its favour. To most people in Great Britain, the glowing mass of fuel in an open grate presents a cheerful appearance. A further advantage of the open fire is that it ensures a liberal amount of ventilation.

The main disadvantage of the ordinary type of open fire from the standpoint of comfort is that the distribution of heat throughout a room is very uneven. If persons within a few feet of the fire are comfortable, those more remote from it are almost certain to be uncomfortably cool. The fire heats only one room, and that inadequately. Another serious disadvantage is that much labour is involved in lighting and tending the fire and in removing ashes. Yet a further drawback is that burning raw coal in the usual open grate is a wasteful way of using fuel, for commonly the useful heat derived from such a fire is below 25 per cent of the total calorific value of the fuel. A substantial part of the waste is due to incomplete combustion of the coal, with subsequent emission of smoke to the detriment of health and of the fabric of buildings.

Stoves

The closed stove, whether of the simple types used, for instance, in army huts, or of the type commonly used in living rooms, emits much less radiation than the open fire, for its surface temperature is far lower. A substantial proportion of the heat given off to the room is transmitted to the air and there is on the whole a better distribution of heat throughout the room. Slow-combustion stoves of the type used for burning anthracite need less labour than open fires, for they can be kept alight overnight, and the work of rekindling is avoided. They do not present the cheerful appearance of the open fire, however, and probably largely on that account they have never achieved the same popularity in Great Britain.

Warm-air Furnaces

The stove as ordinarily used only warms the room in which it is placed and if it is desired to warm the whole of a house several stoves are required, with consequent multiplication of labour. A device much used for domestic heating in Canada and the United States and one which is used for warming some buildings in Great Britain is the warm-air furnace. Air is passed over the surface of a furnace placed in the basement and is then circulated through the building. In some instances the air is passed into the room above the furnace and is then allowed to find its way into other rooms; sometimes ducts are provided through which the warmed air rises to pass into the various separate rooms; and in yet other instances where gravity circulation would be inadequate, fans are used to circulate the air.

Heating by Steam and Hot Water

In many houses, and commonly in larger buildings, steam or hot water is used as the medium by which heat is conveyed from a central boiler to any part of the building. When the steam or hot water has been brought to the room it may be used primarily to heat the air or to heat some radiating surface. Even in large buildings hot water seems to be more popular than steam as a means of heating.

Steam-heating systems are classified into two main groups according to the method of returning the condensate to the boiler. Gravity systems are those in which the condensate gravitates back to the boiler owing to the static head of water in the return pipes, while in mechanical systems the condensate flows to a reservoir by gravity and is then forced into the boiler against pressure. In a gravity system the height of the waterline must be sufficiently below the level of the lowest radiator to allow enough static head to ensure the return of the water to the boiler.

Various systems of piping are in use for the circulation of steam. The gravity systems include one- or two-pipe systems with air vents, and one- or two-pipe vapour systems. Of these, the one-pipe air-vent system is the most widely used. It is cheap and simple. One connection only is made to each radiator and the condensate makes its way back to the boiler along the same pipe. Each radiator is provided with an air valve or is connected to an air exit system. Gravity two-pipe systems are now regarded as obsolete, although many are still in use (A.S.H.V.E. Guide). Vapour systems operate at pressures at about that of the atmosphere. Special automatic air valves which readily allow air to escape are used, and the radiator valves allow the free passage of water. While two-pipe vapour systems operate at pressures only slightly above that of the atmosphere, they may be made to operate at more than 10 lb pressure if they are converted to mechanical systems by the provision of some arrangement for the return of the condensate.

In vacuum steam-heating systems, a partial vacuum is maintained in the return line by the operation of a pump. Systems in which a controllable partial vacuum is maintained on both the supply and return sides of the circuit are known as sub-atmospheric systems. By varying the pressure and volume of steam in circulation the temperature of the heating surface can be regulated. In these systems the partial vacuum is often about 10 in of mercury but it may reach 25 in.

Hot-water pipe installations are arranged on the one-pipe or the two-pipe system. Either of these may be run as a gravity or a forced circulation system. In the one-pipe system the whole of a section of the installation is supplied by one main which also acts as the return. Each radiator has two connections to the main, the upper one serving as the supply connection and the lower one as return. The system has the disadvantage that the radiators at the beginning of the circuit are supplied by the hottest water, and those on the lower floors, or at the return end of the circuit, have the cooler water. Thus, the distribution of heat is uneven. With forced circulation the difference in temperature can be reduced.

The two-pipe system has separate flow and return pipes. Each radiator is connected to both flow and return so that the temperatures in the different radiators are more similar. To ensure equal distribution of heat throughout a building, it is desirable in some large buildings to arrange the circulation on the reversed return principle. In the direct return system the radiator nearest to the boiler has the shortest run of piping on the flow side and also on the return side, while that farthest away has the longest flow and the longest return. With reversed return, while the radiator nearest to the boiler has the shortest length of flow pipe it has the greatest length of return pipe, and *vice versa,* so that the friction losses to all radiators are nearly the same.

Expansion tanks are necessary in hot water systems to allow for the thermal expansion of the water. These may be open to the atmosphere or closed. Closed tanks need to be supplied with relief valves. Hot water systems are run at temperatures of from 150° F to 250° F. The higher temperatures in this range can only be attained with closed systems, for at atmospheric pressure water boils at 212° F.

The water for a heating installation is most commonly heated in a boiler, but sometimes in industrial plants, or where there are district heating schemes, it is convenient and economical to use steam for the purpose of heating the water. This is done by passing the steam through an apparatus known as a calorifier. The steam passes through a heating coil which is immersed in the water to be heated.

Heated Floors and Ceilings

In recent years there has been a considerable increase in the use of panel heating, in which the floor or ceiling of a room is heated by the circulation of hot water through embedded pipes or by electrical heating elements. Electric floor heating has been installed in many houses and office blocks, and in such installations it is commonly arranged that the concrete floors are warmed during the off peak hours, thus taking advantage of the lower tariffs that apply by night, and often also for a period in the middle of the day.

Heating by Gas and Electricity

Gas and electricity can be used for any type of heating. Incandescent gas and electric fires are familiar to all, and both gas and electricity can be used as the source of heat for panel heating or for various types of convectors. They can both be used also for heating the water used in hot-water installations.

Circulation of Hot Air

We have already noted that in many plenum ventilation systems the entering air is not heated merely to the temperature which it is desired to maintain in the room, but that it is often raised to much higher temperatures. Frequently, especially in industrial installations, the circulation of hot air is the only means provided for warming the building. The air may be heated by being passed over surfaces heated by electricity or gas, but more commonly steam or hot-water batteries or heating coils are used. Sometimes air-heating is separated from ventilation, in that the air of the building is re-circulated without any fresh air being taken in by the fans.

Warm-air heating is being used increasingly in dwellings in Great Britain.

Combination of Methods

In many instances two of these methods are employed in combination. Air heating and radiant heating are often combined. Thus, in large buildings in which warm air is circulated, the air is often heated to a temperature very similar to that which it is desired to maintain in the building, and additional heat to compensate for the losses through the structure is supplied by extended heating surfaces such as radiating panels or hot-water pipes. In dwellings, along with open fires, hot-water radiators are often used, and in recent years gas and electric convectors have also been employed.

DISTRICT HEATING

In Great Britain the term 'central heating' is applied to installations in which the heat requirements of a building are provided by the circulation of steam or hot water from a central plant situated within the building. In the United States, the term refers to instances in which a number of buildings closely grouped together, as in a large industrial plant, are supplied from one power plant.

District heating is a large scale extension of central heating in the latter sense. Instead of supplying the heat requirements of a single building, or a small group of buildings from one point, a district heating system supplies the requirements of a whole district. District heating was begun in the United States fifty years or more ago. The district heating service in Detroit is described by Walker (1929) in a booklet issued by the Detroit Edison Company. The distribution of steam was begun in 1903 with 3,000 ft of mains. In 1929 the distribution system covered an area of about 3¼ miles long with a maximum width of nearly one mile. The heat to this large area is supplied by four boiler plants which are to some extent inter-connected, and the distribution system is made up of about 25 miles of underground mains and two miles of tunnels.

Steam, rather than hot water, was chosen as the medium for distributing heat, for the following reasons:

1. Steam lines are cheaper to install, since only a single pipe system is required instead of the double set of mains needed for hot water;
2. Since no circulation pumps are required for steam there is a saving in station equipment;
3. Buildings piped for hot water can be served by steam if calorifiers are used;
4. Steam consumption can be metered;

5. Steam can be circulated to any elevation;
6. The user can control his own supply with steam;
7. It costs less to equip a building for steam heating than for heating by hot water; and
8. Cooking apparatus and water heaters can be served by steam but not by hot water.

The steam is conveyed by a system of underground pipes varying from 20 in to 4 in in diameter. These mains are mostly buried to a depth of 6 to 10 ft below the surface of the street, and they are insulated with asbestos with an outer jacket of waterproof roofing felt. The distribution losses, as indicated by the difference between the steam delivered by the plant to the distributions system and the condensate metered, consist of condensation drained from the mains, leakage of steam from the mains and from consumers' piping, leakage of condensation, and errors in meters. During three years, 1926 to 1928, with an annual output of steam in the neighbourhood of 3,000,000,000 lb, the condensation loss ranged from 17.4 to 18.4 per cent. There has been a steady increase in the number of district heating undertakings and the practice is spreading to other countries.

A district heating system was begun in Manchester in 1911, and the station supplies 72 million lb of steam/year (Smith, 1945). On the continent of Europe, many district heating undertakings were in operation before 1939. In Hamburg, for instance, steam could be delivered over a distance of about 6 miles. The district heating station is commonly built in connection with the local electricity power station. In Dundee, there are two district heating plants, each supplying between 200 and 300 houses; at Staveley there is an undertaking supplying hot water to 2,000 houses, and in Wales and Scotland there are large installations supplying industrial estates. Great strides have been made in Russia. By 1938 there were in that country over 160 district heating stations. In Moscow alone, heating mains extended for over 200 miles throughout the city. Between 1922 and 1939, there were large district heating installations in over 200 cities (Smith, 1945).

Early technical difficulties of district heating now appear to have been very largely overcome, and it is claimed that a heat supply for a whole town can now be provided which will be as reliable as the supply of water, gas or electricity (Margolis, 1935). During the past few years district heating has been used in various areas in Great Britain.

CHAPTER XV

DOMESTIC HEATING

IN recent years there has been a trend towards the more adequate heating of British homes. The Egerton Committee on Heating and Ventilation (Min. of Works, 1945) considered that the whole house should be kept at a temperature of at least 45° to 50° F, and that the equivalent temperature should be between 62° and 66° F in occupied rooms, between 50° and 55° F in bedrooms during the day and during periods of dressing and undressing, and about 60° F in kitchens.

Many houses are being equipped with central heating designed to heat all or most of the rooms to a comfortable level; and in many others background heating is being installed to maintain a minimum temperature of about 50° F, additional local heating being used to bring any room up to a comfortable temperature.

Yet, much domestic heating in Great Britain is still by the open fire burning solid fuel, although since the enactment of legislation giving local authorities powers to establish smokeless zones the use of smokeless fuels has increased.

The Parker Morris Report (1961)

The appliances which are considered in the Egerton Report are somewhat out of date and the main emphasis at that time, 1945, was on solid fuel. The modern type of open fire, the gas-fired warm-air, the small-bore central heating and the electric floor heating systems were not available then.

The Parker Morris Report, 'Homes for Today and Tomorrow', laid down the standards for the 1960-1970 decade but the standards of temperature advocated do not differ very greatly from those of the Egerton Report. The Parker Morris standards are as follows: for an outside temperature of –1.1° C (30° F), the minimum temperatures for the kitchen and circulation areas are 12.8° C (55° F), and for living areas 18.3° C (65° F). Munro and Chrenko (1949) showed that, for sedentary persons, the preferred temperature was nearer 70° F than 65° F and it would appear that 20° to 21° C (68° to 70° F) would be a more comfortable temperature for the living room.

In view of the ever-increasing cost of fuel to the domestic consumer, the matter of thermal insulation of a dwelling house becomes more significant than ever. The Report gives some data for the possible saving of heat obtainable for different types of insulation, *e.g.* the heat saved per annum expressed in Giga-joules, GJ and therms for:

Weather stripping	9.495 GJ	= 90 therms
Roof insulation	9.389	89
Wall insulation	4.958	47
Double glazing of whole house	4.747	45
Double glazing of living room only	1.371	13

It is quite clear from these data that double glazing is of doubtful economic value compared with insulation of the roof.

OPEN FIRES

Fire Grate Design

Modern fire grates are generally designed more or less in accordance with the rules enunciated by Teale in 1883. Teale advocated that in the construction of the fire grate as much firebrick and as little iron as possible should be used. The back and sides of the fireplace should be of firebrick, and the back should lean or arch over the fire so as to be heated by the flame. The bottom of the fire grate should be wide (9 in to 11 in) from back to front, and the slits in the grate should be narrow. The bars in front of the fire should be narrow so as not to obstruct much heat. The space beneath the fire should be closed in front by a fender, in order to stop all currents of air passing through the fire from beneath, thus to reduce combustion and to keep the space under the fire at a high temperature.

Later, fires began to be constructed with a solid firebrick base, thus abolishing the chamber beneath the fire; and in many patterns of grate now in use this principle is adopted. In modern grates front bars are generally absent, or, if present, they are very low. With the considerable depth from back to front much fuel can be put on the fire even when no bars are provided.

The horizontal fire emits much of its radiation in an upward direction, and Harcourt (1915) who studied the efficiency of open fires urged that the fire should be narrow from back to front, and that its face should be vertical so that a large part of the radiation would be emitted horizontally. He agreed with Teale's other principles.

Distribution of Radiation

Fishenden (1925) studied the distribution of radiation from barless fires, and from fires burning in grates with bars. Figure 78 shows the horizontal component of the radiation directly opposite such fires at distances of 1, 2 and 3 yards. With either fire, at a distance of 1 yard the intensity of radiation is strongest at one's feet and legs and there is a steady diminution with height, until at 6 ft above the floor the intensity is only 15 per cent of the maximum. Three yards away from the fire there is little variation with height. At the sides of the fire the radiation intensities are much lower. The radiation intensities at the lower levels were found to be greater with the vertical-fronted grates than with the bar-less ones, and that is a distinct advantage from the standpoint of comfort.

While, without extra radiation, an air temperature of 65° was required for comfort, Fishenden and Willgress (1925) found that with an open fire the mean horizontal components of the additional radiation required for comfort were 30, 75, 120 and 170 Btu/sq. ft/hour, when the air temperatures were respectively 60°, 55°, 50° and 45°F. Although a temperature of 65° without radiation, or the lower temperatures with the appropriate radiation, gave a general feeling of warmth which was

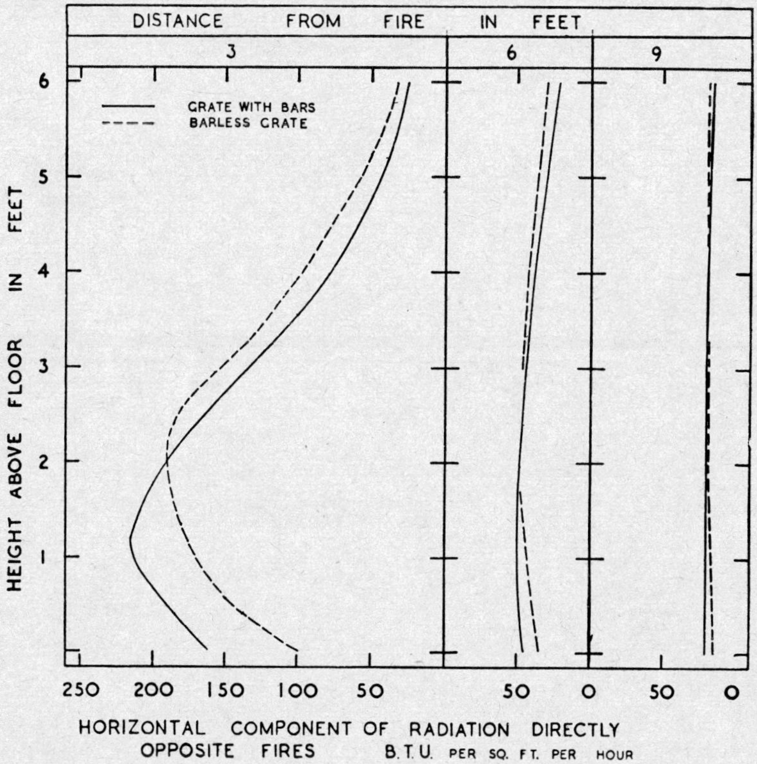

Fig. 78. Horizontal component of radiation directly opposite an open coal fire (Fishenden).

described as comfortable, these investigators preferred the combination of an air temperature of 55° with the radiant heat from the fire (75 Btu/sq. ft/hour). When the temperature was about 62° the room began to feel rather stuffy, even when it was slightly chilly without extra radiation.

With warmer conditions than these the observers tended to feel too warm, and with cooler conditions too cool. Without extra radiation they felt very slightly chilly with an air temperature of 62°, and slightly too warm at 68° F. The combinations of air temperature and radiation which were described as 'comfortable', and those representing the limits of the comfort zone, are plotted in Fig. 79.

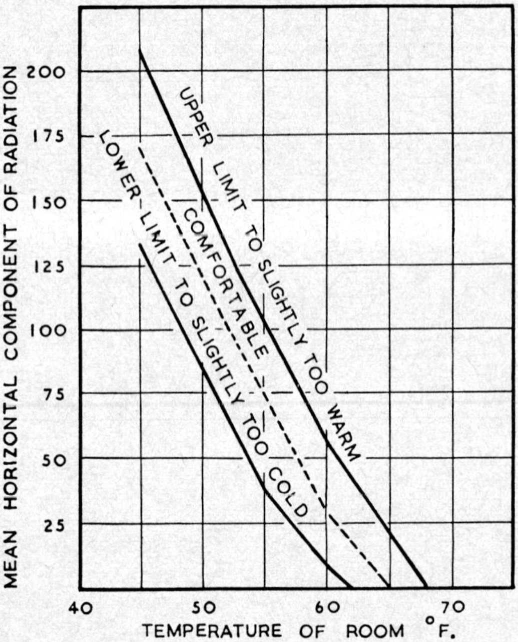

Fig. 79. Radiation from open fire required for comfort with various air temperatures (Fishenden and Willgress). (Fuel Res. Bd. Tech. Paper No. 12, H.M.S.O.)

The intensity of radiation declines rapidly with increasing distance from the fire, so that the distribution of warmth throughout a room warmed by a coal fire is very uneven. The inequality of distribution is the more striking when the air temperature is low and much radiation is required for comfort. This is clearly shown in diagrams given by Fishenden and Willgress. Figure 80 shows the zone in which thermal comfort can be achieved by sitting persons in a room heated by an open fire when the air temperature is 60°, 55° or 50° F. With an air temperature of 60° F one can be comfortable in most parts of the room, but when the temperature is lowered to 55° F the comfortable area is much reduced, and with the still lower temperature of 50° F comfort is only possible when sitting quite close to the fire. The room shown in the diagrams is 14 × 15 ft in plan—its floor area is 210 sq. ft. The zones in which comfort can be attained in such a room, *i.e.* the areas between the lines representing the upper and lower limits for comfort, are

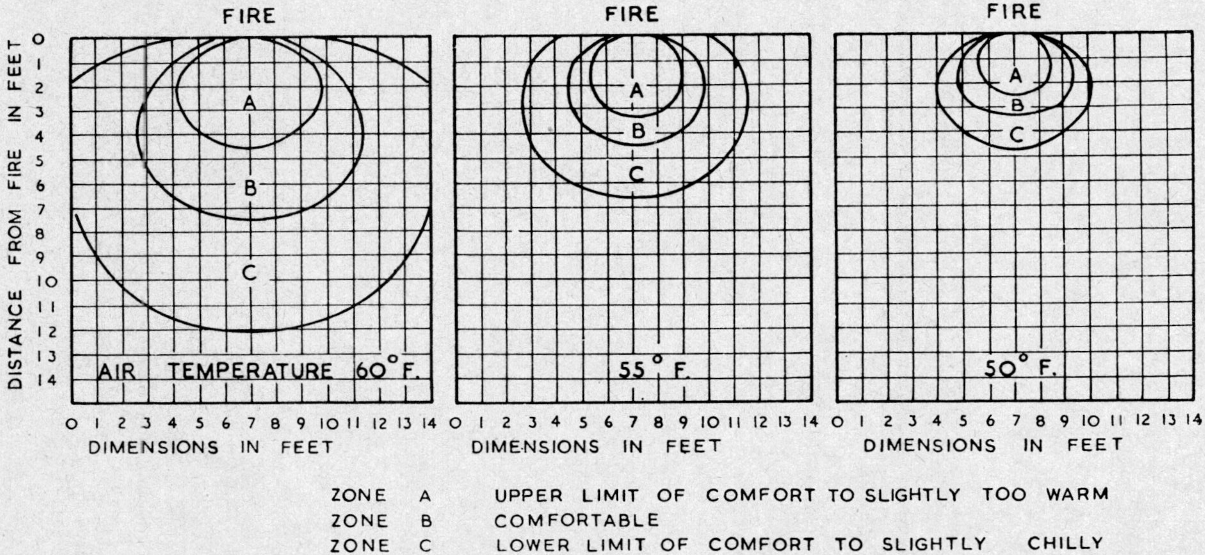

Fig. 80. Comfortable zones in a room warmed by an open fire burning 2 lb of coal/hour (Fishenden and Willgress). (Fuel Res. Bd. Tech. Paper No. 12, H.M.S.O.)

127 sq. ft, 40½ sq. ft and 15 sq. ft with air temperatures of 60°, 55° and 50° F respectively. Thus, with these air temperatures, the proportions of the floor area over which one can be comfortable are 60, 19 and 7 per cent.

In Great Britain it is customary to have a fire burning in only one living room, especially when fuel is scarce or dear, and yet of that one heated room only a small portion is comfortably warm in cold weather. Clearly, comfort would be greatly enhanced by the combination of some air-heating with the radiation from the fire. A smaller fire would then suffice, and comfort would be attainable over a greater part of the room.

Radiation Efficiency

The radiation efficiency of the ordinary sitting-room grate is generally from 17½ to 22½ per cent of the theoretical heat of combustion of the coal used, whatever the rate of burning (Fishenden, 1925). The lower values were given by a grate with a canopy which cut off a considerable proportion of the radiation thrown upwards, and with a projecting curb which prevented much radiation from being emitted downwards. Generally, grates which are narrow from back to front give a higher radiation efficiency than wider ones.

The radiation efficiencies just referred to were observed with ordinary bituminous coal. When anthracite or coke is used higher efficiencies can be obtained. A Welsh anthracite with a calorific value of 14,300 Btu/lb gave an efficiency of 27 per cent in an open grate, against one of 24 per cent obtained with a high-grade bituminous coal of about the same calorific value in the same grate, while with gas coke a radiation efficiency of 28½ per cent was observed.

Much of the heat from an open fire is expended on the gases which pass up the chimney. Some of this heat, sometimes as much as 25 per cent of the heat of the fuel burnt, is conducted through the flue walls. If the flue is an internal one this conducted heat serves to warm the house, but if the flue is built into an external wall much of the heat is lost. When, with an internal flue, 25 per cent of the heat of the fuel passes through the flue the total useful heat, including that radiated into the room, may be as high as 50 per cent of the heat of combustion of the fuel consumed.

In an earlier chapter we have seen that in a living room of ordinary size the ventilation rate when a fire is alight is often of the order of 4 to 5 air changes/hour. This represents an air flow of, say, 5,000 to 8,000 cu. ft/hour. Since this air has traversed the room from air inlets such as doors or windows, and has therefore been warmed to room temperature, the ventilation caused by the open fire is only achieved at the cost of a considerable reduction in comfort.

The ventilating effect of the open fire is often a source of discomfort, in that in many rooms the gap beneath the door is the main air inlet, and the air supply to the fire is drawn along the floor. Sometimes these floor draughts have a velocity high enough to cause acute discomfort. As Reid wrote in 1844: 'Unless an appropriate channel be provided for the ingress of air, its movements cannot be controlled'. In a room heated by an open fire, Fishenden and Willgress observed air velocities of 30 to 90 ft/min near the floor, and in tests made at the Building Research Station (1939*b*) floor level velocities of about 70 ft/min were observed near the hearth.

In a room heated by an open fire the temperature gradients are very slight. Campbell (1857) reported observations in which the temperature at head level was virtually the same as that near the floor. Vernon, Vernon and Lorrain-Smith (1928) found that the temperature 4 ft 9 in above the floor was only 1.3° F higher than that at a height of 6 in, and in the Building Research Station (1939*b*) tests, the temperature was about 2°F higher at head level than at floor level.

Air Heating by Open Fires

Various fireplaces which permit of a useful amount of air heating have been contrived from time to time. One of these was the cheminée de Nancy described by Bernan (1845). Above the firegrate is a metal trumpet, tapering backwards, which leads into the flue into which it is sealed by a plate. The chimney throat is thus constricted to about 15 sq. in. The air in the hearth recess comes into contact with the heated surfaces of the metal trumpet, and is caused to circulate in the room. The room is thus heated more effectively than by radiation alone, and a saving in heat is effected by the restriction of ventilation due to the narrow throat aperture.

Dufton (1942) has experimented with an adaptation of the

Fig. 81. Cheminée de Nancy (Dufton).

cheminée de Nancy in his own home, with notable success (Fig. 81). The room in which the fireplace was installed was formerly not so well heated as the rest of the house, but with the experimental fireplace in use there was no difficulty in obtaining a temperature of 70° F, even 20 ft from the fireplace, when snow was lying on the ground outside. Further experiments were made at the Building Research Station in which an experimental fire of this type was compared with an ordinary coal fire on a stool grate, when the air entering the room was at 45° F. The average experimental results over a 6-hour period are shown in the table.

Comparison of experimental fire (Cheminée de Nancy) with ordinary fire

Type of fire	Fuel consumption lb/hr	Equivalent temperature, °F		Air temperature, °F	Ventilation rate, air changes/hr
		8 ft from fire	12 ft from fire		
Experimental fire	2.0	62.4	58.6	59.4	1.5
Ordinary fire	3.0	61.9	57.4	56.3	3.9

Although the fuel consumption of the experimental fire was only two-thirds of that of the ordinary fire the equivalent temperature up to 12 ft away from the fire was higher with the experimental fire than with the ordinary one, and the air temperature throughout the room was also higher.

The equivalent temperature isotherms in Fig. 82 show that with the experimental fire a comfortable equivalent temperature of between 60° to 70° F was reached over a distinctly greater part of the room than with the ordinary fire and that the equivalent temperature in the parts most remote from the fire was also higher.

From Fig. 83, which gives the isotherms in a vertical section of the room, it can be seen that the efficient air heating with

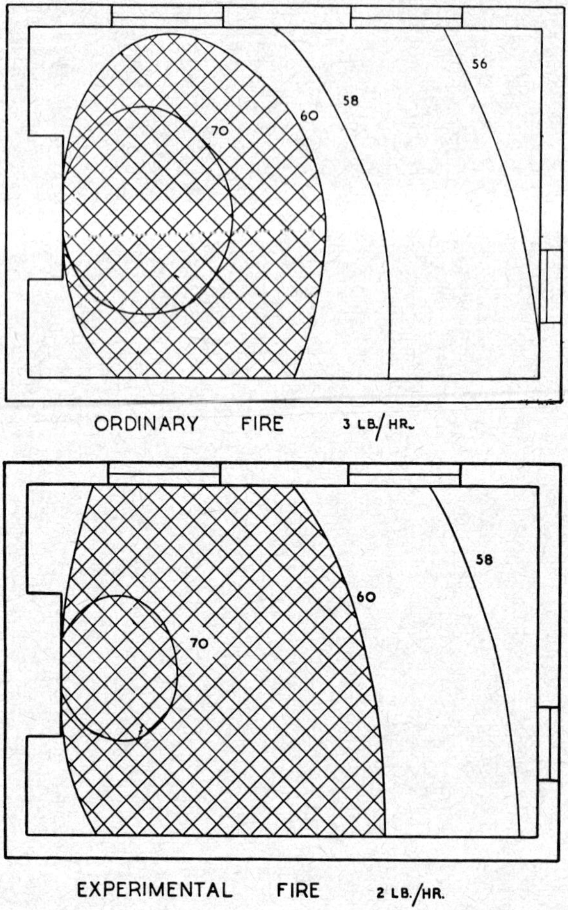

Fig. 82. Equivalent temperature isotherms for (a) *ordinary fire burning 3 lb of coal/hour, and* (b) *experimental fire (cheminée de Nancy) burning 2 lb/hour.*

the cheminée de Nancy was achieved at the expense of somewhat steeper temperature gradients than occur with the ordinary fire. When the room was heated by an ordinary fire there was some difference between the air temperatures in various parts of the room, but the differences with change in height from the floor were slight. With the experimental fire the horizontal distribution of temperature was more even, but there was a difference of 4° or 5° F between the temperatures at floor level and at head level.

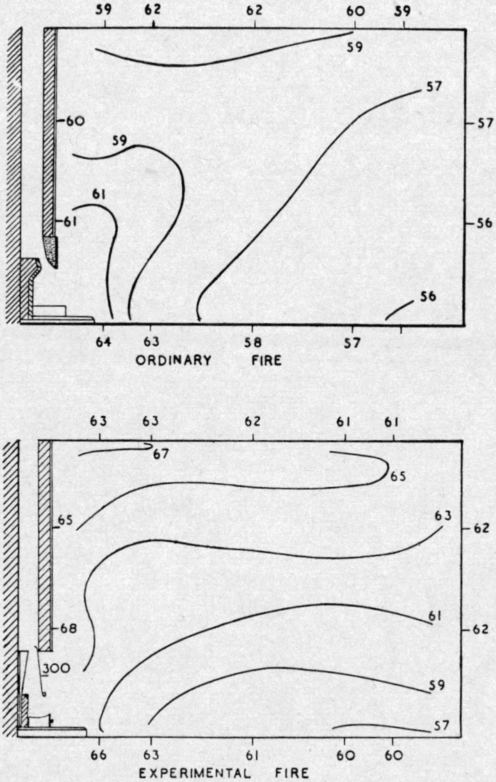

Fig. 83. Isotherms in vertical section of room.

The Committee on solid fuel installations (Min. of Works, 1944*a*) advocated the design of convector fires which would burn for 10 hours without refuelling. One such appliance is the 'Hales' open fire space heater.

The 'Hales' Open Fire Space Heater

This appliance is an open fire which also acts as a space heater. The fire can be kept alight continuously, and when adjusted for overnight burning it has a consumption of only about one-third of a pound of bituminous coal an hour. Provision is made for controlling the rate of burning by means of a damper operated from the hearth, and in this way the rate

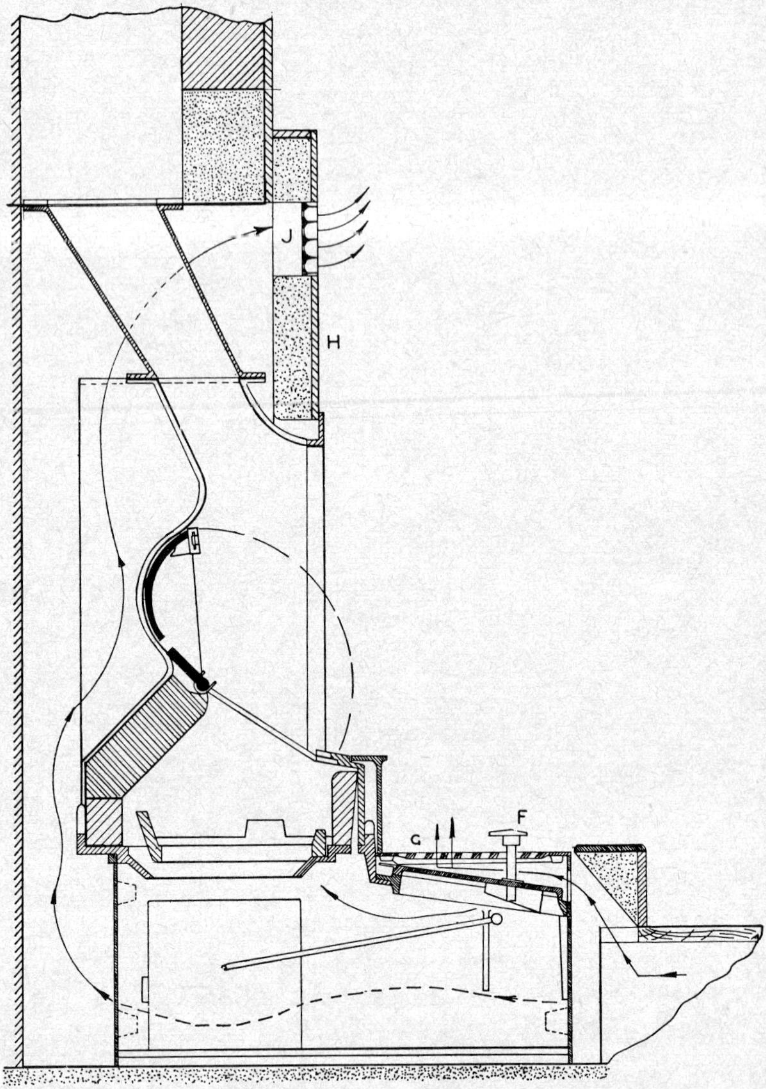

Fig. 84. Section of 'Hales' open fire space heater.

of combustion can be increased rapidly after the fire has been damped down. A portion of the air required for combustion and for feeding the flue is drawn under the floor direct from outside, so that floor draughts are reduced. Ignition and partial

Comparison between continuously-burning space heater and ordinary stool-bottom grate

	Continuously burning space heater		*Ordinary stool-bottom grate*	
Hours of operation/day	12 as open fire 12 shut down overnight		12 as open fire	
Average rate of coal consumption (lb/hr)	2.1 as open fire 0.3 shut down		2.4 as open fire 2.4 as open fire	
	As open fire		*Open fire*	
Thermal efficiency per cent	Radiation	25	Radiation	25
	Convection	19	Total	25
	Total	44		
	Closed fire			
	Radiation	25		
	Convection	37		
	Total	62		
Heat output (Btu)				
Per 12 hours open fire	140,000		92,000	
Per 12 hours closed fire	31,000			
Per 24 hour period	171,000		92,000	
Smoke emission/24 hour period, as percentage of weight of fuel	0.7		2.2	

combustion of the smoke given off by bituminous coal are achieved by means of a special form of fireback which provides a preheated secondary air supply. The ash box is said to be large enough to hold a week's accumulation of ash. A convector system is incorporated so that warmed fresh air is admitted to the room, and a portion of this air can be piped to a bedroom. The thermal efficiency of this unit is about 45 per cent, whereas that of the conventional open fire is about 25 per cent.

The construction of the fire is shown in the sectional drawing in Fig. 84. The ashpit is partly sunk below floor level and partly contained in the raised hearth. The firebox unit terminates in a flue connecting piece, and there is a closure lid. The damper control is shown at (*F*), the louvred inlet in the hearth is at (*G*), and the inlet to the room for warmed air is at (*J*) above the front surround (*H*). By manipulation of the damper control the air can be shut off from the louvred plate and the whole of the available air pressure applied to the fire. The heating surface available for air warming is about 14 sq. ft, and heat expended on warming the air in the convector system amounts to nearly 20 per cent of the total heat available in the fuel. A comparison of the heat yield of this space-heater burning continuously with that of an ordinary stool-bottom grate burning 12 hours/day is given below. It is assumed that the coal consumption of each unit is 28 lb/day.

GAS FIRES

Formerly in Great Britain the gas fire was regarded primarily as an appliance for use for short periods. It had the advantage of producing warmth quickly, almost entirely by radiant heat.

Commonly, the products of combustion from the gas fire have been vented into the conventional flue of 9 in x 9 in cross-section. Such a flue is needlessly large, and when the gas fire is alight this ventilation is excessive, thus removing much warmed air from the room and reducing the practical heating efficiency of the fire. For this reason some form of restriction, either of the flue itself or to the spigot of the gas fire, is now used. Such restriction improves the performance of the fire by 30 to 35 per cent while still ensuring sufficient ventilation (Andrew and Wills, 1961).

Within the past few years a variety of convector gas fires has been available. Typically these have a radiant efficiency of about 32 to 35 per cent, with an overall efficiency of 60 to 67 per cent when the flue is suitably restricted (Berry and Taylor, 1960).

Berry and Taylor describe experiments with a totally enclosed convector gas fire, in which the radiants are placed behind a front screen of heat-resisting glass. Even when fitted to a normal flue the efficiency of such a fire is 70 per cent or rather more, and of the heat output about one-half is radiant and one-half convected heat.

The fitting of thermostats to convector gas fires is now established practice.

CONVECTORS

Amongst the appliances used for warming buildings by direct convection may be included hot-water pipes and radiators, and electric and gas air-heaters. Heating by hot-water pipes will be dealt with in a later chapter.

Electric heaters are of various types. One of these consists of a rectangular casing in which is housed a heating coil. The casing is open at the top and bottom, so that the air heated by passing over the elements leaves at the top and is replaced by cooler air entering below. Thus a circulation is set up. Another type, referred to as a unit heater, is intended to be placed on the floor, and comprises heating elements over which air is forced by a fan incorporated in the unit. Tests on this latter type of heater were made at the Building Research Station (1939*b*). The heated air discharged from the appliance remained near the floor for about 3 ft. It then rose almost vertically to the ceiling, drifted across the room, and then descended down the walls, and passed back to the heater along the floor. With this appliance the temperature gradient between head level and floor level was of the order of 5°F over a considerable part of the room.

Gas convectors take the form of rectangular casings similar to those described above, but inside there are a number of small gas jets. The heat liberated by the combustion of the gas sets up convection currents, and thus a stream of air flows through the casing. These appliances are not provided with flues, so that all the products of combustion pass into the air of the room. The use of these and other flueless gas heaters has been considered by a Committee of the Royal College of Physicians of London (1936). The concentrations of various products of combustion, which, in the view of the Committee, should not be exceeded are:

Sulphur oxides	0.0001 per cent
Nitrogen oxides	0.0006 per cent
Carbon monoxide	0.005 per cent

These limits were specified with the objects of avoiding smell and providing for hygienic purity.

The concentration of products of combustion depends on the gas-rate of the heater, the size of the room, and the ventilation rate, and a further factor is the sulphur content of the gas. If gas is the sole means of heating available in the room, a continuous gas rate of 1 cu. ft/hour/100 cu. ft of room space is usually adequate, while a rate of 1 cu. ft for 50 cu. ft of space may be used for intermittent heating. Gas used for domestic purposes may contain 10 or 20, or even 30, grains of sulphur/100 cu. ft. A sulphur content of 30 grains is now rare. The content is commonly below 20 grains, and there is a general trend towards much lower sulphur contents. Both sulphur dioxide and sulphur trioxide are present in the products of combustion of coal gas. For the mixed oxides, as they commonly occur, the threshold concentration for perception is of the order of 0.0001 per cent, *i.e.* the limiting value recommended by the Committee. A substantial proportion, 50 per cent or more, of the sulphur oxides liberated by the combustion is absorbed by the walls and furnishings. Assuming 50 per cent absorption of the sulphur oxides, the Committee calculated the maximum gas rates for 20-grain sulphur-content gas burnt continuously in flueless heaters in rooms of 500 cu. ft capacity, together with the resulting concentrations of products of combustion in the air of the room, with various ventilation rates. These are shown opposite.

From the table it can be seen that if the gas rate is to be 1 cu. ft/100 cu. ft of room space, as stated above to be the amount usually adequate for continuous radiation, a ventilation rate of 1.7 air changes/hour is necessary if the sulphur content is 20 grains/100 cu. ft.

With the commonly lower sulphur contents the minimal

Concentration of products of combustion in relation to gas rate and ventilation rate

Ventilation rate (air changes/hour)	Gas rate Cu. ft/hour in room of 500 cu. ft	Gas rate Cu. ft/100 cu. ft room volume	Sulphur oxides assuming 50 per cent absorption (per cent)	Carbon monoxide (per cent)	Nitrogen oxides (per cent)
0.5	1.5	0.3	0.0001	0.0006	0.0003
1.0	3.0	0.6	0.0001	0.0006	0.0003
1.5	4.5	0.9	0.0001	0.0006	0.0003
2.0	6.0	1.2	0.0001	0.0006	0.0003
3.0	9.0	1.8	0.0001	0.0006	0.0003

ventilation rates on the above basis will be proportionately lower. With the former high sulphur contents it was undesirable to use flueless gas heaters as the sole means of heating flueless rooms, but when, as with some gas supplies, the sulphur content is well below 10 grains/100 cu. ft such heaters will probably be unobjectionable.

CLOSED STOVES

One of the main disadvantages of the ordinary open fire is the uneven distribution of heat throughout the room. Closed stoves provide a more even distribution of heat, but they have the disadvantage that they tend to produce much steeper temperature gradients than are found with an open fire. This is especially true of the old-fashioned coke stove which is still often used for heating hutments. The stove and its flue-pipe are sometimes heated to a dull red during cold weather, and a freely-exposed surface at such a high temperature is a powerful convector. Hot air rushes rapidly towards the ceiling, and there is a rapid increase of temperature with increasing height from the floor. Observations made in rural schools in the United States show that with such stoves, while the temperature at floor level was generally at between 55° and 65°F, the temperature near the ceiling was usually between 80° and

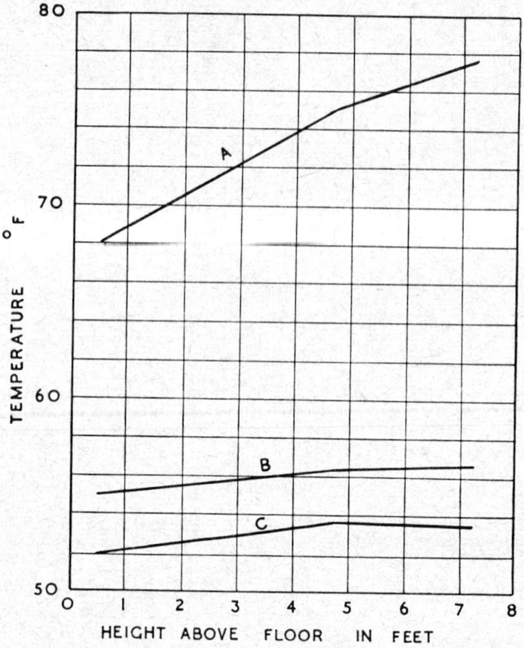

Fig. 85. Comparison of temperature gradients induced by anthracite stoves and by open fire. A. Anthracite stove; B. Gas fire; C. Coal fire.

100° F. There was commonly a difference of roughly 30° F between the temperatures at the two levels (Cole *et al.*, 1931). Such a steep temperature gradient is highly undesirable.

While these old stoves are crude and unsightly, and need a good deal of attention, the modern anthracite stove has a neat appearance and is labour-saving. Dampers enable the draught to be controlled so that the rate of combustion is easily regulated.

The opening from the anthracite stove to the chimney flue is small, and, in consequence, the amount of ventilation caused by the passage of air up the flue is also small. Whereas the rate of ventilation in a room of about 1,600 cu. ft capacity was $4\frac{1}{2}$ air changes/hour with an ordinary open fire, it was only 0.7 air changes when the room was heated by an anthracite stove (Building Research Board (1939*b*)). This latter ventilation rate is a low one, and would not give very satisfactory conditions for more than two occupants. Some additional ventilation, as by

Fig. 86. 'Courtier' stove.

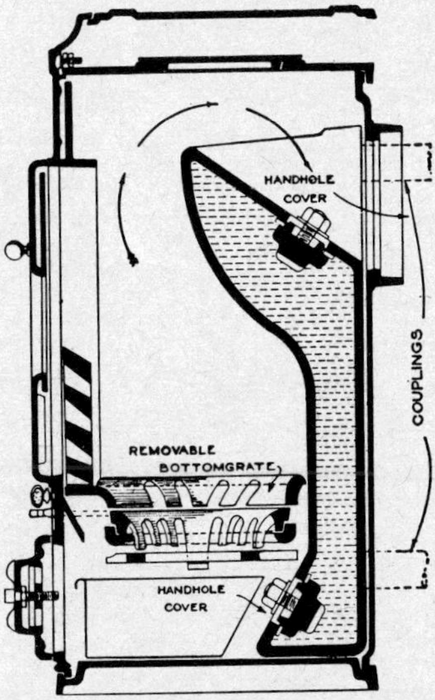

Fig. 87. Section of the 'Courtier' stove.

the opening of a window, is desirable when a room is heated by a stove.

With the anthracite stove placed in front of the chimney breast, as ordinarily fitted, there is free convection, and although the horizontal distribution of heat is much better than with the open fire, undesirably steep temperature gradients may occur. Thus Fig. 85 shows that in observations made by Vernon, Vernon and Lorrain-Smith (1928) the excess of temperature 4 ft 9 in from the floor over that 6 in from the floor was 7.1° F with an anthracite stove, while it was only 1.3° F with a coal fire, and 1.7° F with a gas fire. In the Building Research Station (1939*b*) experiments the temperature gradient between floor level and head level was about 6° F with an anthracite stove.

The Openable Stove

The 'Courtier' stove is another example of an appliance which heats by convection as well as by radiation. The front view of the stove is given in Fig. 86, while Fig. 87 shows the stove in section.

Cool air enters through the air inlets along the bottom of the outer side casing, it is warmed by contact with the inner body of the stove, and passes out through openings near the top of the outer casing. During the day the fire doors can be opened, so that the stove serves as an open fire. By night the fire doors and the ashpit door are tightly closed, and with the draught suitably regulated, by means of the ventilator in the ashdoor, the stove will remain alight overnight. Any kind of solid fuel can be used with this stove, but good quality, hard kitchen nuts are recommended.

The sectional drawing (Fig. 87) shows the 'Courtier' stove fitted with a high-pressure boiler, so designed as to expose a large heating surface to the burning fuel. When a boiler is fitted to the stove the amount of heat available for room heating is

Fig. 88. Siesta stove No. 3.

reduced, and care should be taken that a stove large enough to deal with both services is installed.

This stove is one the appliances which can be used in the heat-service unit designed by the Ministry of Works (*vide infra*).

Another openable stove which is suitable for the heat-service units is the Siesta Stove No. 3. This is illustrated in Fig. 88.

HOT-AIR HEATING

In North America hot-air furnace systems are much used for the heating of dwellings. The furnace, which is enclosed in a casing of metal or brick, is situated in the basement of the building. In the usual form of installation, hot air, taken from the upper part of the furnace casing, is distributed to the various rooms through pipes. The air supply to the furnace may be taken from the inside of the building or from outside, or the supply may be drawn partly from inside and partly from outside.

The simplest form of hot-air system is the pipeless furnace. It delivers air only to the room directly above the furnace, and a recirculating duct leads air back to the furnace for re-heating. This system is only applicable to small dwellings.

Pipe systems may be of the gravity type, or fans may be used to increase the circulation. The temperature of the air discharged from the registers of a gravity system may be about 175° F. With fan systems lower register temperatures—90° to 120° F— are used. It is to be expected that with these lower inlet temperatures there will be less striking temperature gradients than with higher inlet temperatures.

Pre-fabricated Heat Service Unit

In 1945, when the pre-fabrication of low-cost houses was being undertaken in Great Britain, a highly efficient heat-service unit was designed. This unit provided for the supply of warmed air to bedrooms.

A note on the heat-service unit is included in a 'Technical Note on Some Types of Permanent Pre-fabricated Houses' (Min. of Health Circular 182/45 of 16th October 1945).

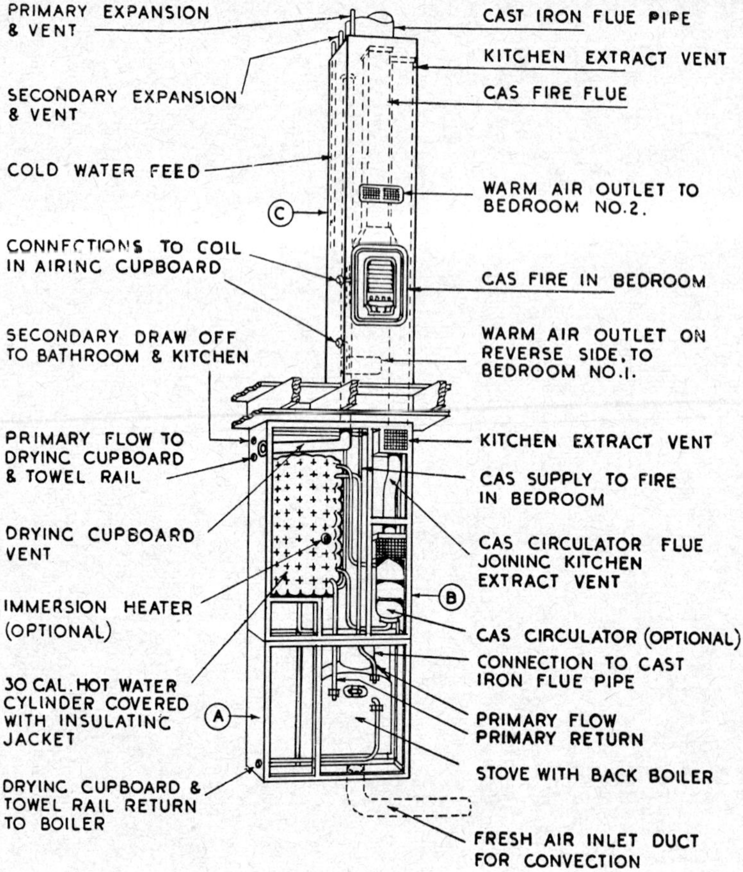

Fig. 89. Diagrammatic sketch of heat service unit. (Ministry of Works.)

Modern Use of Warm-Air Heating

Recently there has been a rapid increase in the use of warm-air heating in houses. An installation may provide for heating the whole house or only two or three selected rooms. The heating unit consists essentially of an air delivery unit, a heat exchanger, and a fan for circulating the air. Any fuel may be used to provide the heat for the heat exchanger. In many installations the fuel used is gas. The design of gas-heated units is discussed by Chance and Dance (1961), who also give valuable guidance based on experience in the laboratory and in the field.

With any type of warm-air heating, whether for a small house or for a factory or office block, vertical gradients of air temperature are always a problem. Chance and Dance cite observations showing that in order to keep such gradients to a minimum the warm air should be admitted at low level, preferably only a few inches above the floor, and the air should be discharged from the inlets in a downward direction, say at an angle of 10° to 25° below the horizontal. The temperature of the entering air should be kept as low as is practicable. In one installation, as the temperature of the air entering was raised from 140° to 190° F the gradient of temperature from floor level to a height of 5 ft increased from 7° to 10° F.

With any form of central heating where mechnical power is used, whether for a fan as in warm-air installations, or as a circulating pump in small-bore hot-water systems, much care is needed to keep noise to a minimum. Chance and Dance review the noise problem as it arises with warm-air systems, and discuss the methods by which noise can be reduced. Care is needed in the choice of the fan and its motor, in the connection of the ducting to the fan housing, and in the construction of ducts and grille boxes.

Improved standards of insulation and lighter building construction have made central heating by electricity a practical proposition from the economic point of view. The system known as Electricaire is a form of central heating by which cheap off-peak electricity is used to store heat in a centralized unit. Heat from this unit is transferred to the various rooms of the dwelling by means of air circulated through ducts. Air is drawn from the heated room through return-air grilles by means of a fan and filtered before being sent to the heating unit. The temperature of the air is raised to a predetermined level and its final temperature is automatically controlled by a mixing device. The heated air is then sent back to the room by ducts.

This system may be supplemented with 1 kW radiant fires in each bedroom for rapid and selective heating from the 55° F recommended general bedroom temperature to 65° to 68° F; a

750 W radiant fire in the bathroom; a 3 kW immersion heater on unrestricted tariff and a 1 or 2 kW radiant fire in the main living area together with some fifteen 13 A socket outlets.

CHAPTER XVI

PANEL HEATING

WE have seen that a serious drawback to open fires is the uneven distribution of heat. The amount of heat radiated from a surface depends partly on the temperature of the surface and partly on its size. The area of an open fire is small and its temperature is high. If, however, a large surface such as a wall or ceiling is heated, only to a comparatively low temperature, it may radiate as much heat as a small but intensely hot surface, with the advantage of a more even distribution of heat. The Romans achieved such a result with their hypocaust, by which the floor and sometimes the walls also were heated.

LOW-TEMPERATURE RADIATION

A system of low-temperature radiant heating which is being used to an increasing extent is that known as panel heating. The system of burying hot-water pipes in the plaster of walls was introduced by Barker in 1908. In the newer installations of this type of heating, the tendency is to place the heating panels in the ceiling (Fig. 90). The temperature of a large area of ceiling is maintained at roughly 100° F.

These panels radiate heat downwards towards the occupants of the room as well as the floor, walls and furniture. Since the panels can be placed in the ceiling wherever required, large rooms can be heated in this way.

It will be evident that unless there is remarkably good insulation above the embedded hot-water pipes there will be some transmission of heat to the floor above. This is to some extent an advantage, for a warm floor is very pleasant to the feet unless its surface temperature exceeds 75° F, or perhaps 80° F, when discomfort tends to arise.

Fig. 90. Heating panel in course of construction.

Such embedded panels have the advantage that they are invisible and take up no floor space. A disadvantage is that any alteration or repair is a serious matter, and to minimize trouble from these sources, exposed panels are sometimes used. These panels consist of large iron or steel plates, on one side of which pipes for the circulation of hot water are closely attached. The plates can be joined together so as to form panels of the required size.

For some years past floor heating has been used in many buildings. Widdows (1921) applied floor heating to schools which he designed for the Derbyshire County Council. Hot-water pipes are laid beneath the floor, and the floor surface is thus maintained at a temperature of 70° or 75° F. This method of heating seems to be especially applicable to nursery schools, for with it children can play on the floor without being chilled.

There is an increasing tendency to use electric heating for floor-warming installations in dwellings and office blocks. The heating cables may be buried in ducts so that they can be withdrawn, or directly embedded in concrete. Commonly, the floors are heated in the off-peak hours.

A modification of the hypocaust system is in use in Liverpool Cathedral. Beneath the floor there is a system of shallow ducts through which hot air is blown, and the surface of the floor attains a temperature of about 70° F.

In another system, known as border-heating, the floor near the skirting, or the skirting itself, is heated to a temperature of from 80° to 140° F. Good results have been claimed for this method, but no part of the floor where persons must stand or sit should be heated to such high temperatures.

MEDIUM-TEMPERATURE RADIATORS

A system of heating by means of small electrically-heated panels is described by Hallet (1931). The panels, 2 ft x 1 ft in area, were operated at a temperature of about 500° F and had a radiation emission of nearly $\frac{1}{2}$ kW/sq. ft. They were mounted high on the wall at an angle of 60° or 70° with the horizontal, and consisted of fire-clay tiles in which were embedded ribbon layers of graphite to serve as the conductors (Vernon, 1934).

In recent years, in some lofty industrial buildings, large panels heated by high-pressure hot water have been hung vertically.

PANEL HEATING AND COMFORT

We shall see that panel heating has advantages over some other forms of heating, but in the early days of its use claims were made for it which could not be substantiated. Thus, one firm of engineers claimed that 'the radiant warmth from a ceiling approximates closely to the effect produced by sunshine' (*vide* Vernon *et al.*, 1928).

Vernon and his colleagues (1928) made a careful examination of the thermal environment in many buildings with panel heating. They found that in panel-heated rooms the temperatures of the walls, ceiling and floor were about 3° F higher than the air, while in rooms heated by convection methods the boundary-surface temperatures were about 3° F lower than the air. It can be estimated from these results that with a given air temperature, and an air velocity of about 20 or 30 ft/min, the equivalent temperature in a panel-heated room would be roughly 3° F higher than that in a room heated by convection. In some later observations made in other rooms heated by ceiling panels, Vernon, Bedford and Warner (1930) found that their sensations of warmth in the panel-heated rooms were equivalent to those experienced in convectively heated rooms in which the air temperature was rather more than 1° F higher. This difference can scarcely be likened to the effect of sunshine.

Many hundreds of observations of their own sensations of warmth and those of factory workers were made by Vernon and his colleagues (1928). Although the influence of heated panels on the mean radiant temperature of the room must have had its effect on sensations there was little evidence of localized effects due to the direct radiation from panels. The only instance in which proximity to a panel caused a sensation of appreciably greater warmth was when the observers sat only 3 ft from a wall panel 5 ft in width, and from the data given below it will be noticed that in that position the mean radiant temperature was 7° F higher than the air temperature.

Comparison of mean radiant temperature with air temperature in office heated by wall panels

Position		*Air temperature °F*	*Excess of mean radiant temperature over air temperature, °F*
A.	3 ft from panel	65.1	7.1
B.	6 ft from panel	64.6	3.6
C.	33 ft from panel	63.9	2.8
D.	5 ft from panel	64.4	0.9

Ceiling Heating

Panel heating gives a very uniform distribution of heat. After studying various forms of heating used in factories, Vernon, Bedford and Warner (1926*a*) remarked that ‘too much stress cannot be laid on the principle of starting the source of heat as near the floor as possible', and at first sight it might appear that heating by ceiling panels is wrong in principle. This conclusion concerning the position of the sources of heat referred to convection heating. With convection heating the warm air rises to the upper parts of the room, and the temperature of the air at the higher levels is always higher than at the lower ones. When ceiling panels are used, however, the source of heat is part of the upper boundary surface of the enclosure, and any air which is heated by contact with the surface, and so rendered less dense, is unable to rise. Thus convection is reduced, and while there is a stratum of heated air near the ceiling the layer does not extend much below the ceiling, and below it there is remarkable uniformity in the temperature of the air at different levels. The air temperature may vary no more than 1°F at different heights, except that a few inches below the ceiling there may be a rise of about 2°F.

Figure 91 shows the temperature gradients observed by Vernon and his colleagues in an office heated by ceiling panels. The surface temperature of the floor was higher than that of the air, probably owing to leakage of heat from panels in the room

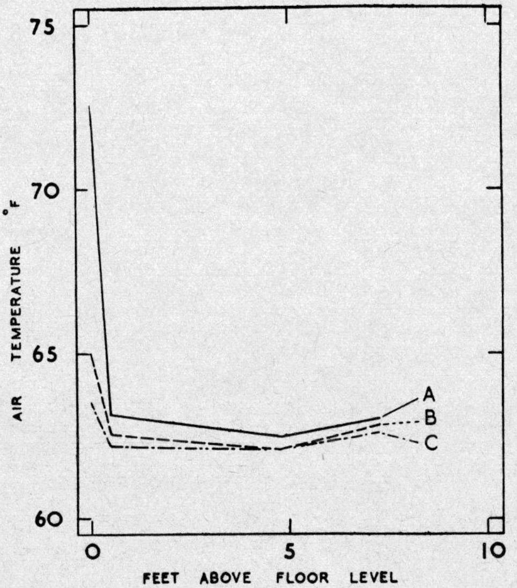

Fig. 91. Temperature gradients in office warmed by ceiling panels.

below, but from 6 in to 7.25 ft above the floor the maximum temperature variation observed was less than 1° F, and at 4.75 ft above the floor—roughly the head level of a person at work—the temperature was lower than that near the floor. Dufton and Marley (1942) found that with ceiling panels the air temperature 5 ft 6 in above floor was 2.5°F higher than that at the 6 in level. These observers noted than when the equivalent temperature at the centre of the room was 65° F, the equivalent temperatures in the four quarters of the room ranged only between 64.1° and 65.2° F. The mean radiant temperature in these observations, which were made in an unventilated room, was 66.3° F when the air temperature was 65.4° F. When observations were made in the panel-heated room with ventilation at the rate of 2.4 air changes/hour, and the equivalent temperature was maintained at 65° F, the air temperature was 63.9° F, and the mean radiant temperature 69.8° F. Vernon and his colleagues (1928) had observed that in one panel-heated office the mean radiant temperature was 1.4° F higher than the air temperature when windows were shut, and 3.3° F higher when the windows were open to some extent. Other observations in another panel-heated room showed that when the windows were shut the excess of the mean radiant temperature over the air temperature was 3.6° F, while it was 5.6° F when the windows were open to a moderate extent.

One of the requirements for a satisfactory thermal environment is that the temperature at head level should not be appreciably greater than that near the floor; the temperature at head level should be little, if any, in excess of that at floor level, and there should not be excessive radiation towards the heads of the occupants. In rooms heated by low-temperature ceiling panels, the gradient of air temperature is small and unobjectionable, but it remains to be considered whether or not the radiation from the panels to the heads of the occupants produces disagreeable results. After prolonged study of panel heating Vernon (1934) concluded that the radiation from low-temperature panels is so mild that local heating of the head is not perceived unless the panels are less than 10 ft from the floor.

Hallet (1931) claimed that when medium temperature radiators 2 sq. ft in area and emitting 500 W/sq. ft are hung at an angle of 60° or 70° with the horizontal the vertical intensity of radiation is less than if the same total heat load were spread uniformly over the whole ceiling, while with the small radiators at the higher temperature the horizontal radiation will be considerably greater than if the whole ceiling were warmed. It was claimed by Hallet that these small medium-temperature radiators, properly distributed, project the maximum of direct radiation on to the occupants of the room, and as little as possible on to the walls where it is largely wasted. He assumed that the intensity of radiation given by such an installation would be enough to compensate for a reduction of 10° F in the air temperature. Dufton (1931*b*) studied the thermal environment in a school heated by means of these radiators. In ventilated rooms so heated the equivalent temperature was almost identical with the air temperature, while in similar rooms heated by electric tubular heaters (a form of convector) the equivalent temperature was about 3° F below the air temperature: thus in these instances the effect of radiators was to permit of a reduction of the air temperature by 3° F. In closed rooms heated by the radiators the equivalent temperature was 2° to 3° F higher than the air temperature, while in others heated by convectors the air temperature exceeded the equivalent temperature by about 3° F. Hence in closed rooms the effect of radiators was to permit of a reduction of the air temperature by perhaps 6° F. Crowden and others (1933) made observations on the children in classrooms heated by these radiators. Their data are summarized in the table.

In the panel-heated room the temperature of the air was lower than in the other room and the air velocity was higher, and the radiation from the electric panels by no means compensated for these differences. In the panel-heated room the average temperature of the children's fingers was 5° F below the average in the other room, and the children were too cold to work efficiently. Vernon (1932*a*) had already shown that in a similarly heated classroom the globe thermometer temperature was only 2.6° F higher than the air temperature.

Comparison of conditions in classrooms warmed by
(a) *electric radiating panels, and*
(b) *hot-water radiators*

	Average condition in room heated by	
	Electric panels	*Hot-water radiators*
Air temperature, °F	57	61.5
Air velocity (ft/min)	29	13
Mean skin temperature of fingers of children, °F	66.7	71.7
Outside air temperature, °F	43.5	39

Since these medium temperature heaters are suspended below ceiling level they give off a substantial amount of heat by convection, and because of their high surface temperature the convection currents are rapid. The heated air rushes towards the ceiling and sets up considerable temperature gradients. Figure 92 (from Vernon, 1934) illustrates the magnitude of these gradients. From floor to ceiling there was a rise of 13.6° F in the air temperature, but it may be noted that the gradient over the height of a standing man was only 1 or 2°.

Vernon deprecated the use of these heaters fixed horizontally 10 ft or so above the floor, for the radiation on the heads of persons standing beneath them is most unpleasant. He found that a cloth surface supported horizontally above the floor and

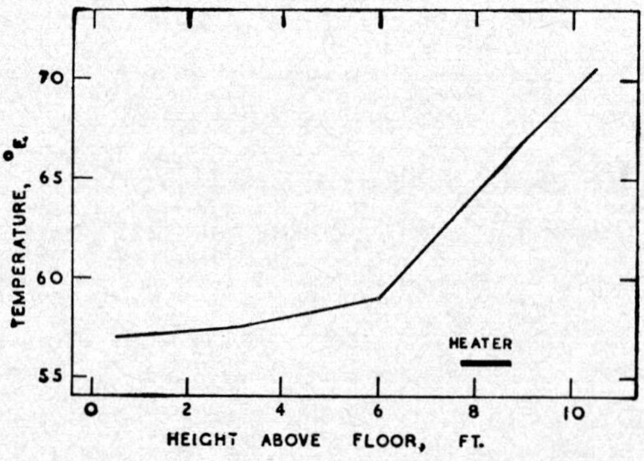

Fig. 92. Temperature gradients in a room heated by medium temperature panels.

4 ft below the heater attained a temperature of 19° F higher than that of the air.

Various claims have been made that panel heating is much more economical than heating by hot-water radiator systems. Experimental tests in which rooms heated (*a*) by ceiling panels, and (*b*) by convection, were maintained at an equivalent temperature of 65° F, are described in the reports of the Building Research Board for 1930 and 1931. When the rooms were closed the panel-heated room required a heat input 14 per cent greater than that needed for the room heated by convection, but when each of the rooms was ventilated by means of a small wall-ventilator the room heated by convection required 16 per cent more heat than before, while the panel-heated room required no extra heat.

Dufton's (1931*b*) observations in the classroom heated by medium-temperature radiators showed that on a mild day the electricity consumption was 6 per cent less than in the room fitted with tubular heaters, while a slightly higher equivalent temperature was maintained; and on a colder day the radiators maintained an equivalent temperature 1.5° F higher than that in the room with tubular heaters, and yet consumed 15 per cent less electricity.

It was evident that, in some circumstances, the radiant heat from the ceiling could be quite unpleasant. Chrenko (1953*a, b*) undertook an extensive investigation of the problem. His 150 subjects were exposed to radiation from a heated ceiling, or from an experimental panel overhead, at different heights for periods of 30 min. All the experiments were made in calm air, and in most instances the boundary surfaces of the room, apart from the heated portion of the ceiling, were maintained at a uniform temperature.

The subjects were questioned as to their sensations of warmth with reference to the head and to the feet as well as to their general feeling of warmth. They were asked to say whether they found the environment pleasant or not, and those who found it unpleasant were invited to say in what way it was unpleasant. In addition they were questioned as to their impressions of freshness.

It was found that the percentage frequency of discomfort due to excessive warming of the head was more closely associated with the elevation of mean radiant temperature at head level, due to the overhead radiation, than with the intensity of the radiation received either by the forehead or the top of the head. The table shows the incidence of uncomfortable warming of the head in relation to the elevation of the mean radiant temperature due to the heated ceiling.

Percentage incidence of uncomfortable warming of the head in relation to elevation of mean radiant temperature at head level due to the heated ceiling (Chrenko, 1953a)

Increase in mean radiant temperature at head level °F	*Percentage frequency of occurrence of uncomfortable warming of the head*
1	5
2	7
3	8
4	11
5	13
6	16
7	19

Even when the subjects could not say that their heads were uncomfortably warm they often reported unpleasant sensations usually referred to the head. Uncomfortable warming of the head accounted for about 40 per cent of the total number of complaints of unpleasantness. Stuffiness and general thermal discomfort accounted for about 10 per cent of the symptoms, and gradients of sensation, *e.g.* a feeling of warmth of the head while the feet were chilly, accounted for 20 per cent. The remainder of the symptoms, sometimes ill-defined, were referred to the head.

The table shows that even when the elevation of the mean radiant temperature was only 1°F some subjects reported that their heads felt too warm. It does not follow, however, that

Maximum desirable temperatures for ceiling panels 48 sq. ft in area, at different heights

Height of ceiling (ft)	*Maximum desirable temperature (°F) of panel of dimensions*		
	2 ft x 24 ft	*4 ft x 12 ft*	*8 ft x 6 ft*
7	101	88	85
8	112	95	92
9	124	104	100
10	135	113	109
11	146	124	120
12	158	135	130

under those conditions they would have made unsolicited complaint. Yet, as the intensity of radiant heat is increased and a greater number of the occupants experience some discomfort, some will be acutely uncomfortable, and strong complaints will arise. So as to keep the incidence of discomfort at as low a level as possible, Chrenko recommended that the elevation of the mean radiant temperature at head level should not exceed 4° F.

The intensity of radiation from a ceiling panel impinging on the heat of an occupant of the room is a function of the temperature of the panel and of the solid angle subtended by the panel at the head. The maximum desirable temperature of the panel will depend on the height of the ceiling and on the dimensions of the panel. The radiation from a panel of a given area at a point below it will be more intense if the panel is square than if it is long and narrow. Chrenko gives the following table showing the maximum desirable temperatures for ceiling panels 48 sq. ft in area in ceilings of various heights, on the assumption that the ceiling contains only the one heated panel.

Chrenko (1961) made a further examination of the results reported in his earlier paper to estimate the effects of altering the air temperature. The average air temperature in his experiments was 67° F, and the recommended limit of 4° F for the elevation of mean radiant temperature applies to that air temperature. It appears that if the air temperature were only

61° F overhead heating could raise the mean radiant temperature by 10° F with only slight risk of discomfort, while, on the other hand, with an air temperature of 70° F if the elevation of the mean radiant temperature were 4° F there would be a 20 per cent incidence of discomfort.

When ceiling heating is used in multi-storey buildings it is important to ensure that the transmission of heat from the ceiling to the surface of the floor above is kept to a minimum. otherwise, occupants of the upper storeys may experience discomfort from overwarming of the feet. Pearse (1961) made observations in hospitals warmed by ceiling panels, and in some instances members of the nursing staff complained of hot and tired feet.

Valance Heating

A method of heating which, though primarily by convection, is regarded as being essentially a form of radiant heating is the so-called valance system, which has been used in houses and hotels in the United States of America (Bailey, 1962). Convectors are fixed to the walls near the ceiling and concealed by valances. The convectors produce a thin layer of very warm air across the ceiling, which is thus heated and in turn heats the room and its occupants by radiation. Tests carried out by the John B. Pierce Foundation in cold weather showed that the gradient of air temperature over the height of a standing man was very small, even when the air 3 in below the ceiling was at a temperature of 96° F. The valance system can be used for cooling in hot weather.

Floor Heating

Floor heating is commonly regarded as one type of panel heating. Vernon and his colleagues (1928) measured the temperature gradients induced by the floor heating in the nave of one of the transepts of Liverpool Cathedral. The air temperature at head level was nearly 2° F lower than that at foot level.

In some schools in Derbyshire heated by hot-water pipes running beneath the floor, Vernon, Bedford and Warner (1930) measured more temperature gradients. When the floor temperature was 76° F and the air temperature about 62° F, the temperature 6 in above the floor was 2° F higher than that at head level.

Marley (1937) reports observations made in a room the floor of which was heated to an average temperature of 73° F, and when the equivalent temperature in the centre of the room was maintained at 65° F. From 6 in above the floor to a height of 6 ft 6 in the increase in air temperature averaged only about 0.5° F, and even within 2 in of the ceiling the temperature was less than 1° F higher than that 6 in from the floor.

In schools where the floors were heated to a temperature of about 75° F the author and his colleagues found the conditions agreeable, but when on a few occasions the floor temperature reached or exceeded 80° F the floor was tiring to the feet. It seemed from those observations that the temperature of a warmed floor should not much exceed 75° F. Later, Herrington and Lorenzi (1949) concluded that floor temperatures above 75° F were physiologically undesirable, and that with floor temperatures exceeding 80° F there were clearly noticeable effects on the temperature of the feet.

Chrenko (1957) carried out experiments on men and women wearing their normal footwear who, on different occasions, sat with their feet on a heated floor or walked about on it. The air temperatures in the different experiments ranged between 58° and 76° F, averaging 68° F.

In the four sets of observations, women sitting or walking and men sitting or walking, the relations between the skin temperature measured on the sole of the foot and local thermal sensations were so similar that the results could be pooled. The coefficient of correlation between (*a*) the skin temperature and (*b*) the sensation of warmth on the foot expressed on an arbitrary numerical scale, in a total of 801 experiments, was +0.72. Some subjects experienced discomfort when the skin temperature on the sole exceeded 86° F, and when it reached 93.5° F 50 per cent were uncomfortable.

No uncomfortable warming of the feet occurred when the temperature of the floor surface was below 77° F, and it was concluded that in a room maintained at a temperature such as is usually regarded as satisfactory in Great Britain the temperature of the floor surface should not exceed 77° F or, better, 75° F.

Local skin temperature is influenced by the general thermal state of the body as well as by local exposures, so that the temperature of the soles of the feet is affected by the air temperature as well as by the temperature of the floor. Thus, a higher floor temperature can be tolerated as the air temperature is lowered. From his multiple regression equations Chrenko estimated the floor-surface temperatures which, with various air temperatures, would evoke sensations of warmth on the soles of the feet similar to those caused by air temperature of 68° F and a floor temperature of 77° F. The various combinations are set out below:

Experimental group	*Estimated floor temperature with air temperature (°F) of*		
	60°	*65°*	*70°*
Women sitting	92°	83°	75°
Women walking	84°	79°	75°
Men sitting	87°	81°	75°
Men walking	79°	78°	77°

The regression equations make it clear that when the subjects walked, and thus made firmer contact with the floor, the effect of the floor temperature on the temperature of their feet was greater than when they sat. In the sitting subjects a change of 1° F in the air temperature had a greater effect than a similar change in the temperature of the floor, but when the subjects walked the floor temperature exerted the greater effect.

PANEL HEATING CALCULATIONS

Mean Radiant Temperature, and Mean Enclosure Temperature

When a room is heated primarily by radiation, as is the case with panel heating, the temperature of the air in the room depends on the mean temperature of the boundary surfaces.

The mean temperature of those surfaces, which may be termed the mean enclosure temperature, will be higher than the air temperature by an amount which depends on the dimensions of the room, the loss of heat through walls and floors, the outdoor temperature, and the ventilation rate (Bruce, 1936). The mean enclosure temperature is the arithmetic average of the temperatures of the various boundary surfaces.

The American Society of Heating and Ventilating Engineers Guide for 1945 gives a simple example of the calculation of the mean surface (or enclosure) temperature:

Surface	*Area*	*Temperature (°F)*	*Product (area x temperature)*
Internal walls	480	70	33,600
External walls	360	60	21,600
Glass	216	30	6,480
Floor	480	70	33,600
	1,536		95,280

The weighted mean surface temperature is obtained by dividing the sum of the products by the sum of the surface areas: thus, 95,280/1,536 = 62.93°F.

It is said that this figure is the mean radiant temperature. That is an unfortunate expression. The term 'mean radiant temperature' has been used as a measure of the mean radiation intensity at any point. The uniform temperature of a black surface which would radiate with that intensity is known as the mean radiant temperature. Unless the boundary surfaces of an enclosure have a uniform temperature, which very rarely occurs, the mean radiant temperature will vary at different points inside that enclosure. We have already seen that in an office heated by wall panels the mean radiant temperature at four observation points varied between 65.3° and 72.2°F, but it may be that none of the observed mean radiant temperatures was identical with the mean enclosure temperature calculated as indicated above. In the calculation of the mean enclosure temperature, the temperature of any element of the boundary surfaces carries weight in proportion to the area of that element, but the influence of any particular area of surface on the mean radiant temperature is proportional to the solid angle subtended by that portion of the surface at the point at which the mean radiant temperature is being determined.

Values of the mean radiant temperature derived from globe thermometer observations as described in Chapter III denote the mean radiant intensity at a point, or on the surface of a small sphere. Such measurements are of much value in the study of the thermal environment. They give equal weight to the radiation received from all directions, whereas a standing man resembles a long cylinder rather than a sphere, and the direction from which he receives radiation, as well as the intensity of that radiation, has its influence on him. Yet, as has been pointed out earlier, in ordinary circumstances the mean radiant temperature referred to a point at about 45 in from the floor gives a good indication of the radiant heat as it affects a man.

The eupatheoscope, the standard instrument for the measurement of equivalent temperature, is a vertical cylinder, and it likewise is affected by the direction of the incident radiation. Dufton and Marley (1942) have derived a method of calculating the mean radiant temperature of an enclosure as it influences the eupatheoscope. The various surfaces of the room are portrayed on a plane chart so that each area of the diagram gives a measure of the effect of the corresponding surface upon a body shaped like the eupatheoscope.

The surface of a sphere can be shown on a chart by means of an equal-area projection in which angles of azimuth are marked along the horizontal scale, and sines of angles of altitude along the vertical scale. Thus the position of any point on the surface of the sphere can be shown, and the area of any zone is shown by a proportional area on the chart. On such a chart the vertical boundaries of a rectangular plane wall are shown as parallel straight lines, and the upper and lower boundaries as curves.

In adjusting such a diagram so that corresponding areas give a measure of the effect of the surfaces they represent upon a eupatheoscope, the ordinate of any point is made proportional to

$$\frac{a}{2}\left\{\frac{\sin^2 W}{\tan W} + W + \frac{\sin^2 W}{a}\right\}$$

where W is the angle of altitude above the horizontal plane through the centre of the eupatheoscope, and a is the value 3.73. For angles of altitude below the horizontal plane the expression is modified to allow for the fact that the wooden base of the eupatheoscope is substantially non-conducting. It becomes:

$$\frac{a}{2}\left\{\frac{\sin^2 W}{\tan W} + W\right\}.$$

The area of projection of each of the surfaces in a room so obtained represents the radiation effect on a eupatheoscope per degree of temperature excess, and measurements of surface temperature may be readily set out on such a diagram. Figure 93 shows the projection of a room constructed by the method described, and the surface temperatures measured by a revolving thermopile are also shown.

Tredre (1960) used the method of Dufton and Marley to construct a projection of her test room so that each area of the diagram represented the effect of the corresponding surface on a standing cylinder of height 68 in and diameter 12 in. In her experiments surface temperatures were measured and weighted according to the projected areas of the various elements so as to ascertain the mean radiant temperature as it affected such a cylinder. The temperatures thus calculated, when the small experimental room was warmed by a panel either suspended horizontally from the ceiling at a height of 8 or 10 ft or fixed to a wall, were then compared with the mean radiant temperatures measured by using metal models of men. When the overhead panel was at a height of 8 ft and was heated to about 120° F the calculated values referred to the cylinder underestimated the mean radiant temperature as measured by means of the metal men by 1.6° F, while when the panel was at a similar temperature but fixed vertically on a wall the cylinder values overestimated the mean radiant temperature by more than 3° F.

The average error was substantially greater than that incurred by assuming that the mean radiant temperature obtained from a single observation of the globe thermometer made at a height of 45 in above the floor measured the radiation effect on the metal

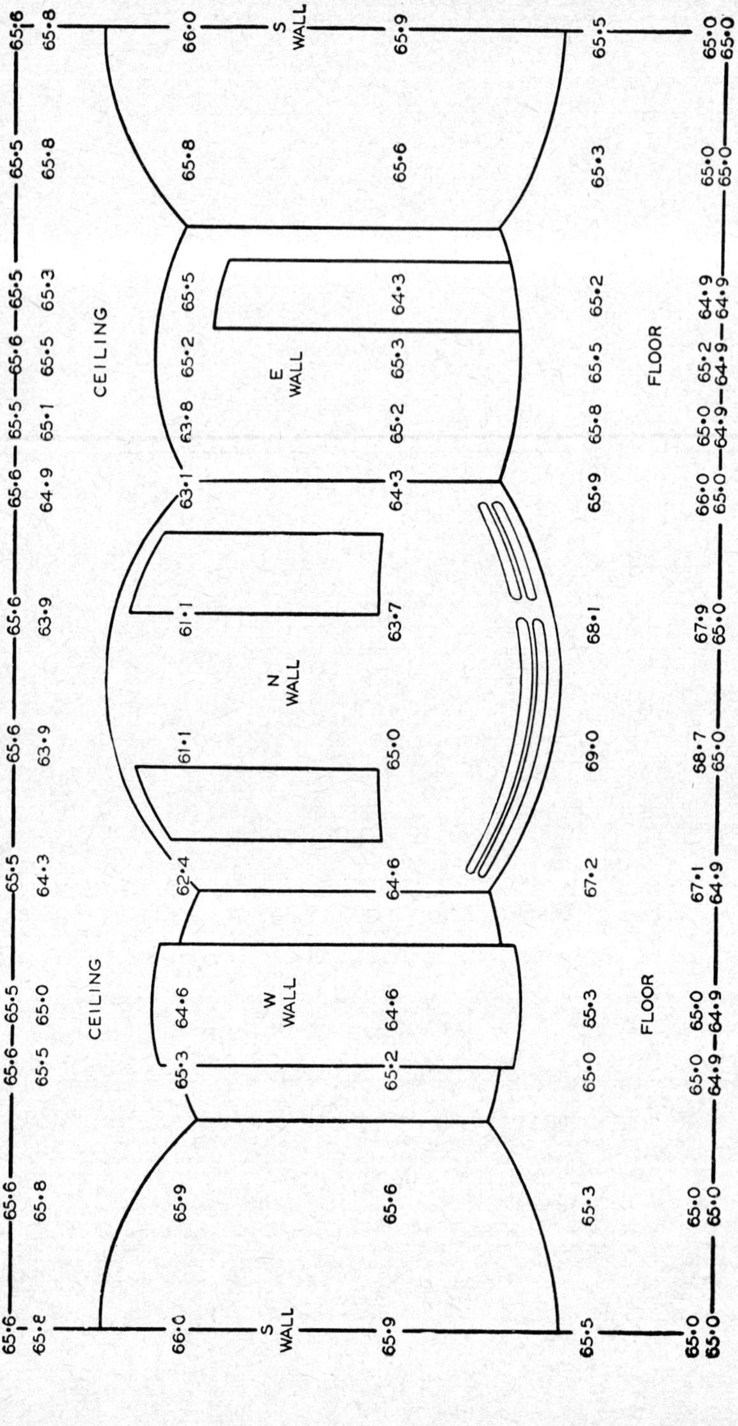

Fig. 93. Projection of a test room showing surface temperatures measured by revolving thermopile. (Dufton and Marley.)

men. On the other hand, when the mean radiant temperatures at a point 45 in from the floor were estimated by measuring the temperatures of various areas of the boundary surfaces and weighting them according to the solid angles subtended at that point, the values so obtained agreed very well with those measured by use of the metal men.

DISTRIBUTION OF RADIATION FROM PANELS

An important point in the design of a system of panel-warming is that the radiating surfaces should be so distributed that the amount of radiation received by the occupants directly from the panels will be reasonably uniform. Raber and Hutchinson (1943) have studied the distribution of radiation from hot panels. A full-sized model of a man was used, and the proportion of the radiation emitted from a small element of panel surface reaching the model was measured. Determinations were made of the energy received from ceiling panels by a standing man, (*a*) when facing the horizontal direction of the radiation, (*b*) when in the profile position, and (*c*) when half way between full-face and profile positions, *i.e.* at an angle of 45°; and also by a sitting person in front- and rear-view positions, and in a position 45° from the full rear-view position. Besides this comprehensive study of ceiling panels, observations were made of the radiation received from wall panels when the subject was in such a position that the horizontal line between him and the element of the wall panel was normal to the panel. The radiation received from elements of a wall panel to the right or left of the normal was calculated.

In the study of ceiling panels, ceiling heights of 8, 10 and 12 ft were investigated. Even with a ceiling as low as 8 ft the maximum fraction of the energy leaving a small element of a panel which is received by an occupant of the room is less than 5 per cent. The fraction contributed by a panel surface at a horizontal distance of 6 ft from the subject is less than 2 per cent while at a distance exceeding 8 ft, less than 1 per cent of the energy leaving the panel passes directly to the occupant. The maximum energy exchange between the occupant and the ceiling panel does not occur when the occupant is directly under the panel, for in that position his projected area is only that of the plan view of his head and shoulders. As the occupant moves from directly under the panel area, the projected area of his body rapidly increases, and with an 8 ft ceiling the maximum energy exchange occurs when the horizontal distance from the subject to the panel area is 1.5 ft to 2 ft. With a 12 ft ceiling the maximum energy received by a standing subject is only a little over 1 per cent of that radiated by the panel element, and the horizontal distance at which this maximum occurs is 5 ft.

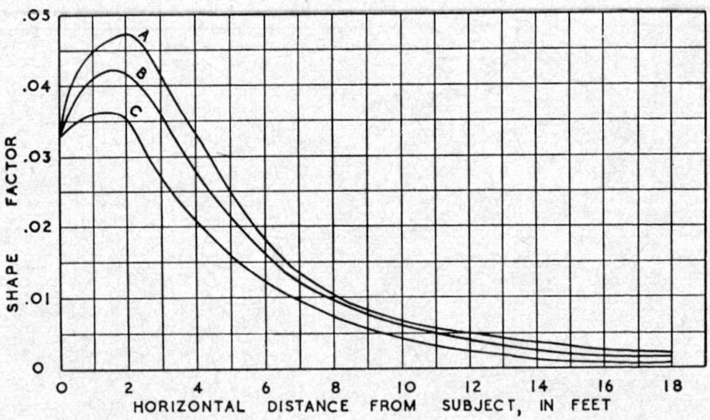

Fig. 94. Fraction of energy radiated by an elemental panel area in an 8 ft ceiling which reaches a standing subject. (Raber and Hutchinson.)

In Fig. 94 Raber and Hutchinson's curves showing the fractions of energy reaching a standing subject from a panel in an 8 ft ceiling are given, and Fig. 95 gives the curves for a person facing a panel in a ceiling 8 ft, 10 ft or 12 ft high.

The radiation exchange between an occupant and a ceiling panel above him is much greater when the occupant is standing than when he is sitting. The greater average distance of the subject from the panel more than offsets the increased area projected by the body when seated. At horizontal distances of more than 6 ft, however, the differences in radiation exchange for seated and standing persons are small.

From data such as these, the overall fraction of the radiation from a panel of any shape and size which will be received by a occupant in a given position can be calculated. The plan view of the panel is drawn to scale and divided into small elements, *e.g.* squares of 1 ft or 2 ft sides. The location of the subject is fixed on the plan view, and his distances from the centres of the various elements of the panel then measured. Then the radiation which will be received from each element is ascertained. Such computations have been made for various distributions of panels with different ceiling heights. From these studies, Raber and Hutchinson conclude that in a ceiling 8 ft high the most desirable arrangement of panels is in a square pattern, with the centre lines spaced at intervals of 10 ft. Wherever possible,

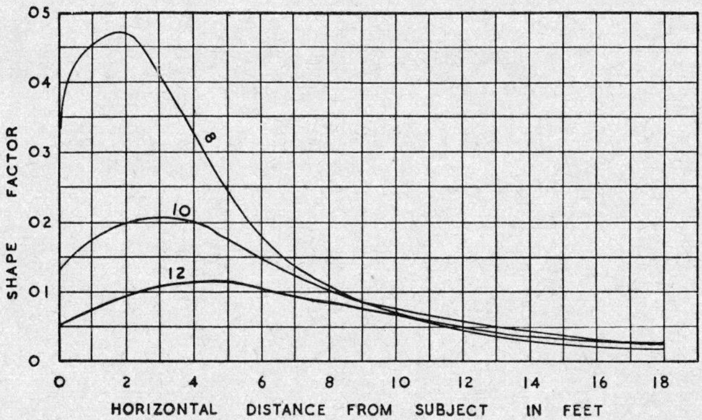

Fig. 95. Fraction of energy radiated by an elemental panel area in a ceiling 8 ft, 10 ft or 12 ft high which reaches a standing person facing the panel area. (Raber and Hutchinson.)

spacings of less than 9 ft or more than 14 ft should be avoided, and the use of a single block of panel in the centre of the room is not recommended.

For a 10 ft ceiling a similar square pattern is desirable, but the spacing of the centre lines should preferably be 14 ft, and neither less than 10 ft, nor more than 15 ft. For a 12 ft ceiling a square pattern is again recommended. The optimum spacing of the lines is 20 ft, and the lines should not be less than 17 ft, nor more than 21 ft apart. In a room smaller than 14 ft square, a cross-pattern with the centre lines running along both axes of the room is satisfactory. Although with ceiling heights of 8 ft and 10 ft it is undesirable that the whole ceiling should be warmed, with a ceiling height of 12 ft there is a satisfactory uniformity of radiation exchange when the whole ceiling is used as a panel.

Although Raber and Hutchinson have shown that with a ceiling as low as 8 ft panels can be arranged to give a distribution which is satisfactorily uniform from the standpoint of the total radiation received by occupants in different positions, it does not follow that heating with such a low ceiling will be satisfactory. We have noted Vernon's conclusion that with low-temperature panels local heating of the head is not perceived unless panels are less than 10 ft from the floor, and it is possible that warming by panels only 8 ft high may lead to undesirable local warming of the heads of the occupants.

The curves of the radiation received from wall panels are very different from those relating to ceiling panels. One can stand much closer to a wall panel, and the body area projected towards it is much greater than that projected towards a similar panel in the ceiling. The projected area is greater when the subject faces the panel than when he stands with his side to it. In Fig. 96 some of Raber and Hutchinson's curves for a standing person facing a wall panel are reproduced. The curves show the fraction of radiation from an element of wall panel reaching the subject, for various heights of the panel element, and separate curves are given for different horizontal distances from the wall.

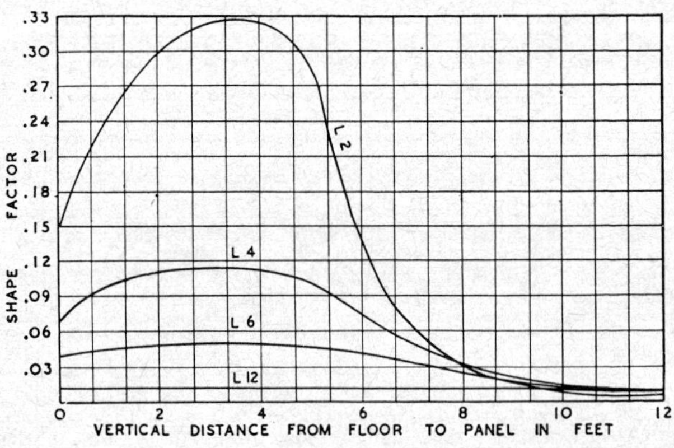

Fig. 96. Fraction of energy radiated by an elemental area of a wall panel which reaches a standing subject facing the panel. (Raber and Hutchinson.)

When standing only 2 ft from the wall the subject receives nearly 33 per cent of the radiation from an element of panel 3 ft 6 in to 4 ft above the floor, but the amount of radiation received falls off rapidly as the distance of the subject from the panel is increased. When the subject is in the profile position the maximum fraction of radiation received at a distance of 2 ft is 21 per cent.

In a room heated by ceiling panels, or by wall panels at high level, the occupants are warmed not only by the radiation from the panels more or less above them, but by all the panel surface from which radiation can reach their bodies. This fact is of importance when panel installations are being designed. Hallett (1931) gives calculations of the radiation intensity in two rooms, one a large drawing office, and the other room only about 16 ft square. Both these rooms were warmed by means of small moderate-temperature panels suspended at high level. The radiant energy supplied per cu. ft of space was approximately the same in the two rooms, but at a height of 3 ft from the floor the radiation intensity in the larger room was approximately 40 Btu/sq. ft/hour, whereas in the smaller room the intensity was only 10 Btu. The difference was due to the effects of distant radiators. Hallett refers to an instance in which a building 80 ft × 20 ft was heated by means of these moderate-temperature radiators. A partition was erected 20 ft from the wall at one end, and it was then found that the larger portion was noticeably cooler, while the smaller portion was considerably too cool to be comfortable and further radiators had to be installed, although the partition had divided the radiators so that the load/unit of space was the same in both portions.

The influence of the size of the room on the radiation intensity received from ceiling panels can be illustrated by a simple example. Consider two circular rooms, one 10 ft in diameter and the other 30 ft in diameter, both rooms being 10 ft high. Assume that the walls and floors of these rooms are uniform at a temperature 65° F, and that the whole of each ceiling is warmed to 100° F. The ceiling in the larger room subtends an angle at the centre of the room equal to slightly more than one-third of the surrounding sphere, while the ceiling of the smaller room represents only about one-seventh of the sphere. It can be calculated that the mean radiant temperature in the larger room will be 77° F, whereas that of the smaller room will be 70° F.

Design Calculations

While with panel heating most of the heat supplied to a room is transmitted by radiation, there is an appreciable amount of convection. A ceiling panel delivers approximately 70 per cent of its heat by radiation and the remaining 30 per cent by convection. A floor panel gives about 55 per cent by radiation, and 45 per cent by convection, while a vertical wall panel gives 65 per cent by radiation, and 35 per cent by convection (A.S.H.V.E., 1945).

In calculating the heat losses from a panel-heated room it is necessary to include the heat transmitted to the other boundary surfaces by direct radiation from the hot panels. Bruce (1936) pointed out that in considering the heat lost through any one wall, the mean enclosure temperature as viewed by that wall is higher than the mean temperature of the enclosure reckoned as the mean temperature of the six boundary surfaces.

In designing a panel-heating system calculations should be made of the mean radiant temperature and the air temperature to which an average occupant will be exposed, the heat lost by ventilation must be allowed for, and the heat balance of the unheated room surface must be taken into account. Hutchinson (1945) has reviewed the complex rational design procedure, and has shown that it can be reduced to the substitution of numerical design data in a single equation. The equation is by no means a simple one, but, it is claimed, its solution involves no more than a routine substitution followed by arithmetical computation.

In those days computers were not available. Fanger (1970) gives an example of a thermal analysis of a room heated by a panel which completely covers the ceiling. He assumes that the temperature of the air and mean radiant temperature will be unequal and gives the following expression:

$$t_{\mathrm{comf}} = \frac{t_a + \dfrac{dt_a}{dt_{mrt}}}{1 + \dfrac{dt_a}{dt_{mrt}}}$$

in which t_{comf} is defined as that uniform value of air t_a and mean radiant temperature t_{mrt} which under the conditions stated will give optimal thermal comfort, and dt_a/dt_{mrt} is the partial differential coefficient in the comfort condition.

Given that the room is to be occupied by sitting persons wearing clothes with an insulating value of 1 'clo', in which the air velocity is less than 0.1 m/sec, his graphs give 0.93 for the differential coefficient, and $t_{\mathrm{comf}} = 23.0$°C. The equation above thus becomes

$$\frac{t_a + 0.93t_{mrt}}{1 + 0.93} = 23.0$$

Subsequent analysis is complicated and in the end some ten equations are solved by iteration on a digital computer. The result includes a value for t_a of 20.5°C and t_{mrt} of 26.1°C. It seems, however, that for all practical purposes, in rooms with a low air velocity, the comfort temperature is the arithmetic mean of the air and mean radiant temperature. But suppose one divides $t_a + 0.93t_{mrt}$ by 1 + 0.93. The answer is

$$0.52t_a + 0.48t_{mrt} = t_{\mathrm{comfort}}.$$

The student will recall that Bedford's equation for equivalent temperature was

$$t_e = 0.522t_a + 0.478t_w - 0.01474\sqrt{v}(100 - t_a)$$

and if the air velocity is very low and the convection term is omitted,

$$t_e = 0.522t_a + 0.478t_w$$

which is virtually identical with Fanger's expression. It is true that the 'clo' unit had not been invented in 1936, when Bedford's researches were done, and he did not measure the insulation of the clothing of his subjects who were female factory workers. One might conclude that, by inference, the insulation of the clothes of these people was about 1 'clo'.

CHAPTER XVII

HEATING BY PIPES, RADIATORS AND CONVECTORS

WHILE the open fire was the most favoured means of heating dwellings in Great Britain, in buildings of other types hot pipes or radiators are very widely used. In industrial buildings, rooms are often heated by long lines of pipes carrying steam or hot water, but in most modern buildings of the non-industrial type, and often in industrial buildings also, radiators are used in preference to simple pipes when direct air heating is required. Radiators have the advantage of compactness; a radiator of given length has a much greater heating surface than that of a similar length of pipe.

Fig. 97A. Column radiator.

Fig. 97B. Tubular radiator.

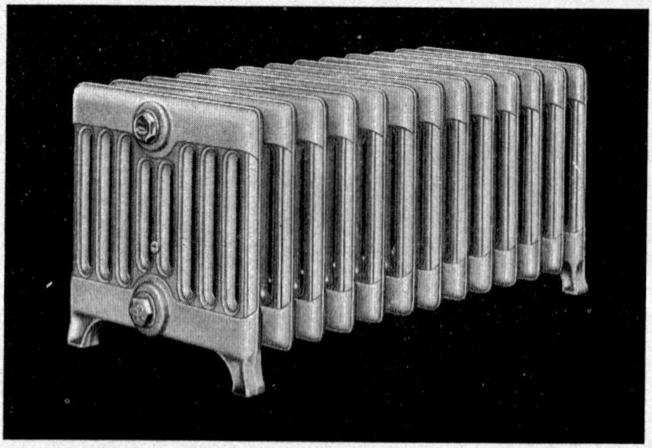

Fig. 97C. Window radiator.

Fig. 97D. Wall radiator.

Radiators are of four main types, *viz.*: column, tubular, window, and wall radiators. The column radiator, once very widely used, is now being superseded by the other types. It consists of cast-iron columns built in sections which are connected together by screw nipples. The illustration shows a single-column radiator of seven sections, but in a column radiator, each section may consist of more than one column.

Tubular-type radiators are similar to column radiators, but they have smaller columns. Window radiators are of the tubular type, but they are of smaller height, and to compensate for this each section comprises more tubes. Wall radiators have the advantage that they are only about 2 in in width, so that their use does not result in appreciable loss of space. A further advantage is that these wall radiators can be joined together so as to make up any desired area of heating surface.

HEAT LOSS FROM RADIATORS

Although much of the heat given off by radiators is in fact emitted as radiation, a considerable fraction of it is transmitted to the air by convection. The proportions of radiated and convected heat vary with the type and size of the radiators. Allen and Walker (1931) give the results of tests of radiators heated by steam at 215° F and with a room temperature of 70° F. The radiators were of the column type with one, two, three, or four columns, and of various heights.

The curves to the left of Fig. 98 show the relationship of the heating surface to the height and number of columns of a 10-section radiator, and at the right of the figure the rates of heat emission/sq. ft of rated surface are shown.

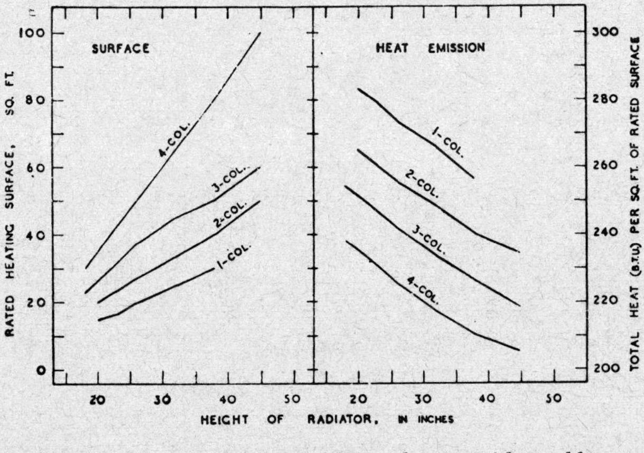

Fig. 98. Heating surfaces of 10-section column radiators, with total heat emission at room temperature of 70° F and steam at 215° F.

The table shows the increase in the proportion of heat transmitted by convection as the number of columns in the radiator was increased. With the single-column radiators, roughly 60 per cent of the heat was convected and 40 per cent radiated, while with the 4-column radiators 70 per cent was convected and 30 per cent radiated.

Proportion of convected heat from column radiators 10 sections in length (room at 70° F and steam at 215° F)

Height of radiator	*Convected heat as percentage of total heat emitted*			
	One column	*Two columns*	*Three columns*	*Four columns*
38 inches	58.6	62	65	71.5
26 inches	59.4	58	69	69

The amount of heat radiated by a heating surface, and consequently the total heat emission, is influenced by the finish of the surface. As we noted in Chapter II, surfaces coated with ordinary paint emit practically maximal radiation, and the emission is about the same whatever the colour of the paint, while when metallic paints are used the radiation is reduced. The total heat output from a surface covered with metallic paint is about 10 per cent less than that from an ordinary painted surface. The radiation from a bright metallic surface is still less than that from a surface treated with metallic paint.

The heating capacity of radiators is now commonly expressed in terms of what is called the *equivalent radiator surface.* For a steam radiator the total heat emission in Btu with steam at 215° F, and the room temperature at 70° F, is divided by 240, and the quotient is the rated surface area in equivalent sq. ft. For hot-water radiators the basis is 150 Btu/sq. ft (A.S.H.V.E., 1945). There is now a tendency to refer simply to the rate of heat output from the radiator without quoting any figure in terms of surface area.

CONVECTORS

Although from long usage the appliances which have been discussed in the preceding paragraph are known as 'radiators', we have seen that they emit one-half or more of their total heat output by convection. The term 'convector' is generally reserved for appliances which emit nearly all their heat by convection. Radiators are often concealed in some form of enclosure for the sake of appearance. In such instances virtually all the heat supplied to the room is convected, and the amount of radiation is negligible.

The arrangement of the radiator enclosure has a substantial influence on the heat output from the radiator. Most enclosures reduce the output below that of an unenclosed radiator. The mere placing of a shelf 3 in above the radiator hinders convection, and may reduce the output of the radiator by 20 to 35 per cent according as the width of the shelf is the same as, or 50 per cent greater than, the width of the radiator. An enclosure which does not affect the heat output is illustrated in Fig. 99. Dimension *B* equals 80 per cent of dimension *A*, *C* is 150 per cent of *A*, and *D* is equal to *A*.

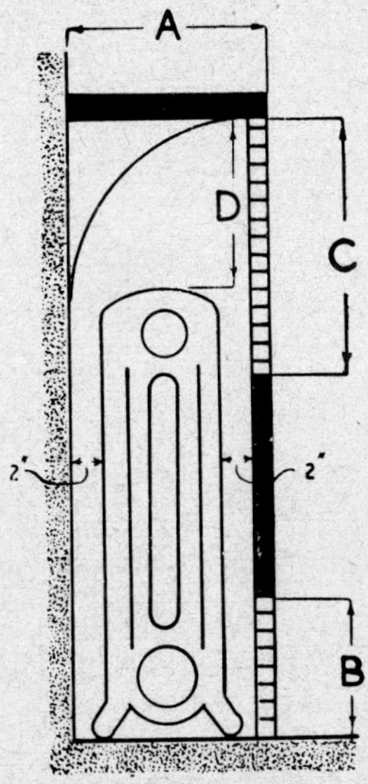

Fig. 99. Radiator enclosure which does not reduce heat output.

The placing of a screen in front of the radiator, with no obstruction above the radiator and with an adequate opening for the entry of air below the screen brings about a stack effect, and increases convection to the extent of about 10 per cent (Allen and Walker).

LOCATION OF RADIATORS

The location of radiators in a room is a matter of much importance. In cold weather the temperature of the air which is in contact with the surface of a window is lowered and the air tends to fall with considerable velocity, and if this downward current is not checked, a person near the window is likely to suffer discomfort from draughts. If a radiator is placed beneath the window the upward current of heated air tends to neutralize the down-draught from the window and thus promote comfort. One the other hand, if the radiator is placed against an internal wall the down-draught from the window is unchecked. It may be increased to some extent by the convection effect of the radiator, for the warm air from the radiator placed against the opposite wall will rise to the ceiling and will tend to travel across the room and then, as it approaches the cold external wall in which the window is situated, it will fall towards the floor. This effect is shown in Fig. 100, which illustrates some of the results obtained by Dufton and Marley (1942).

The diagrams show the direction of air currents across a section of the room. The experiments illustrated in diagram (*a*)

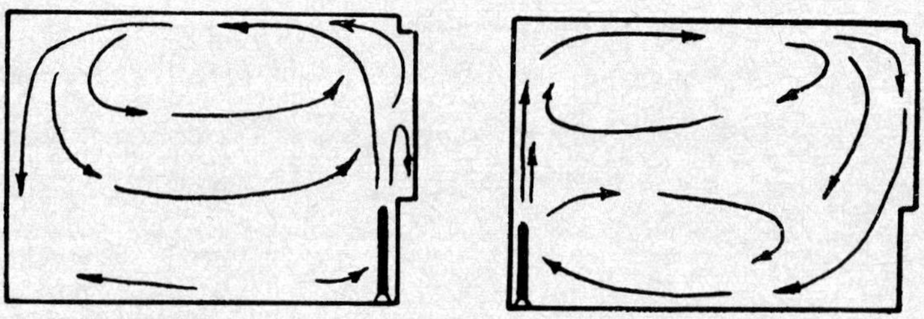

(*a*) (*b*)

Fig. 100. Direction of air movement in radiator-heated room with one cold wall: (a) *radiators under window,* (b) *radiators against opposite wall. (Dufton and Marley.)*

were made when the radiators were placed under the windows, and those in diagram (*b*) when they were placed against the wall opposite the windows.

The temperature distribution throughout a room is also much affected by the position of the radiators. It is perhaps self-evident that when radiators are placed under windows the mixing of cool and warm air currents which occurs will tend to produce a greater equality of temperature than will prevail when the sources of heat are placed against internal walls. This effect was clearly shown by Marley (1937). A test-room was heated by hot-water radiators placed sometimes under the windows and sometimes against the opposite wall, and the temperature of the air at the centre of each of the four quarters of the room was measured. Each series of tests was run for several days. When the room was unventilated the mean deviation of the individual temperatures from the average for the room was only 0.18° F when the radiators were placed under the windows, whereas it was 1.45° F when the radiators were placed against the opposite wall. When the ventilation was at the rate of 2.4 air changes/hour the corresponding mean deviations of temperature were 0.50° and 1.7° F.

The placing of radiators or other heaters under windows to prevent down-draughts is no recent innovation. Some years ago, Bedford was asked to investigate the ventilation of the Debating Chamber of the House of Lords. Amongst other things it was complained that cold draughts were experienced in one part of the chamber. Fume tests showed that in cold weather there were strong down-draughts from the lofty windows. Over the greater part of the length of the chamber these currents tended to veer horizontally some distance above head level, but in the position about which complaints were made the currents swept right down on to the occupants. Bedford proposed that heaters should be placed under the windows to counteract these downward currents, and then learned that when the building was erected in the middle of the last century heaters had in fact been placed there. They were fed by steam and were sometimes noisy, and they had, therefore, been unused for many years. Arrangements were made to feed them with hot water.

It is often of great advantage to place a run of hot pipes along the lower edge of skylights.

TEMPERATURE GRADIENTS

The kind of heater and its position both influence the vertical distribution of temperature in a room. We have earlier remarked that the source of heat should be as near the floor as possible. It is stated in the Annual Guide of the American Society of Heating and Ventilating Engineers (1945), that when radiators are placed near the ceiling there is usually such a large difference in temperature between the floor level and the ceiling that it becomes difficult to heat the living zone of the room satisfactorily. In that Guide it is also said that smaller temperature gradients can be maintained with long, low, thin radiators than is possible with tall radiators, and that tall column-radiators placed at the sides of window openings do not produce as comfortable heating effects as long, low radiators placed beneath the window openings.

Tredgold (1824) described a steam heating system in a Watford silk mill, in which a steam pipe was suspended just below the ceiling along the length of each room. This course was necessary because the factory was already occupied with the machinery in position, and, as Tredgold remarked, 'the power of making a good arrangement was in this case extremely limited'. Discussing the question of steam pipes he mentioned that they could be arranged in whatever manner was thought most convenient, provided they were near the floor.

Since Tredgold's day, many factories have been heated by steam or hot water pipes suspended 8, 12 or more feet above the floor. The air heated by contact with these pipes rises and tends to pool at the higher levels, and steep temperature gradients are thereby caused.

The slight temperature gradients associated with long and low wall radiators placed near the floor were noted by Vernon, Bedford and Warner (1926*a*). Many hundreds of observations were made in a factory heated by hot-water radiators, 7 ft 3 in long, 1 ft high, and 3 in wide, placed beneath the windows and 10 in above the floor. In one room the average rise in temperature from 6 in above the floor to a height of 4 ft 6 in was only 0.2° F, and in another room where some heat was given off by heated machines the temperature gradient between foot level and head level was 1.7° F. Temperature gradients tend to be greater in cold weather than in mild, for when the outside temperature is low fuller use is made of the heating apparatus, and ventilation is often restricted. The table shows the frequency of occurrence of various differences between the temperatures at head level and foot level in this factory.

Even when the outside temperature was only slightly above freezing point the average temperature gradient was only about 1° F. With external temperatures of 60° F or more, when the heating apparatus was not in use, the temperature gradients averaged 0.2° to 0.3° F, and 62 per cent of the observations indicated gradients of less than 1° F.

Marley (1937) describes experiments in which a test room was heated by hot-water radiators when the wall containing the windows was exposed to a cold surface at 25°F, and the remaining surfaces were maintained at 60° F. In a study of the effects of placing the radiators (*a*) under the windows, and (*b*) against the wall opposite the windows, some observations were made with the room unventilated, and others when it was ventilated at a rate of 2.4 air changes/hour with air which entered at a temperature of 40° F. The equivalent temperature in the room was maintained at 65° F. In the unventilated room the temperature gradient from 3 in above the floor to the ceiling was 4.9° F when the radiators were under the windows, and 5.8° F when they were against the opposite wall. When the room was ventilated the corresponding gradients were 4.2° and 5.2° F.

In a further paper, Marley (1938) gives the temperature gradients set up in the test room, under similar experimental conditions as to external temperature and ventilation, and with the equivalent temperature maintained at 65° F, when the heating was (*a*) by radiators, 32 in high and 18 in long, placed under the windows, and (*b*) by electric tubular heaters placed a few inches above the floor level. Figure 101 shows the results obtained with these appliances.

Frequency of occurrence of various differences between temperature at head and foot levels (Factory A)

Outside temperature	*No. of observations*	*Mean temperature difference*	*Per cent of observations when difference was*							
			−2.1° or more	*−2.0° to −1.1°*	*−1.0° to −0.1°*	*0° to 0.9°*	*1° to 1.9°*	*2° to 2.9°*	*3° to 3.9°*	*4° or more*
30° to 39.9°	108	0.9°	3	4	13	40	21	15	2	2
40° to 49.9°	162	1.0°	1	1	7	41	33	12	4	1
50° to 59.9°	147	0.7°	–	1	9	58	29	3	–	–
60° to 69.9°	207	0.2°	1	1	25	62	10	1	–	–
70° to 79.9°	63	0.3°	–	2	24	63	9	2	–	–

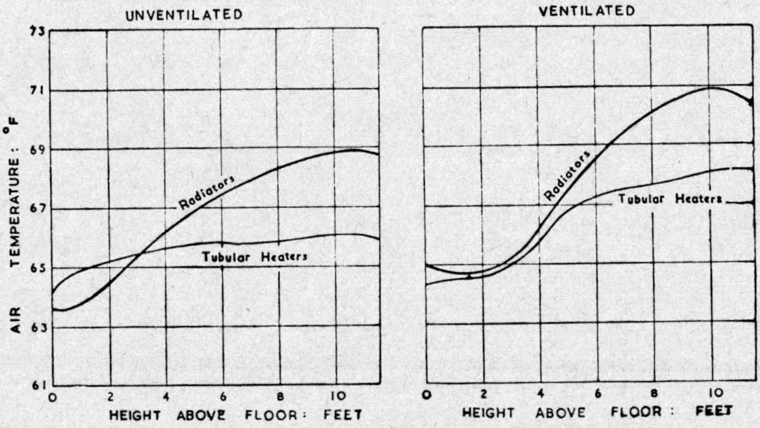

Fig. 101. Temperature gradients in test room, ceiling height 11 ft 6 in.

The tubular heaters caused only slight temperature gradients. The rise in temperature between heights of 6 in and 5 ft 6 in from the floor was only 1.1° F when the room was unventilated, and 2.6° F when it was ventilated. The gradients with radiators were somewhat greater. Between 6 in and 5 ft 6 in from the floor the temperature rise was 3.5° F in the unventilated room, and 3.1° F with ventilation. In other tests, made when the test room was warmed by electrically-heated convectors, the temperature gradient up to a height of 5 ft 6 in in the unventilated room was 6.2° F (Dufton and Marley, 1942). It should be noted that in these observations the ceiling height was 8 ft 6 in instead of 11 ft 6 in as in the tests referred to above.

Figure 102 illustrates the temperature gradient observed with convectors, and for contrast the negligible gradient observed with floor heating is also shown.

The radiators so far mentioned in this discussion of temperature gradients were heated by hot water, and their surface temperatures were not very high. When radiators are heated by steam their surface temperatures are higher, the convection currents arising from them are more rapid, and the heated air rises more swiftly towards the ceiling. When the external temperature is far below freezing point, as so often happens in North America, a greater heat input is required, and this also

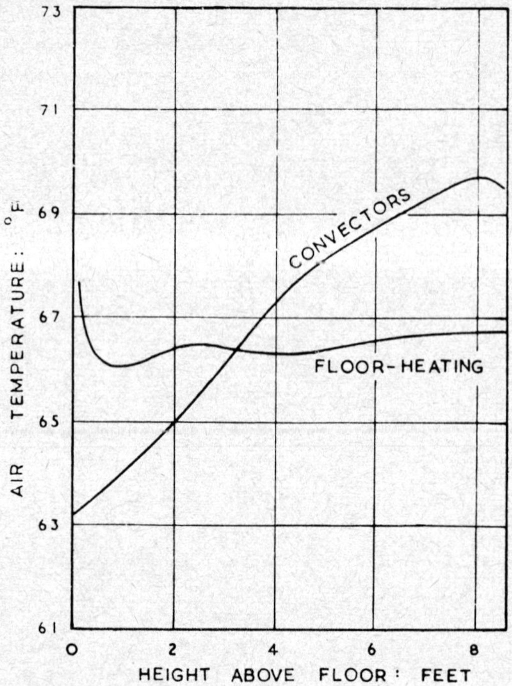

Fig. 102. Temperature gradients in unventilated test room 8 ft 6 in high.

tends to cause greater variations of temperature at different heights above the floor. These facts are illustrated by the observations of Willard and others (1929). A small test room 9 ft high was heated by a steam radiator placed under the window. The external temperature was below 0°F, the temperature of the external wall surfaces was –1.3° F, and the steam temperature was 216.5° F. The air temperature near the floor was 53° F, and that near the ceiling 76° F. Five feet above the floor the temperature was 68.7° F. Thus over the height of a man the temperature gradient was about 16° F. Such gradients are distinctly greater than will ordinarily be expected in the milder climate of Great Britain.

Willard and Fahnestock (1930) studied the effects of two different types of steam radiators on the vertical distribution of temperature in a room 10 ft high and 13 ft × 12 ft in plan. In

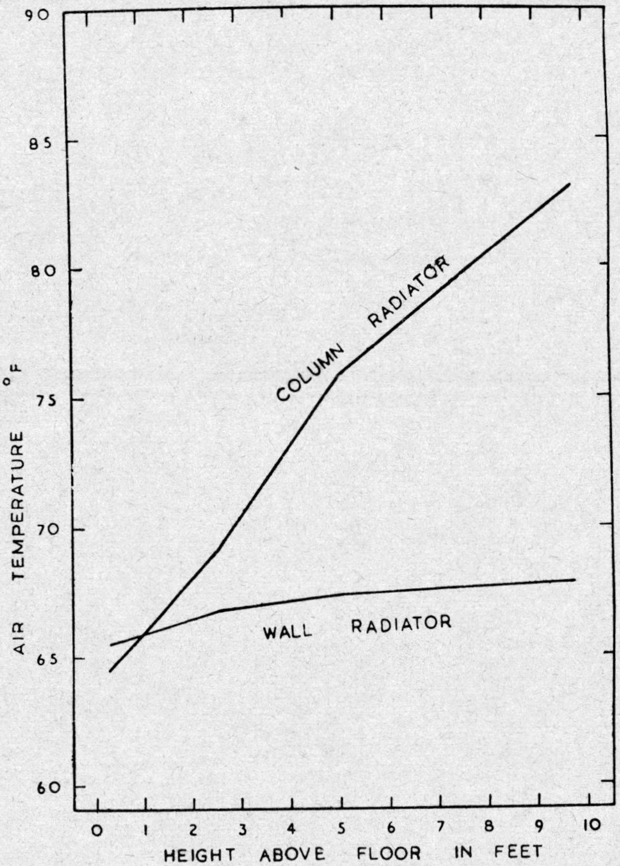

Fig. 103. Temperature gradients with column- and wall-type steam radiators.

one series of tests the room was heated by a two-column radiator 38 in high, and with a surface area of 36 sq. ft, placed against the external wall opposite the window. In the second series this radiator was replaced by a long, low, narrow, wall-type radiator with a surface area of 21 sq. ft, placed below the window. The temperatures observed at various heights from the floor in the middle of the room are shown in Fig. 103. The tests were made in very cold weather when the external temperature was in the neighbourhood of zero on the Fahrenheit scale. In both instances the radiators were supplied with steam at about 230° F.

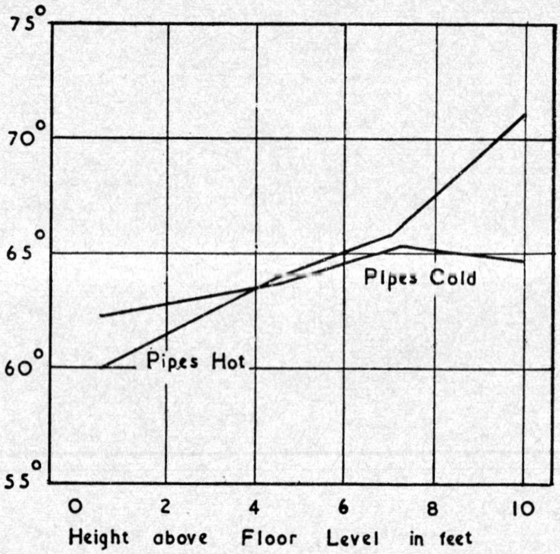

Fig. 104. Temperature gradients with overhead steam pipes.

When the column radiator was in use the average difference between the temperatures 3 in from the floor and 3 in from the ceiling was 18.7° F, while with the wall radiators the difference was only 2.2° F. The differences between floor level and head level were 11.3° F with the column radiator, and 1.9° F with the wall radiator. With the wall radiator far more comfortable conditions were maintained, and with a smaller heating surface and a lower steam consumption. The steep temperature gradient produced by the column radiator was undesirable from the standpoint of comfort, and it resulted in much needless loss of heat through the ceiling.

Whereas Vernon, Bedford and Warner noted only very slight gradients in the factory warmed by low hot-water radiators, they found steeper gradients in factories warmed by overhead steam pipes. Figure 104 shows the temperature gradients in a single-storey factory where low-pressure steam pipes were fixed 8 ft above floor level. When the steam pipes were in use the rise in temperature between foot level and head level was 4° F, and from foot level to a height of 10 ft from the floor it was 11.1°F. When the steam was turned off and the pipes had cooled the difference between the temperatures at foot level and head level was reduced to 1.2° F.

CHAPTER XVIII

UNIT HEATERS

INTRODUCTION

WHEN the plenum system of mechanical ventilation is employed it is customary to warm the entering air in cold weather. With some types of installation, the temperature of the air supplied by the fans is raised to about the level which it is desired to maintain in the building, and other means of heating are employed to compensate for the heat losses through the structure. More commonly, however, the hot air supplied by the fans is the only means of heating provided, and the air is discharged from the supply system at a temperature which may be far above that which it is desired to maintain in the working zone; then the system is truly an air-heating installation in which heating and ventilation are combined.

Formerly, in order to ensure an even distribution of heat it was generally necessary to install either a plenum ventilation system or overhead pipes in extensive industrial buildings. We have seen that overhead pipes are very undesirable on account of the temperature gradients which they set up, and the consequent discomfort and waste of heat, while the installation costs of elaborate plenum systems, with their long runs of ducting, are often heavy. The unit heater is an appliance which fulfils the heating functions of a plenum system at a more economical cost, and it can also be used as a ventilating mechanism.

Essentially, a unit heater consists of a fan and a heater enclosed within a casing. The fan draws air into the casing, where it passes over the heater, and then propels it forward into the room. The heater may be a steam or hot-water coil, or an electric element, or it may be of the direct-fired type incorporating a gas burner. The fan may be of the disc or propeller type or of the centrifugal type. In some unit heaters the heating element is on the intake side of the fan, and in others on the outlet side.

The larger unit heaters are mounted on the floor, and are designed to take air in from near floor level and discharge it horizontally at a height of 7 or 8 ft above the floor. These units are equipped with centrifugal fans, and can handle large volumes of air, which they discharge at a high velocity.

The suspended type of unit heater is intended to be hung from the ceiling or from the roof members. It is usually provided with a disc fan and handles smaller volumes of air than the floor-mounted type, and it also has a lower outlet velocity. This kind of heater comprises a rectangular box-like casing in which the fan and heating element are housed. It is provided with an adjustable louvred outlet so that the air can be discharged either horizontally or downwards at an angle of about 45° F with the horizontal.

A further type of heater, designed for suspension overhead, is known as a vertical-blow unit heater. Heaters of this type are mounted on the ceiling or the roof members, and discharge their air vertically, or almost vertically downwards. Unit heaters of these three types are shown diagrammatically in Figs. 105, 106 and 107.

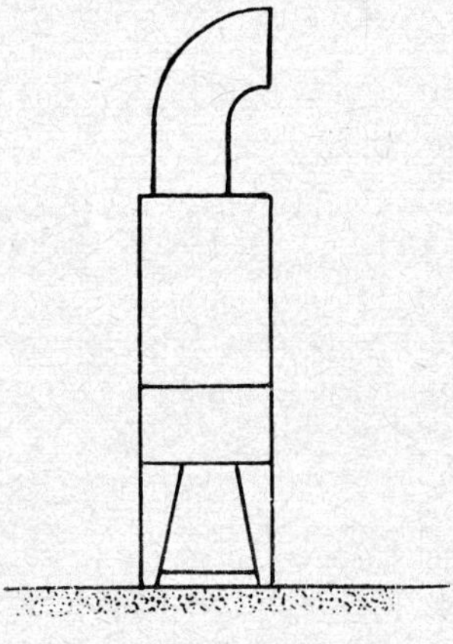

Fig. 105. Floor-mounted unit heater.

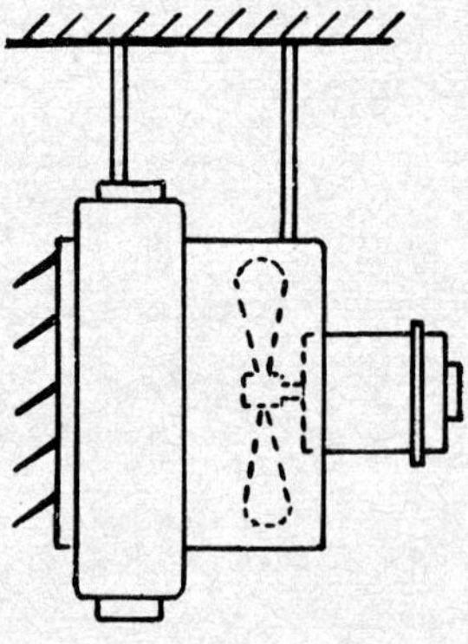

Fig. 106. Suspended unit heater.

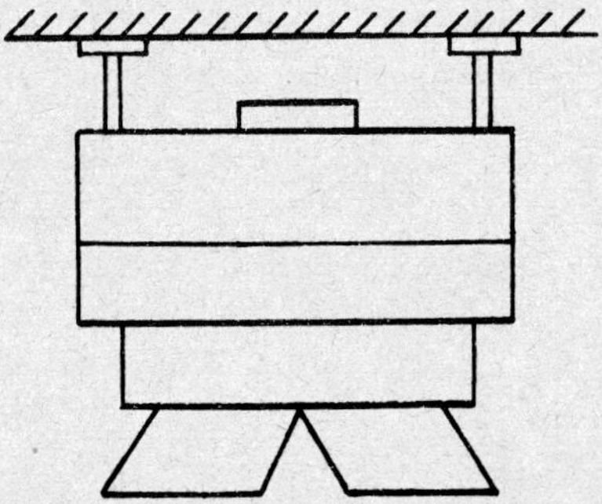

Fig. 107. Vertical-blow unit heater.

Although such a unit is primarily meant to heat and recirculate air, it can be converted into a ventilator by fitting to the intake side a duct which communicates with the outside air.

Units of the suspended type discharge air at velocities of from 400 to 1,200 ft/min, floor-mounted units with centrifugal fans have outlet velocities of about 1,500 to 2,000 ft/min, while units of the vertical-blow type have discharge velocities of 1,200 to 2,200 ft/min.

OUTPUT OF UNIT HEATERS

Unit heaters of the various types and sizes which are available vary greatly in heating capacity. Small heaters of the suspended type can be obtained with a heating capacity of only about 20,000 Btu/hour, while a larger unit of the same type may yield about 200,000 Btu when the heater is fed with hot water, and about twice as much when high-pressure steam is the heating medium. Floor-type units may give up to 300,000 Btu with hot-water feed, and higher outputs with steam.

The amount of heat given off by a unit heater depends on the quantity of air passing over the heater as well as on the temperature of the heating medium. Many heaters are supplied with two-speed fans. When the fan is run at low speed the heating effect of a unit is much lessened, for with the heating element at a given temperature, the reduced velocity of the air passing over it results in a diminished convection transfer. The reduction in the amount of air passing over the coils results in an increase in the temperature of the air as it leaves the unit, but this increase does not compensate for the reduction in the quantity of air passing, and the total heat output of the unit is diminished. Thus when one kind of suspended unit heater is run at low speed the heat output is reduced by about one-third, but the temperature of the air leaving the unit is raised by about 20° F.

The output required to heat a building can be obtained by circulating a relatively large volume of air raised to a relatively low temperature, or a smaller volume heated to a higher temperature. Since in order to obtain a particular heat output with a low outlet temperature it is necessary to deal with larger volumes of air, larger fans and power units are required than are necessary with higher outlet temperatures. Larger fans and motors entail increased costs, and therefore, on the score of economy, there is a temptation to use the cheaper units with their higher outlet temperatures.

TEMPERATURE GRADIENTS AND DRAUGHTS

We have already seen that when hot air is discharged horizontally from overhead ducts undesirably steep temperature gradients may be caused. The heated air from unit heaters suspended overhead is usually discharged downwards at an angle of about 45°, with the intention that the stream of hot air will approach the floor and heat the lower levels of the room. A stream of air moving at high velocity entrains some of the surrounding air, and it may be expected that as the air discharged from a unit heater is projected downwards it will become mixed with some cooler air, and that the temperature of the moving stream will be, to some extent, reduced. On the other hand, it should be remembered that air heated to perhaps 40° or 70° F above room temperature is much lighter than the air into which it is discharged, and it tends to rise very rapidly to the higher parts of the room. The desirability or otherwise of a unit heater installation must depend on the ability of the installation to heat the working zone without causing steep temperature gradients, and without exposing the occupants to unpleasant draughts.

The circulation of air discharged from unit heaters was studied by Lewis (1926). When a disc-fan unit suspended 12 ft above the floor was made to discharge downwards at an angle of 45°, with an outlet velocity of 400 ft/min, the air stream descended to within 8 ft of the floor and then began to ascend. The stream had again reached the 12 ft level by the time it had traversed a horizontal distance of 25 ft. When the same unit, still suspended 12 ft from the floor, was made to discharge vertically downwards the air stream only descended through a distance of 7 ft, turning upwards again 5 ft from the floor. When a unit provided with a centrifugal fan, housed at the same height and provided with 60° deflectors, was made to discharge downwards with an outlet velocity of 1,500 ft/min, the air stream reached the floor and travelled along it in a turbulent manner for a distance of 30 or 40 ft and then began to rise. Some minutes later the smoke-laden air used in the tests had diffused along the length of the building, but beyond the immediate range of the heater the diffusion took place only at the higher levels.

With the discharge velocity of 400 ft/min, none of the air from the heater was able to mix with the stratum of cold air lying along the floor, and although the air discharged downwards at 1,500 ft/min reached the floor, a downward blast with such a high velocity from a height of only 12 ft would be objectionable in most situations. Neither type of unit was satisfactory.

Lewis made other tests with a floor-mounted heater of the centrifugal-fan type discharging horizontally at a height of 8 ft, with an outlet velocity of 1,500 ft/min. The air stream maintained a horizontal direction for about 150 ft, while there was some spread above and below the horizontal line of discharge. There was much diffusion of air from the heater into the warm zone. Lewis concluded that floor-mounted unit heaters with high outlet velocities are able to move the heat they deliver into the working zone, and allow as little of it as possible to escape into the higher parts of the room. Since they accomplish this without blast in the working zone they can be used to heat industrial buildings, producing only minimal temperature gradients, and with the greatest heating economy.

There seem to be few actual measurements of the temperature gradients in buildings heated by floor-mounted unit heaters, but a few observations which Bedford made some years ago lend support to Lewis's contention. In a large factory with heaters of this type, when the outside temperature was 40° F, a temperature of 64° F was maintained at a height of 4 ft from the floor, and the temperature 6 in above the floor was, on the average, only 2.7° F below that figure. In ten sets of observations the temperature gradients between the two levels varied from 1° to 4.8° F.

It is sometimes said that the high discharge velocities from units of the floor-mounted type must create objectionable draughts at head level near the line of the air stream, and that since these heaters draw their air supply from near floor level there are likely to be undesirable draughts along the floor. In Bedford's investigations the air velocities close to the heaters were not specially measured, but in observations made at scattered points in the factory no undesirably high velocities were observed either at head level or near the floor. The mean velocities at these two levels were respectively 28 and 30 ft/min—velocities which are almost identical with the averages of a large number of observations made in well-ventilated factories. The highest velocity observed 6 in from the floor was 51 ft/min and the highest at 4 ft above the floor was 45 ft/min.

Lewis's observations of the direction of flow of the air discharged from a suspended unit heater of the disc-fan type, with a 45° deflector, were made with an outlet velocity of only 400 ft/min. Many modern units are made with substantially higher discharge velocities, and although in Lewis's observations the hot air from the unit only descended to a height of about 8 ft from the floor, the air discharged from units with higher outlet velocities can reach working level, as was demonstrated by Lewis's tests on the high-velocity unit discharging downwards. When these units are installed great care should be taken to ensure that the downward stream of hot air does not traverse positions where people must work. One may occasionally encounter such a stream of air when passing through a building with unit heaters, and even a short exposure to a blast of hot air is unpleasant.

Some of my colleagues had occasion to visit a factory office fitted with overhead unit heaters which were supplied with hot water. The building was 150 ft long by 20 ft wide. The height from the floor to the springing of the roof was 8 ft and the apex of the roof was 6½ ft above the springing. Unit heaters were mounted at a height of 9 ft from the floor. The office staff disliked the hot air being blown directly towards their heads, and the louvres had been adjusted so that the heated air was discharged horizontally. Whatever possibility there might have been of a reasonable mixing of the hot air with that at the lower levels when the air stream was deflected downwards, there was no chance of such mixing when the air was discharged horizontally, and steep temperature gradients resulted. The gradients observed in the middle and at the two ends of the room, on a day when the outside temperature was 49° F, are shown in the table.

Temperature gradients with overhead unit heaters

Position	*Air temperature °F, at height above floor of*				*Temperature at 10 ft level minus temperature at 6 in level*
	6 in	*4 ft*	*7 ft*	*10 ft*	
A.	65.0	72.0	82.0	90.0	25.0
B.	64.0	69.0	78.0	85.0	21.0
C.	60.5	66.5	72.5	80.0	19.5

These temperature gradients averaged 2° to 2.5° F for each foot above floor level. Such a rate of increase in temperature is in close accord with those reported by Smith (1930), who said that in buildings heated by low-velocity suspended units the air temperature generally rises from 1.25° to 2.25° F/ft. Smith also said that with high-velocity units the rise is 1° to 2° F/ft, but in the observations mentioned in an earlier paragraph the temperature-rise observed with floor-mounted units was less than 1° F/ft. A temperature-rise of 2° F/ft over a height of 5 ft represents a difference of 10° F between the temperatures at floor level and at head level. Such a steep gradient is objectionable. It causes considerable discomfort as well as much waste of heat.

Hale (1946) recommends that in a low building used for light industry, small units with fairly high outlet temperatures should be used, and suggests that slow speed fans may be desirable to keep down air velocities and thus avoid draughts. With more lofty buildings, he says, high velocities, larger units and lower outlet temperatures may be used; while in large, high buildings it is absolutely necessary to use large units fixed at a reasonable distance from the floor, fitted with high-speed fans, and having outlet temperatures of 100° to 110° F or perhaps lower. Hale adds that the circulation of greater volumes of air helps to give a better distribution of heat, and to avoid excessive temperature gradients. It seems doubtful whether high outlet temperatures are desirable in rooms with low ceilings. The discharging of gentle streams of very hot air just below the ceiling must, one would expect, result in steep temperature gradients.

It seems to be a reasonable maxim that unit heaters should be arranged to discharge large volumes of air at relatively low temperature rather than smaller volumes of air at much higher temperatures, and that where there is serious danger of draughts a greater number of units with fairly low discharge velocities, but with relatively low outlet temperatures, should be employed.

LOCATION OF HEATERS

There are various views concerning the way in which unit heaters should be arranged. One well-known firm of manufacturers states that as a general rule the heaters should be so placed that the heated air is discharged towards the exposed walls of the building. Allen and Walker (1931) recommend that the heaters should not be placed to discharge in a haphazard way, thus setting up cross-currents, but that instead they should be arranged so that they will assist each other to promote the circulation of air. They express a preference for placing heaters near the wall, so that they will discharge tangentially and thus set up a rotary circulation. Copping (1946) also prefers an arrangement which will set up a rotational movement of the general mass of air, and advocates that the heated air should be directed to the colder parts of the building.

Faber and Kell (1943) refer to the divergence of views as to the location of units, and say that provided the air streams are directed towards the wall surfaces and areas of greatest heat loss, and that they do not blow directly against one another, the layout may be varied within wide limits.

In deciding the placing of unit heaters attention must be paid to the horizontal distances over which the air streams are projected. According to Allen and Walker, air at a temperature of about 135° F can be spread horizontally some 15 to 20 ft for every 100 ft/min of outlet velocity. Increase of outlet temperature will decrease this distance, owing to the lower density of the heated air. Floor-type unit heaters with an outlet velocity of 1,500 ft/min should not generally be placed more than 200 ft apart, while suspended type heaters should not be spaced further than 50 to 100 ft apart when their outlet velocities are 300 to 600 ft/min. Molloy (1945) makes recommendations in terms of the amount of floor area which can be dealt with by one unit. He thinks that the maximum floor area for one floor unit is 8,000 sq. ft and for a ceiling unit 1,000 sq. ft.

ADVANTAGES AND DISADVANTAGES OF UNIT HEATERS

One great advantage of unit heaters is their economy of space. The floor-mounted type takes up considerably less space than would be needed for radiators which would give the same heat output, while the suspended type leaves the whole floor area free. The forced circulation of air through the units makes it possible to raise the temperature of the air to the required level very rapidly, and with this quick response it is also possible to arrange cheap and speedy thermostatic control of the air temperature. Yet another advantage is that in warm weather, by running the units with the heat supply shut off, beneficial air movement can be set up. A further advantage is that the installation costs of this type of heater are low.

Disadvantages of these appliances are their tendency to cause steep temperature gradients, and the possible exposure of occupants to unpleasant air currents. In planning an installation of unit heaters care should be taken to guard against these dangers. It should be ensured that temperature gradients are kept as slight as possible, and that none of the occupants is exposed to the stream of hot air discharged from a heater.

Much skill in design is necessary if these difficulties are to be overcome. Faber (1946) remarks that, except where skill and experience have been used to the full, the heat delivered by the units disappears into the roof space without reaching the working plane, causing a 'tremendous heat gradient'.

CHAPTER XIX

AIR CONDITIONING

INTRODUCTION

MODERN developments in ventilation are largely in the direction of air conditioning. By air conditioning is meant the control of the temperature, moisture content, and purity of the air, so that air conditioning is a combination of ventilation with heating or cooling and with adjustment of the atmospheric humidity. With modern equipment almost any desired condition of temperature or humidity can be maintained. The air conditioning of a building may be desirable in order to create a comfortable environment for the occupants, or it may be called for on account of special industrial processes.

One of the earliest examples of air conditioning in the tropics encountered by Stamp (1964) was in 1934 in the dining room of the Emir of Kano's palace in northern Nigeria, where a huge refrigerator, quite empty, doors open and lights on, was part of the furniture of the royal dining room.

The general method of air conditioning is illustrated in Fig. 108. The air is passed through a humidifying chamber in which water is discharged through nozzles as a very fine spray. In this chamber soluble gases and some suspended solid particles are removed, and the air becomes nearly saturated with water vapour at a temperature which depends partly on the temperature of the water spray and partly on that of the entering air. The moisture content of the air delivered from the plant depends on the temperature of the saturated air as it leaves the humidifier. A thermostatic device is used for controlling the temperature at which the air leaves the spray chamber, thus keeping the moisture content constant.

The temperature of the air in the spray chamber can be raised, and its water content increased, by heating the air before

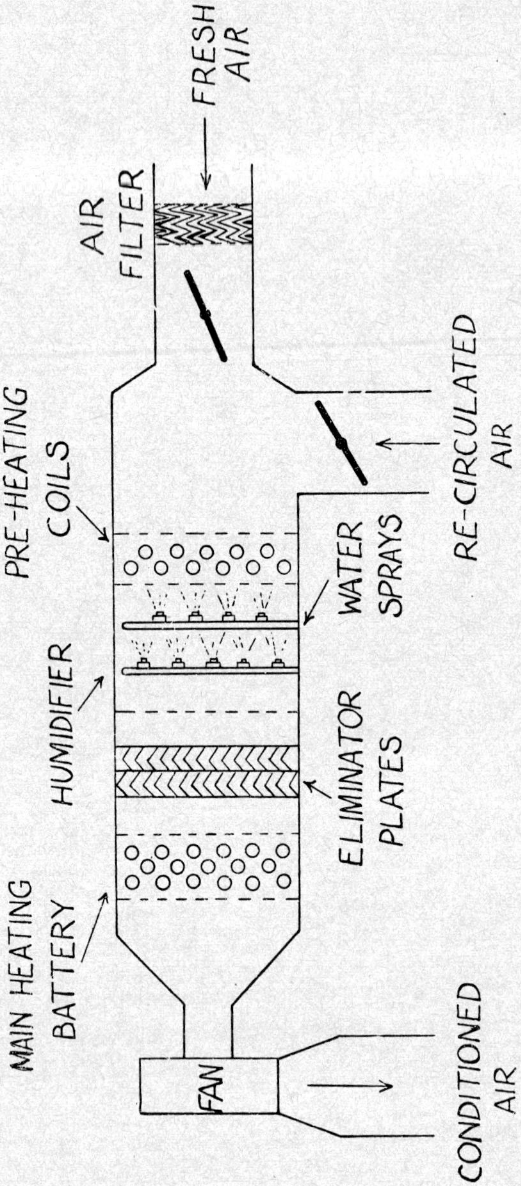

Fig. 108. Diagram of air-conditioning plant.

it enters the chamber, or by raising the temperature of the water supplied to the sprays. If the temperature of the spray water is sufficiently low the air entering the humidifier will be cooled below its dew point, and moisture will then be condensed from it.

As it leaves the humidifier the air passes over a series of eliminator plates, which serve partly to remove dirt and partly to prevent droplets of water from passing out with the air stream. An additional air-cleaning device is often used before the air enters the washer.

After the air has been brought to the desired dew-point temperature in the spray chamber, it is heated to the required temperature and is then propelled through ducts and delivered where required. The temperature to which the air is raised after leaving the humidifier is controlled automatically.

The trend today is to use cooling coils and separate humidifiers. Cooling batteries, which are by no means new devices, do the same job as a spray-type dehumidifier. The idea arose when, in 1930, Thornton Lewis suggested that the heater battery of a unit heater could be economically converted to a cooling battery by the circulation of chilled water through it.

The subsequent development of R12 refrigerant led to rapid advances in the design of cooling and dehumidifying equipment which made such equipment smaller in size and hence cheaper to produce.

AIR CLEANING

The air of towns contains much suspended matter. This consists very largely of the soot and grit particles emitted from chimneys, detritus from roads, and some textile fibres. The total amount of suspended matters varies greatly in different localities and at different times, but even if the solid matter amounts to only a few milligrams/cu. m the amount of dust drawn into a building by a ventilation system is considerable, and in air-conditioning systems and plenum ventilation installations air-cleaning devices are very desirable. The air-cleaning devices employed fall into three main groups, viscous impingement devices, dry filters and electrostatic precipitators.

Dry filters generally consisted of cotton wool, cellulose fibre, or some suitable textile fabric. A difficulty with these filters is that they offer considerable resistance to the flow of air, and a large filtration area is therefore necessary. Further, as the surface layers of the filters become clogged with the filtered dirt the resistance increases substantially, and the filters have to be renewed, or where possible cleaned, frequently.

Viscous impingement devices may consist of a series of metal plates so arranged that the air passing through the cleaner is made to take a tortuous course, or it may comprise a filter of glass or metallic fibres. The metal plates or the fibrous filtering media are coated with a suitable viscous substance—often an oil or grease—so that as the air passing through is made to impinge on a succession of viscous surfaces, dust particles are entrapped. Frequently the air filter is built up of a number of units, and these can be cleaned or renewed in turn.

The choice between dry and viscous filters will be influenced by the type of dust which is to be dealt with. Figure 109 taken from data given in the A.S.H.V.E. Guide for 1945 illustrates the different responses of these two types of filter to two kinds of suspended matter. The curves on the left-hand side refer to tests on a dry filter with a face area of 4 sq. ft, through which air was passed at the rate of 1,000 cu. ft/min, and those on the right to a viscous impingement filter with a face area of 2.8 sq. ft with air passing at the rate of 800 cu. ft/min. The two kinds of dust

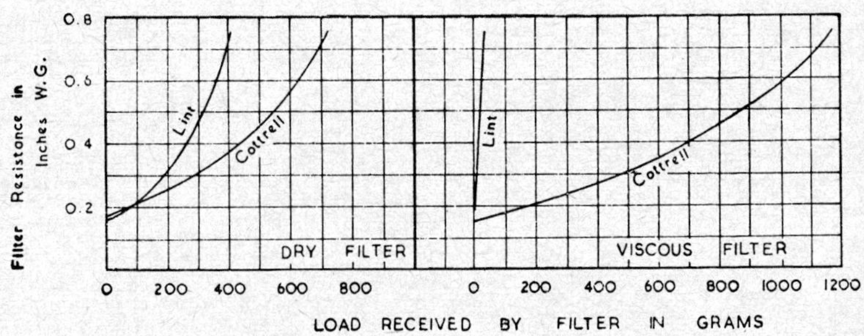

Fig. 109. Dust- and lint-holding capacity of two filters.

were (*a*) smoke particles from a Cottrell precipitator used at a power plant for smoke abatement, and (*b*) cotton linters. The clean viscous filter had a resistance of about 0.17 in of water. A very slight amount of lint raised the resistance to 0.6 in, whereas 1,000 g of precipitator dust were required to raise the resistance to the same level. The resistance of the dry filter when clean was very similar to that of the viscous filter, and its resistance reached 0.6 in when it had become loaded with about 350 g of precipitator dust or 630 g of lint. The lint-holding capacity of the dry filter was greatly superior to that of the viscous filter, whereas the viscous filter was much the better with the fine dust from the precipitator.

It is many years since electrostatic precipitators were first used for the collection of dust in industrial processes, but it is only in recent years that they have been widely used for air cleaning in ventilation. The arrangement of an electrostatic precipitator is shown in Fig. 110.

At the entrance to the precipitator there is an ionizing unit consisting of (*a*) wires supplied with direct current at 12,000 to 15,000 V, and (*b*) earthed cylinders. The plates shown in broken lines are supplied with direct current at about 6,000 V, and those shown by firm lines are earthed. Large numbers of ions pass between the wires and the earthed cylinders and an electric charge is imparted to any particulate matter which may be in the air passing into the precipitator. Then, as the air passes

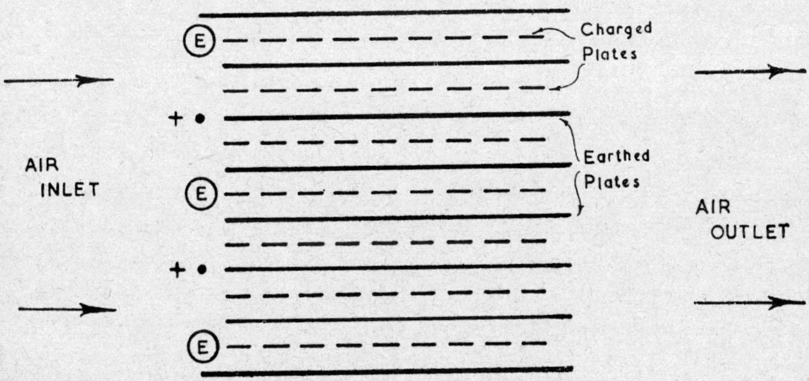

Fig. 110. Arrangement of the electrostatic precipitator.

onwards through the electrostatic field between the charged and the earthed plates of the precipitation chamber, the particles are driven to the earthed plates where they yield up their charges and are deposited. The collecting plates are cleaned at intervals as required. If the electrostatic field is of high intensity, the distance between the plates small (it is only $\frac{5}{16}$ to $\frac{1}{2}$ in in one well-known type of precipitator), and the velocity of the air as it passes through the cell not too high, even very small particles of high mobility will be precipitated. These electrostatic precipitators are made on the unit principle, so that as far as cost and limitations of space allow any required number of cells can be assembled into one unit to cope with the required volume of air and give the desired efficiency of cleaning. It is claimed that 99 per cent of the total weight of dust present in the air can be collected. Particles as small as 0.1 micron can be removed. The resistance to the flow of air is slight, and practically constant: there is no building up of resistance such as occurs in filters.

Because of these characteristics, electrostatic filters not only remove dust and smoke from incoming air but also bacteria and even viruses. In fact, electrostatic precipitators are used for direct sampling of air for the presence of micro-organisms. Artenstein and Miller (1966)* have thus detected meningococci and adenoviruses when using a combination of electrostatic precipitation with a sampling rate of up to 10,000 litre of air/min.

The firm of Honeywell Ltd. have put on the market small transportable unit air-filters including electrostatic precipitators for domestic use.

Tobacco smoke will easily pass through ordinary filters and it is interesting to test one of these domestic filters in this way. As soon as the device is switched on, the smoke in the outgoing air disappears.

Viscous filters are of value for removing coarse particles but they may be inadequate if there is much fine dust in the supply air. Then some kind of fabric filter or an electrostatic

* Artenstein, M. S. and Miller, W. S. (1966). Air sampling for respiratory disease agents in army recruits. *Bacteriol. Rev.* **30:** 571-572.

precipitator may be desirable. In some installations the air is first passed over a viscous filter to remove the coarser particles, and then through a fabric filter. It must be remembered that the resistance of a fabric filter rises very considerably when it is dirty, and the filter must be replaced or cleaned at suitable intervals. In many installations the dirty fabrics are thrown away and replaced by new ones; in others provision is made for the filters to be cleaned automatically.

Sometimes, especially in cases where there is inadequate maintenance, a lazy or incompetent plant operative may not change the filters at the right time, or what is perhaps worse, 'solve' the problem presented by the reduction in the volume of air supplied by the system by the simple action of removing the clogged filters. Unfiltered air together with airborne dust is then carried farther along the ventilating ducts so that, in time, the air outlets become clogged with dirt. The result is that the supply of air to a room may be reduced to 5 per cent of the original designed volume and, in one case known to the present writer, the supply of air was virtually cut off.

REFRIGERATION

For cooling the water used in air washers, or for supplying cold brine or other liquid for cooling coils, mechanical refrigeration is used. The diagram in Fig. 111 illustrates the arrangement of a refrigerator system. The refrigerant, a liquefi-

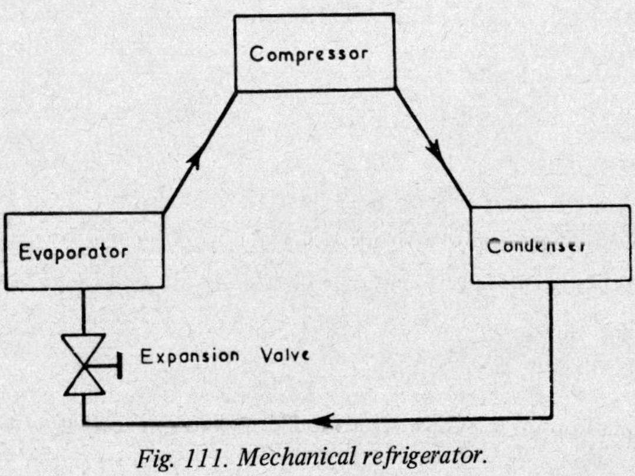

Fig. 111. Mechanical refrigerator.

able vapour, is in a closed circuit. In the compressor the refrigerant vapour is brought under high pressure, and its temperature is raised by the heat of compression. The hot, compressed gas then passes to the condenser. The external surface of the condenser is in contact with a cooling medium—air or water—and the refrigerant gives up heat to this and is thus cooled and condensed. From the condenser the liquid refrigerant passes, still under high pressure, towards the evaporator. Before entering the evaporator it passes through an expansion valve, which reduces the pressure. The evaporator coils are in contact externally with the air or liquid which the refrigerator is designed to cool. In the evaporator heat is exchanged between the refrigerant and the medium which is being cooled, and thus the refrigerant is warmed and evaporated. From the evaporator the vapour passes, under low pressure, to the compressor where the cycle begins again.

Many substances satisfy the thermodynamic requirements for a refrigerant, but in practice the choice is limited by the further considerations of toxicity, cost, and fire risks. The refrigerants most commonly used are monochlorodifluoromethane (Refrigerant 22), dichloro-difluoromethane (Refrigerant 12), and monofluorotrichloromethane (Refrigerant 11). The choice of refrigerant depends also on the type of compressor. Ammonia is the most extensively used refrigerant in industrial and commercial refrigeration.

In many small refrigerators the condenser is air-cooled. The condenser has a large surface area over which air is blown by a fan. This is the typical arrangement in the small domestic refrigerator used for the storage of food. In the larger refrigeratiors the condenser is water-cooled.

The cooling capacity of a system of refrigeration is often expressed as 'tons of refrigeration' and one TR is the rate of cooling required to produce one 'short ton', *i.e.* 2,000 lb of ice from water at a temperature of 32° F over a period of 24 hours. This represents a rate of cooling, or extraction of heat, of 12,000 Btu/hour.

If there is a plentiful supply of water, this may be used for cooling the condenser but, as a rough working rule, some

50,000 gallons of water per annum are needed for every 'ton' of refrigeration.

A guide to packaged air conditioning units prepared by the Electricity Council (1968) gives the quantity and cost of mains water used to cool condensers:

Total cooling expressed in tons of refrigeration (TR)	*Gallons of water per annum*	*Cost per annum at 2/10 or about 14 np/1000 gallons (£)*
1	43,000	6
3	130,000	18
4	198,000	28
5½	290,000	41
8½	433,000	61
13	690,000	97

Air-cooled condensers require something like 1,000 cu. ft of air/min/ton of refrigeration, *i.e.* approximately 2 tons of air/hour for every ton of refrigeration.

The evaporator may be used to cool air directly, or to cool either water or brine. When it is used for cooling water, the water may then be used as the supply to the air washer, and the air thus cooled. Brine cooled by contact with the evaporator may be circulated in pipes and then used for cooling water for an air-washer, or for direct cooling of the air in the building.

THE HEAT PUMP

The refrigerator takes in heat at the evaporator and discharges it at a higher temperature at the condenser. Thus a refrigerator acts as a heat pump because it serves to raise the potential of the heat energy. Ordinarily one is concerned with the ability of a refrigerator to remove heat and disperse it elsewhere, but by suitable arrangements a refrigerator can be used as a means of heating. When it is so used the refrigerator is referred to as a heat pump.

The principle of the heat pump as applied to air conditioning is illustrated in Fig. 112. Whereas in hot weather the evaporator takes in heat from the building, and the condenser discharges its heat to the external air or to water, when the refrigerator is required for warming the process is reversed. The evaporator then takes heat from the external atmosphere, or from well water, and the condenser discharges this heat into the building. The use of a heat engine in this way was discussed in principle by Lord Kelvin nearly a century ago. In 1930, T. G. N. Haldane

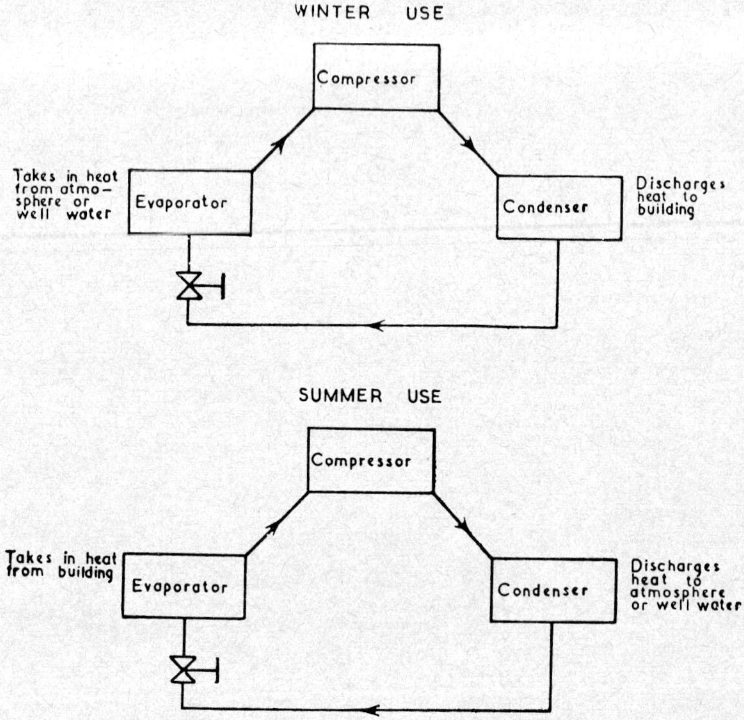

Fig. 112. Refrigerator as heat pump.

described a heat pump installation in his home, and since that time heat pumps have been used for heating water, not only for warming buildings, but also for swimming baths and other purposes. Haldane suggested that one useful way of employing the heat pump would be in connection with a system of panel heating in which very hot water is not required. By simple arrangement of the pipe connections the system could be reversed so that the panels could be used for cooling the building.

Raber and Hutchinson (1945) present data showing that, when a refrigerator is used for the removal of heat, one unit of electrical energy supplied to the system is capable of causing the extraction of 4 or more units of energy from the low-temperature sources. When the machine is used as a heat pump for the supply of heat, the heat equivalent of the energy supplied as work, as well as that extracted from the low-temperature source, is available for warming, so that one unit of energy used in operation will make available 5 units of heat. One unit of electrical energy used for direct heating, say by an electric fire, yields only one-fifth as much heat as is given by a heat pump to which the same amount of electrical energy is supplied. In practice the heat obtained is three to five times the work introduced (A.S.H.V.E. Guide, 1945).

An obstacle to the widespread use of the heat pump for heating purposes is the capital cost of the plant. From a study of costs Onslow (1945) concludes that the full economic benefit of this system is not obtainable unless summer air conditioning is also employed. Raber and Hutchinson mention that, when air conditioning is installed for use all the year round, the additional cost of converting the refrigeration system for use as a heat pump is likely to be not more than the cost of a separate heating plant.

The coefficient of performance of a refrigerating system is defined as the ratio of the desired effect to the energy required to produce that effect. The performance of the heat pump is measured by

$$\frac{T_1}{T_1 - T_2}$$

where T_1 is the absolute temperature of the condenser and T_2 the absolute temperature of the evaporator. The performance is better when the temperature of the condenser is relatively low and that of the evaporator high. Thus if the external air is relied upon as the source of heat better results will be obtained if the winter temperature is high than if it is low. Where the external air is very cold some other source of heat is necessary for efficient performance. Well water is a very suitable source and may be preferable to air, since its temperature is likely to be higher than that of the atmosphere.

With a given temperature of the evaporator the efficiency declines as the temperature of the condenser rises, so that it is uneconomical to work at a higher output temperature than is necessary.

In Great Britain the heat pump has been used in a number of installations in recent years, and various investigations have been carried out by The British Electrical and Allied Industries Research Association.

HUMIDIFICATION AND DEHUMIDIFICATION

The Air Washer

The air washer in an air-conditioning plant serves the dual purpose of removing some of the suspended matter from the air and of adjusting the water content of the air. It may be arranged that the washer adds water vapour to the air or removes it.

Consider what occurs in an air washer in which the water is recirculated. The recirculated water takes up the wet-bulb temperature of the air passing through the washer. When under adiabatic conditions (*i.e.* when no heat is added to the system or abstracted from it during the process) unsaturated air has its degree of saturation increased by the evaporation of water which is already at the wet-bulb temperature of the air, the heat required to vaporize the water—the latent heat of vaporization—is taken from the air, with the result that the dry-bulb temperature is lowered while the wet-bulb temperature remains practically unaltered. Ultimately, if saturation becomes complete, the dry-bulb temperature is reduced to about the original wet-bulb temperature. The total heat of a given mass of saturated air is equal to the total heat of the original air plus that of the water which has been evaporated to bring about saturation. Take the example of air with a dry-bulb temperature of 60° F and a wet-bulb temperature of 50° F which is then saturated by the addition of water at 50° F. On entering the washer the water-vapour content of the air is 38 grains of water/lb of dry air, and the total heat is 12.6 Btu/unit of moist air containing 1 lb of dry air (*i.e.* 7,000 + 38 = 7,038 grains of moist air). When, through the evaporation of water, the dry-bulb temperature is reduced to 50°—the original wet-bulb temperature—the total heat value of 12.6 Btu/lb of dry air represents a water content of 53.4 grains/lb, and the wet-bulb temperature is 49.8°. The relative humidity is 99 per cent.

If the air is heated before it enters the spray chamber, and is then saturated adiabatically in the chamber, it will take up more water vapour, and leave the spray chamber at a higher dew-point temperature, than if it had not been preheated, for the preheating raises the wet-bulb temperature. Thus, if air at a dry-bulb temperature of 60° F and a wet-bulb temperature of 50° F is heated to 90° F dry-bulb without change in its water content of 38 grains/lb of dry air, its wet-bulb temperature will be raised to 62° F, and the total heat content/lb of dry air will be 19.9 Btu. If this air is now humidified, by recirculated spray water at 62° F, until the dry-bulb temperature falls to 62° F, the wet-bulb temperature will be 61.5° F. The amount of water vapour/lb of dry air will be 81.5 grains, as compared with the 53.4 grains given by humidification without preheating. Thus, in this example, preheating will result in an increase of about 50 per cent in the water content of the humidified air.

In the foregoing examples it was assumed that the water used in the spray chamber was recirculated, and thus reached the wet-bulb temperature of the air entering the chamber. If the spray water is heated, the amount of water taken up by the air will be increased, partly because of the warming of the air by contact with the heated spray, and partly owing to the increased heat content of the warm water.

So far we have considered the influence of water sprays in increasing the humidity of the air, but air can be dehumidified by passing through the spray chamber. If the spray water is sufficiently cooled, the temperature of the air will be reduced below its previous dew point and some of its water vapour will be condensed. A spray saturates the air at nearly the same temperature as that of the water used in the spray (Faber and Kell, 1943).

Sometimes only one bank of spray nozzles is used in the washer, and sometimes more than one bank. A double bank gives a greater degree of saturation than a single one.

Steam Humidification

In some air-conditioning installations the air is humidified by the addition of steam. In textile mills, where high atmospheric humidities are required for certain processes, steam has been very widely used for this purpose.

Dehumidification

One method of reducing the amount of water vapour contained in the air is to cool it below its dew point, and thus condense some of the water. One example of this mode of cooling has already been mentioned, namely, cooling in the air washer. Another way of achieving this cooling is to pass the air over blocks of ice, or over coils in which cooled water or brine is circulated.

A further method is to remove the water by adsorption on substances such as activated alumina or silica gel, or by absorption by a solid such as calcium chloride or by aqueous solutions of calcium chloride or lithium chloride.

The adsorbents, silica gel and activated alumina, have a high adsorption capacity. The latent heat of condensation is given up by the water as it is adsorbed, and this heat raises the temperature of the adsorbent. A further source of heat is the so-called heat of wetting of the adsorbent, and the total heat of adsorption amounts to about 1,293 Btu/lb of water vapour adsorbed (A.S.H.V.E. Guide, 1945). As the adsorbent approaches saturation its adsorption efficiency diminishes, and to restore full efficiency it is necessary to drive off the adsorbed water. This 'reactivation' is accomplished by heating the adsorbent.

It will be obvious that when a liquid absorbent is used the solution will be greatly diluted by the water absorbed from the air. The diluted solution is brought back to the desired concentration by heating.

AIR-CONDITIONING UNITS

Air-conditioning installations are commonly designed to deal with the requirements of the whole of a building, or of a large part of it. The various component parts of the installation are assembled at some convenient central spot, and the conditioned air is piped through ducts and delivered where required. In recent years, however, air-conditioning units have been used to an increasing extent. These units are factory made, the necessary appliances being assembled and encased before they leave the factory. They are supplied either complete or in sections, so that the only work that is necessary on delivery is to connect them with the necessary services such as heat, power and water, and, if they are transported in sections, to connect the sections.

These units are of varying types. Some are designed for complete air conditioning for all seasons, and include provision for heating or cooling, and for humidity control and air cleaning. Others are meant only to deal with cooling and humidity control, while yet others are cooling units rather than air-conditioning units, for they are intended only to cool the air. Provision is made for the introduction of fresh air, and for varying the proportions of fresh and recirculated air discharged from the units. Small units with a capacity of $\frac{1}{3}$ to $1\frac{1}{2}$ h.p. and larger units of sizes up to 30 h.p. may be obtained. The smaller units are suitable for conditioning the air of single rooms such as those found in dwellings and hotels, while the larger units are employed to satisfy the requirements of groups of rooms or of other spaces such as those in large offices or stores.

THE NEED FOR AIR CONDITIONING

In the climate of Great Britain there are relatively few days in the year when dwellings and other buildings which can be well ventilated by natural means would really benefit greatly from air conditioning. Except on unusually hot or sultry days most of us would probably prefer to tolerate a fairly high temperature, and enjoy the air movement from widely opened windows, rather than to be enclosed in an air-conditioned room. For the short period when air cooling and dehumidification would be really appreciated air-conditioning appliances would be a costly luxury for most people. However, for other buildings which are less easily flushed with air, and for places such as restaurants, large stores, and places of amusement where many people congregate, air conditioning with its full provision for cooling and drying the air is very desirable. It may also be desirable in large office blocks where windows must be closed to exclude noise.

In hot countries the trying effects of climate can be overcome to a very large extent by the application of air conditioning. Even a small air-conditioning unit in one's office or sleeping quarters can do much to ease the white man's burden in the tropics, while travel in hot climates is made much more comfortable by the air conditioning of railway coaches or other means of transport.

A special field for air conditioning is in the control of conditions in hospitals. In operating theatres air conditioning is used for reducing the explosion risk, for providing a suitable environment for the patient and for reducing the risk of wound infection; and another important application is in the maintenance of suitable temperatures and humidities in nurseries for premature infants.

In hot and deep mines air conditioning has been used. Air cooling and dehumidification have been used in a number of metalliferous mines in India, South Africa, and America. If air cooling is employed in mine ventilation with the object of reducing the air temperature at the working face, it is desirable that the cooling should be done underground as near as possible to the position where the cooled air is required, for, owing to its low specific heat, cold air is rapidly warmed by contact with the warmer rock surfaces if it has to travel for any considerable distance underground. Good results have been obtained by cooling the air at the surface to such an extent that much of its contained water vapour is removed. Then, while the dry-bulb temperature at the working face may not be much affected, the wet-bulb temperature is lowered.

Various experimental trials of local air conditioning have

been carried out in British coal mines (Bedford and Warner, 1931; Hancock, 1927*a*, 1927*b*), but owing to the layout of these mines it has not hitherto appeared to be an economic proposition to employ air conditioning on any large scale. It appears that better results are obtainable by maintaining large airways and circulating large volumes of air.

There are many industries in which air conditioning is desirable, not so much for the comfort of the operatives as for the benefit of the product. Thus, in colour-printing, control of atmospheric humidity is desirable, for changes in humidity affect the size of paper sheets and may therefore cause spoilt work. The strength of textile threads is much influenced by the humidity of the air, and artificial humidification has long been practised in cotton and linen factories. Other industries in which humidity control is desirable include the manufacture of paper, films, rayon, and tobacco, and some processes in food manufacture.

It may interest students of occupational hygiene to note that in 1911 an Act was passed which prohibited artificial humidification in the industry of cotton weaving if the temperature of the wet-bulb thermometer exceeded 75° F. The permissible temperatures varied with the temperature of the dry-bulb thermometer; the latter temperature would, if not controlled by some system of air-conditioning, vary with the time of the day and the season of the year. But the dominant factor for the industrial process which involves the handling of hygroscopic materials like textiles appears to be the *relative* humidity. The breaking strain of 20-denier cotton thread is said to be 74 lb at a relative humidity of 40 per cent and 92 lb at a humidity of 80 per cent. The level of the dry-bulb temperature of the air does not seem to affect the material. Suppose that a relative humidity of 80 per cent is optimal for the material then, at a dry-bulb temperature of 60° F, the actual vapour pressure of aqueous vapour is about 14 millibars but at a dry-bulb temperature of 80° F it is nearly 28 millibars. So far as physiological strain of the human operatives is concerned it is the combination of the high dry-bulb temperature and the high vapour pressure which is so undesirable.

Much of the old conflict between management and operatives in the textile industry might have been avoided by carefully designed systems of air conditioning in the factories.

One may also note that the same Act of 1911 laid down a standard for the permissible amount of carbon dioxide in the air in weaving sheds. It was laid down in 1898 that the carbon dioxide content should not exceed 0.09 per cent, but this was relaxed in 1911 and the limit was fixed at 8 volumes/10,000 in excess of the carbon dioxide in outside air, that is, the upper limit of carbon dioxide in the shed was raised to 12 volumes/10,000 or 0.12 per cent. There is no evidence to show that there was anything harmful in this although some public health inspectors would not, if they had their way, permit this to occur in offices.

In some industrial processes air conditioning may be indispensable in the interests of the product but it is equally important to recognize that so-called 'comfort' air conditioning has made some industrial production possible. For instance, as Dudley Stamp (1964) points out, the artificial climate produced by air-conditioning equipment has made possible the spectacular economic development of Hong Kong in the last decade or so. In the summer season in Hong Kong the temperature may *average* 87° F or 30.5° C with a very high humidity yet Hong Kong factories can do high-quality printing for the European market because of internal climatic control of the environment.

But the need for similar climatic control has increased in Great Britain because of the considerable post-war development in building in general and in particular the high-rise multi-storey block of which the 27-storey tower block of the new London Stock Exchange in the City of London is an outstanding example. It would be impossible to ventilate a structure of this size and shape by natural means. Natural ventilation may be used if the high-rise block is narrow so that air can pass across the comparatively small space separating the long walls; hence the characteristic shape of the high-rise blocks of domestic dwellings. But the designers of the Stock Exchange building have not ignored the assistance to ventilation provided by the outside wind: the operator in the control centre of the building described by Spanswick (1972) is provided with an indicator which enables him to select one of two air inlets for the main plants located on either side of the tower. This enables the supply of fresh outside air to the inlets for the air-conditioning plants to be augmented without cost.

THE IMPORTANCE OF GOOD DESIGN

This is a matter which cannot be discussed in any detail in a single chapter. The guide prepared by the Electricity Council is a simple practical handbook on packaged systems of air conditioning yet it runs to some 80 pages.

The most important factor in the design according to the authors of the guide is the proper assessment of the cooling and heating loads. This leads to the correct sizing of the plant. Only too often the plant is under-sized, conditions are not as expected and the occupants complain. Over-sizing is perhaps less common; it leads to inefficient and uneconomic running of the plant apart from increased capital costs in the first place.

Even if the plant is correctly sized and designed to provide the right amount of air at the desired physical conditions, this conditioned air must be correctly distributed within the space occupied.

Many installations have failed in this respect although much work has been done on the design of side-wall registers, ceiling diffusers and cill-type linear diffusers.

It has sometimes been imagined that the conditions to be aimed at by the designer are those associated with a sunny spring day out of doors, especially when authorities like the late Sir Leonard Hill are alleged to specify such conditions. What Hill actually said was that indoor conditions should be approximated to outside conditions *when hard muscular work is being done.*

CHAPTER XX

HEATING AND VENTILATION OF HOSPITALS

IN the planning of hospitals the provision of suitable means of ventilation and heating is of great importance, and on account of the current hospital-building programme much consideration is being given to the subject. It is necessary to maintain a satisfactorily pure atmosphere, and temperatures that will be at the same time suitable for both the patients and the staff. The requirements for operating theatre suites and for ward blocks are discussed below.

OPERATING THEATRE SUITES

Requirements

The requirements for operating theatre suites are discussed in the Ministry of Health's Hospital Building Bulletin No. 1.

It is no longer regarded as necessary for the patient to maintain a high temperature in operating theatres and anaesthetic rooms. Commonly the required temperature will be between 65° and 70° F, although 75° F may be needed sometimes. In recovery rooms temperatures of up to 75° F may be needed, but in other rooms of the suite a temperature of 65° F is regarded as satisfactory.

Yet Macmillan (1967) reported that as recently as 1960 two patients died from hyperthermia when the temperature of the air was 86° F.

For the prevention of static discharges a high humidity is desirable. It has been suggested in the United States of America that the minimum relative humidity should be 55 per cent, and in the Ministry of Health Bulletin it is said that a humidity of between 50 and 65 per cent will make a valuable contribution to safety.

When certain anaesthetics like cyclopropane and ethyl ether are mixed with air, oxygen or nitrous oxide there is a risk of an explosion. A series of explosions occurred in the early fifties and the subsequent outcry led to an investigation by the Ministry of Health. It was found that about 30 per cent of the anaesthetics used were liable to explosion but, although anaesthetic was used 2,750,000 times in a year, there were only 36 explosions in seven years and the main cause was static electricity which probably caused 22 of these explosions. However, a survey of 230 explosions in the United States indicated that only 30 per cent could be attributed to static electricity.

The amount of fresh air required for an operating theatre is still the subject of investigation. For theatres for general surgery a ventilation rate of 10 air changes/hour was recommended by the Departmental Committee on the Cost of Hospitals in 1937. That figure has been generally accepted, but the results of recent research suggest that a higher ventilation rate is desirable. In other rooms in the operating suite a ventilation rate of 3 air changes/hour is acceptable. There are objections to the recirculation of air in operating suites, and in Great Britain recirculation has not been practised.

The concept of recirculation of air appeals to the designer as he can reduce the size of the heating and cooling plant with subsequent reduction in capital and running costs. It has been suggested that recirculation of air carries with it an increased risk of infection but this has been countered by those who maintain that pathogenic bacteria are trapped by filters.

Explosive anaesthetic gases cannot be removed by filtration and the risk of explosion appears to be the basis for the ban on recirculation. Yet, the National Fire Prevention Association of America will permit recirculation.

Methods of Heating and Ventilation

It is accepted that in the main rooms of operating suites mechanical ventilation is necessary, but although in special cases, *e.g.* theatres intended for cardiovascular and neurological operations, full air conditioning is used, the general use of air conditioning in theatres is at present regarded as unjustifiable in the climate of Great Britain. It is admitted that theatres are liable to become very warm in summer, but it is said that there should be no difficulty in keeping the temperature below 75° F, except on a very few hot days, if all measures are taken to reduce the emission of heat by equipment, if cold water is used for the humidifying sprays and circulated through the heater batteries and the heat gain from solar radiation is minimized. That may be so, but even a temperature of 75° F is rather high for any prolonged task requiring a high degree of skill and concentration, especially when it is remembered that the heat radiated by the lights above the operating table contributes to the local heat; and, furthermore, there may be difficulty in maintaining the atmospheric humidity at a comfortable level. It seems likely that the demand for full air conditioning in operating theatres will rapidly become more pressing, and it may be argued that while a large programme of hospital building is in hand it would be well to anticipate that demand and provide air conditioning.

Heating. It is undesirable that the whole of the heat required should be provided by the mechanical ventilating system. It is better to use central heating to provide for the heat losses through the structure and to admit the ventilating air at approximately the temperature that it is desired to maintain.

Hot water will commonly be used for the central heating, but the desirable method of distributing the heat is still debatable. For use in operating suites there are objections to the conventional hospital-type radiators on the grounds that they harbour dust and also interfere with the direction of the air flow which the ventilating system is intended to provide. Flat wall-surface panels are easy to clean, and if they are fitted at high level they are less likely to circulate dust or interfere with the intended direction of air flow, but from the standpoint of comfort, panels at low level would be preferable.

Heating panels embedded in the ceiling or floor provide an alternative to wall panels. In the Bulletin of the Ministry of Health it is suggested that ceiling panels may be preferable to floor panels because the warmed floor may be tiring to the feet, while its dryness will reduce its ability to remove and neutralize static charges. It has been shown earlier that the floor surface temperature should not exceed 77° F if discomfort is to be avoided. With a ceiling height of only 10 ft, which is that now recommended for operating theatres, care is necessary in the siting of ceiling panels, but it is suggested that if such panels provide only background heat, and are placed round the perimeter of the ceiling, the intensity of the radiation reaching the heads of the occupants will be within acceptable limits.

An alternative method of heating worthy of consideration, is skirting heating.

Ventilation. In order to prevent the entry of contaminated air to operating theatre suites positive-pressure ventilation is employed. The ventilating air is taken from a suitable spot outside and filtered. It is said that fine filtration is not warranted for general use (Min. of Health, 1957). Excellent results have been obtained by using a coarse glass-wool primary filter and fabric filters in series (Blowers and Crew, 1960).

It has been the practice to combine mechanical extraction of air with the plenum input to operating suites, so that the volume of air extracted is about 75 per cent of the input volume. From recent research it seems inadvisable to use mechanical extraction, for with it the outflow of air through an open door is not enough to ensure that no polluted air can enter (Blowers and Crew, 1960). To ensure an outward flow at all levels through a double doorway of 30 sq. ft area the surplus of the input over the exhaust needs to be about 1,200 cu. ft/min. With 75 per cent extraction that figure represents a ventilation rate equal to 80 air changes/hour in a theatre of 360 sq. ft floor area and 10 ft in height.

Blowers and Crew recommend instead that pressure-relief valves should replace the extraction fans, so that when the doors are closed the valves are open, and as soon as a door is opened the valves close and the air passes out through the open door. With this arrangement the volume of 1,200 cu. ft/min is the total required ventilation, and this represents 20 air changes/hour in a theatre of the size mentioned above. This rate is twice the figure suggested by the Departmental Committee on the Cost of Hospitals, but that suggestion was made in 1937 when theatres were constructed with high ceilings. The quantity of air required to prevent the entry of contaminated air is distinctly less than twice that implied by 10 air changes/hour in 1937.

As to the arrangement of the ventilation in operating theatres opinion is divided. There is general agreement that exhaust ports should be at low level and the air inlets at high level, so that any airborne contamination will tend to be carried downwards. The ideal might seem to be that the air should enter through a perforated ceiling and pass steadily downwards to floor level, and then out through the exhaust ports. Some people favour this downward-displacement system. Others are in favour of sending air in through louvres at high level so that some turbulence is created.

A difficulty with the downward-displacement method is the very low air speed that it implies. Even with a high ventilation rate the downward velocity of the air would only need to be about 4 or 5 ft/min, so that the convection currents set up by the warm bodies of the occupants, lights, and other warm surfaces, would upset the air flow and cause turbulence. Any movement of the occupants would add to that effect.

Discussing the relative merits of turbulent and downward-displacement ventilation, Blowers and Crew (1960) considered that the choice depends on the sites from which pathogenic organisms are mainly disseminated during operations On the available evidence they favoured downward displacement. On the other hand, Lidwell and Williams (1960) considered that to prevent local clouds of contamination derived from the activities of the operating team persisting near the operating site there should be appreciable air movement in the centre of the room.

Humidifiers. Many of the humidifiers used are of the air-washer type, with jets spraying cold water into the air stream. The surplus water is drained back into a tank or trough from which it may be recirculated several times. In such recirculating units bacteria may multiply in the water. Blowers,

Lidwell and Williams (1962) remark that while most of the organisms that cause surgical infections do not grow freely in water tanks, one that causes serious wound infection, *Ps. pyocyanea,* multiplies freely in such situations, and has been found growing in humidifiers in operating theatres. Other less notoriously pathogenic organisms have also been found multiplying in humidifying plants, with possible risk to patients. It is suggested that the spray water should be drawn directly from the public supply and that the surplus should be run to waste. Recirculation of the water should be avoided unless efficient disinfection can be achieved. The addition of chemical disinfectants is not always suitable, but ultra-violet irradiation may be satisfactory. Maintenance instructions should stress the importance of the weekly cleaning of cooling coils in refrigeration units.

WARD BLOCKS

For hospital wards the Departmental Committee of 1937 recommended a temperature of 65° F with an external temperature of 35° F, allowing for a fresh air supply of 3 air changes/hour. The B.S. Code of Practice (1949) recommended 65° to 67° F. Such temperatures have been observed in a number of hospitals during the heating season and members of the nursing staff have found the conditions comfortable (Pearse, 1961).

Heating

There is discussion as to the desirable methods of heating wards. Stead and Rose (1961) refer to a dual system, such as embedded floor panels for background heating and locally-controlled flush metal radiant panels for maintaining the exact temperature required; but they consider that the modern hospital with its large window area requires a zoned system. They instance a system of background heating to balance the fabric losses, with a fresh-air plenum system zoned for solar gains and windage losses. This latter suggestion requires the mechanical ventilation of wards, but except in certain situations, or for special purposes, it seems likely that reliance will generally be placed on natural ventilation.

From the standpoint of comfort heating by means of conventional hot-water radiators can give satisfactory results, but the objection is made that such radiators harbour dust. As alternatives there are flush metal heating panels, or embedded floor or ceiling panels. If ceiling panels are used it should be ensured that discomfort will not be caused by excessive radiation to the heads of the occupants or by overheating of the floors above ceiling panels (Chrenko, 1953).

Pearse (1961) made observations in six hospitals where wards were heated by means of embedded ceiling panels. At four of these, few nurses complained of discomfort of the feet due to the heated floors. At two of them, where floor surface temperatures often exceeded 80° F, nurses suffered much discomfort, and some complained that the hot floors ruined their shoes.

Satisfactory results are likely to be obtained by providing flush metal panels in the walls at low level. Such panels arranged to form a heated skirting are worth consideration.

Ventilation

Where the outside air is heavily polluted, or where external noise makes it necessary to keep windows closed, mechanical ventilation of ward blocks may be necessary. It may also be necessary if hospital buildings are built more compactly than has been the recent custom, for then natural cross-ventilation may be impracticable. In other situations, where for ordinary wards natural ventilation is sufficient, mechanical ventilation may be needed to ensure that infected air does not enter a ward, or does not pass from it to other wards.

In the narrow multi-storey blocks that were typical of hospitals adequate ventilation of wards could be obtainable by means of windows if these were suitably designed. With suitable windows there should be no difficulty in ensuring even lavish ventilation in hot weather. With windows placed in opposite walls ventilation rates as high as 20 air changes/hour have been observed in factory buildings, and by the judicious regulation of window openings such rates were obtained with air movement that caused no discomfort on a warm day.

The recommended allowance for ventilation in the design of the heating system is 3 air changes/hour. That amount of ventilation will probably be satisfactory in cold weather. In warm weather it is likely to be insufficient; a higher ventilation rate will be needed to keep the temperature as low as possible, and to cause the more rapid motion of air that is of great importance for comfort when the temperature is high. If windows cannot be opened the ventilating system should be capable of providing at least 6, and preferably 10, air changes/hour.

Measurements of Plant Performance

It is, as Lidwell (1966) says, useless to design a particular system of ventilation to satisfy certain specifications unless there is also a likelihood that the plant which is installed will function according to specification. But the fact is that once an installation has been built, the designer usually ceases to take any interest in the performance of the plant. Moreover, he often fails to check whether or not the plant operates as planned, and, because of this, many plants never function correctly from the date of completion, and others cease to do so soon afterwards.

Lidwell states that the customer is paying for the delivery of certain volumes of air, amongst other things, but the volumes of air actually being delivered are, in his experience, never measured.

Burfoot (1966), writing more in sorrow than in anger, comments on criticisms in newly-built hospitals which arise from failure to check workmanship before handover or to 'commission the engineering services in any scientific manner'. He cites a number of instances in which the systems of ventilation were criticized for not doing the job they were designed to do when the reason for the failure was that the controls which were to operate the systems had never been wired up!

For testing the plant serving the operating theatre suite, Macmillan (1967) recommends not less than 17 groups of measurements given below:

1. The wet-bulb and dry-bulb temperatures of external air.
2. Wet- and dry-bulb temperatures of air in each internal space.
3. Dry-bulb temperatures in main return-air ducts.
4. Dry-bulb temperatures of air before and after any heaters.
5. Dry-bulb temperature of air in main supply ducts.
6. The quantity of fresh air supplied to the system.
7. The velocity of air in the theatre.
8. Pressure-drop across the system.
9. Differences in pressure between the internal spaces with doors open and with doors closed.
10. Noise levels with the plant switched on and off.
11. Output of fans, *i.e.* air flow and pressure.
12. Air flow and resistance across filters in both clean and dirty conditions.
13. Air flow and resistance across heater and cooler batteries.
14. Air flow in main supply and return-air ducts and at all supply grilles and diffusers.
15. Fan and motor speeds.
16. Operation of thermostatic and humidity controls.
17. Operation of control panels including any safety and sequencing arrangements.

As Lidwell says, measurements of temperature and humidity are not difficult and they are usually done—after a fashion. Faults in the system of control of the air temperature such as 'hunting' may evoke complaints of cold draughts when a stream of air emerging during the low-temperature part of the cycle impinges on the neck of the surgeon or the anaesthetist. Or, more bluntly, as the authors of the Guide to Air Conditioning (Electricity Council) put it, 'however effective the equipment, if it results in a blast of cold air at someone's left ear it isn't doing its job properly.'

To measure the volumes of air actually being delivered, the only satisfactory method for initial measurements and continual monitoring throughout the life of the plant involves the use of devices which are built into the plant. Such devices could include the provision of safe tracer gas such as helium for studies similar to those conducted by the Building Research Station for monitoring the rate of ventilation in occupied houses. Better known is the orifice plate which also needs long straight sections of ductwork. Pitot-tube traverses can be reliable and accurate but they are also laborious. Ma (1965) has described a simple system consisting of two tubes at right-angles placed across the diameter of a duct: the tubes are pierced with a row of small holes on the upstream side. The four ends of the tubes are connected together and provide an averaged value of the velocity head which is measured against the static head of pressure obtained from four commoned holes in the circumference of the duct placed between the ends of the Pitot averaging tubes. For reliable readings, one needs six duct diameters of straight run of duct.

The direction of air flow may be ascertained by the use of small smoke generating cartridges or small swabs dipped into liquid titanium tetrachloride and mounted on thin sticks, and the velocity of turbulent air flow can be measured with the silvered kata-thermometer.

For the measurement of differences of pressure between various spaces a simple sloping manometer used with 1 millimetre bore plastic tubing will suffice and Medical Research Council workers suggest the use of a CF Casella portable 'draught gauge, type T 10804', which is sensitive to 1/200 in water gauge.

It is important to understand that the equipment records and test points provided and used during the phase of commissioning of the plant should form an actual part of the installation so that similar patterns of test can form part of the programme of maintenance.

In the discussion so far it has been assumed that the design of the hospital has been satisfactory from the beginning. In a necessarily brief chapter one cannot discuss faults in design but it should be pointed out that these have occurred in newly-built hospitals. One of the chief criticisms is a lack of flexibility in control and performance; for instance, the staff of a number of hospitals noticed that the temperatures of rooms facing north and south were subject to one common control with the inevitable discomfort to patients and staff.

As for poor performance, as Burfoot says, 'it is hardly credible that the winter temperature of a delivery room in a hospital built only a year or so ago should be under 50° F, of a babies' nursery 44° F, that a pipe should freeze in an anaesthetic room and that steam should rise from a patient's abdomen when it was opened up in an operating theatre.' Examples could be multiplied but it only emphasizes that a hospital is a complicated building calling for the best possible team approach from those concerned with design, commissioning and maintenance of all its services.

CHAPTER XXI

WORK AT HIGH ENVIRONMENTAL TEMPERATURES

WHEN wearing ordinary clothing and doing light work, most people in Great Britain are comfortable at temperatures of about 60° to 68° F. At higher temperatures an increasing number of people feel too warm, and when the indoor temperature is above 75° F most people experience discomfort.

At temperatures up to 70° or 75° F the humidity of the air has a comparatively small effect on the rate of heat dissipation from the body, but at higher temperatures, when people are sweating, humidity becomes increasingly important. In moist atmospheres the evaporation of sweat is hindered, and for that reason hot moist air is much more unpleasant than dry air at the same temperature.

When hard work is done in hot and moist atmospheres the evaporation of sweat may be insufficient to remove the heat produced by the body. Then the body temperature rises and in extreme cases heat-stroke may occur. It is practically certain that the victims of the Black Hole of Calcutta died from heat-stroke.

With regular exposures to hot conditions, whether in hot climates or in hot industries, people become accustomed to heat. There is a psychological effect in that one becomes inured to heat, and accepts as commonplace conditions which would have evoked complaint or resentment otherwise, and there are physiological adjustments by virtue of which the body is enabled the better to withstand heat. It is primarily these physiological adjustments that are referred to when one speaks of acclimatization to heat.

ACCLIMATIZATION TO HEAT

With successive daily exposures to hot conditions circulatory adjustments bring about increased stability of the blood pressure, and there is a substantial reduction in the heart rate during work. With the same level of stress due to heat and work the rate of sweating increases and the onset of sweating occurs at a lower body temperature, thus helping to maintain the lower body temperatures that are associated with acclimatization. Along with these changes there is a progressive improvement in the apparent physical condition of men at work, and in severe conditions the incidence of the various manifestations of heat sickness declines.

After a month's experiments during which he worked in a hot room Vernon (1923) could tolerate a wet-bulb temperature 6° F higher than at the beginning of the month. Vernon and Warner (1932), step-climbing at the same rate at the same air temperature and humidity, had pulse-rates of several beats a minute less in summer than in winter, thus even in the British summer they had acquired some acclimatization to heat.

In the deep gold mines of South Africa hard physical work has to be done at high temperatures, and in such conditions acclimatization to heat is of great importance. For instance, Mavrogordato and Pirow (1927) reported four deaths, presumably from heat-stroke, of native 'boys' working for the first or second time in hot places, and their sudden collapse was probably due partly to lack of acclimatization. Further cases of heat-stroke amongst native mine-workers stimulated investigations of heat-tolerance and acclimatization, and as the result of the experimental work undertaken at the Rand mines, native recruits were examined and their heat-tolerance estimated before they went underground. They were put through tests in a hot chamber, and according to their reactions to those tests they were allowed varying acclimatization periods before being put to work in the hottest parts of the mine (Dreosti, 1935).

Vernon achieved a very substantial measure of acclimatization to heat during the course of a month's experiments. But a considerable measure of acclimatization can be achieved in much less than one month. Dreosti's mine labourers were acclimatized after an exposure of only a few days, indeed some of them gained sufficient acclimatization as the result of one day's heat-tolerance tests.

The rapidity of acclimatization during the first few days of exposure is illustrated by Fig. 113 which is taken from the work of Robinson and others (1943) who carried out experiments in which men walked on a treadmill for 1 to $1\frac{1}{2}$ hours a day in a room where desert conditions were simulated (air temperature 104° F and relative humidity 23 per cent). At the beginning of the experiments, which were done in the winter months, the work was hard enough, and was continued for long enough, to bring on symptoms of heat-exhaustion. During the first seven days the men rapidly became acclimatized to work in the heat, and thereafter their acclimatization improved more slowly up to 23 days. After acclimatization the men performed the standard task more comfortably, their pulse rates were lower, and their rectal temperatures and skin temperatures were also lower, than at the beginning of the experiments. Before acclimatization the

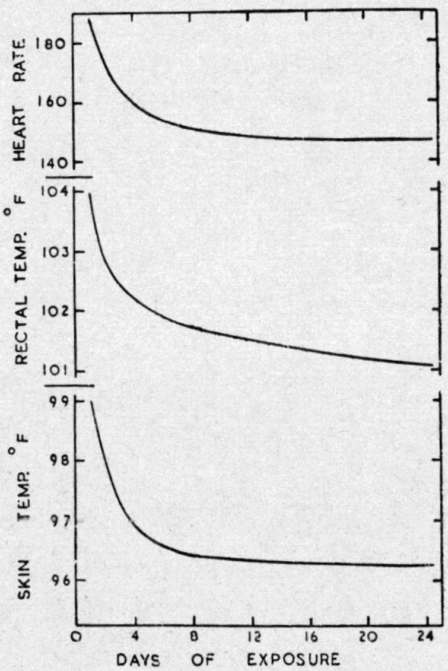

Fig. 113. Acclimatization to heat, as shown by changes in heart rate and body temperature (Robinson et al.*).*

body heat-production while performing the standard task was substantially greater than in the cool room, but after acclimatization the rate of heat-production had declined to the same value as in the cool room.

At the beginning of the experiments walking for 1½ hours proved to be exhausting, but after acclimatization two of the subjects walked at the standard rate in the hot room for 4½ hours with ease.

The increase in the rate of sweating with acclimatization is shown in Fig. 114, taken from the work of Ladell (1951). The Fig. shows the average amounts of sweat lost by 17 men when they were carrying out a routine of step-climbing in a room at 100° F dry-bulb temperature and 93° to 94° F wet-bulb temperature, with an air speed of 50 ft/min. Since the men

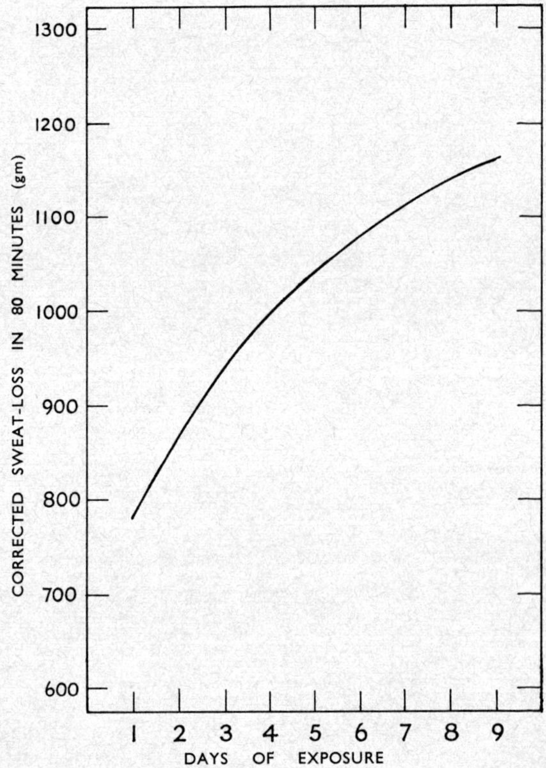

Fig. 114. The influence of acclimatization on the rate of sweating (Ladell, 1951).

varied in weight the figures were 'corrected' to a standard body weight of 65 kg (143 lb). Under the same conditions of stress there is wide variation in the rates at which men sweat. On the various days of exposure the standard errors of the averages through which the curve is drawn were from 6 to 8 per cent of the average values.

Once acquired, acclimatization is not permanent. The ability to endure severe heat stress diminishes after only a few days in cool surroundings, and after one month's sojourn it will have almost disappeared. Caplan (1943–4) remarked that European workers on the Kolar Gold Field felt that they had less stamina after a week-end's absence from work, and that Indian workers felt the heat more after two or three days' leave. Many patients who suffered from heat collapse attributed their collapsed condition to an absence from work of two or three days, and Caplan considered that there was enough evidence to support the contention that loss of acclimatization predisposed to collapse after such a short absence.

One becomes acclimatized to a certain level of stress due to heat and work. A person who has been acclimatized to only moderate stress will need further acclimatization before he can perform satisfactorily in severe conditions, but one acclimatized to severe stress will endure less severe conditions the more readily for that reason. A measure of acclimatization is gained by merely resting in hot surroundings, but that will not enable one to do hard work in such conditions. Work in the heat is necessary for the attainment of fuller acclimatization.

There is general agreement that maximum acclimatization which will enable one to work strenuously in severe heat is most rapidly attained by performing graded, progressively increasing work in the heat. This procedure is convenient for research purposes, but it may not be easy to carry out under industrial conditions. Thus, Wyndham and others (1954), while agreeing with this principle, consider that in the training of African mine labourers it is better to acquire the full rate of working from the outset and to grade the exposures to heat.

It is commonly thought that the native of a tropical country is endowed with a greater capacity to withstand heat than is a white man. In the tropics the native is acclimatized to some measure of heat and can cope with hot conditions better than the unacclimatized European, but when tropical natives and Europeans are both similarly acclimatized to work in severe heat there is little to choose between their responses. Adam, Ellis and Lee (1953; see also Macpherson (1960)) subjected Asian men to artificial acclimatization and then compared their responses with those of similarly acclimatized Europeans. When allowance was made for differences in physique, nutrition and motivation the two groups behaved similarly. The main difference was in the amount of sweat lost, for the Asian subjects adjusted to the stress of the hot environment with a marked economy of water turnover. Given similarity of age, nutrition, physique and experience of work and climate, any effect of race *per se* appears to be very small.

INDICES OF HEAT STRESS

Both work and heat impose stresses on the body, and the wearing of clothing may aggravate the thermal stress in hot surroundings. It is the total stress due to work, clothing and the thermal environment that must be taken into account when limiting conditions are considered. Scales of warmth such as those described earlier (Ch. IV) indicate the stress imposed by the environment, and in so far as equal values on a scale, *e.g.* effective temperature, connote equality of sensory response to the environment they may be regarded as based on equality of physiological strain. Others, more obviously based on equality of strain are the index of physiological effect and the predicted 4-hour sweat rate. As Macpherson (1962) points out, in designing these indices the approach has been to expose subjects to a wide range of conditions and then, by analogy with Hooke's law, conclude that conditions of equal environmental stress are those which produce an equal physiological strain. Yet other indices have been based on the calculation of heat exchange, and of these the best known is probably that devised by Belding and Hatch (1955).

Index of Physiological Effect (E_p)

This index was devised by Robinson, Turrell and Gerking (1945). It is the sum of the effects of heat on the heart rate, skin temperature, rectal temperature and rate of sweating, E_h, E_s, E_r, and E_w, where these are calculated thus:

$$E_h = \frac{100(H_3 - H_1)}{H_2 - H_1}$$

where H_1 is the individual subject's heart rate when engaged in a given form of activity and either nude or wearing a given assembly of clothing in a cool environment;
H_2 is the heart rate when engaged in that activity and with the same conditions with respect to clothing in the most severe environment that he can tolerate over a 2-hour period; and
H_3 is the heart rate during similar exposure to the environment which is being evaluated;

and similarly with *S, R* and *W,* the skin and rectal temperatures and the rate of sweating.

The values of heart rate and of temperature used in these calculations are the averages of all observed values during the second hour of exposure, and the rates of sweating are the averages of the first and second hours.

The maximum score for each of the four components is 100, so that in the worst conditions that can be tolerated for 2 hours the maximum attainable value of E_p will be 400. The index is intended to evaluate the stress imposed by heat over and above the stress due to the performance of the same task in a cool environment. Roughly speaking the subjects were able to maintain themselves in heat balance until the E_p value exceeded 200. Robinson and his colleagues give for clothed and unclothed men expending energy at different rates, psychrometric charts on which are drawn contour lines showing the combinations of dry- and wet-bulb temperature at which various values of E_p occur.

This method of assessing strain is arbitrary. As Yaglou (1949) pointed out, the four physiological characteristics that are taken into account are normally interrelated and the relative importance of each as a component of the total heat strain may vary considerably according to conditions and according to the individual subject. Thus, no simple summation of component strain ratios can be taken as a true index of the total physiological strain. Yaglou also remarked that the index does not apply to unacclimatized subjects since no consistent measurement of the responses under maximal conditions can be obtained with such persons.

Although it is possible, as Robinson and his associates have done, to take a particular response to heat and ascertain the ratio of the actual strain to the maximum tolerable there is, as Yaglou pointed out, the difficulty that one characteristic may approach a limiting value while others do not. It might appear that a composite index such as the E_p figure should be related to medical appraisals of the physical state of the experimental subjects, so that the validity or otherwise of the method of combining and weighting the various strain ratios could be established. Unfortunately, however, such appraisals which depend on the judgment of the observer necessarily lack precision, and they tend also to lack consistency. There are no present means of judging the accuracy of such measures of bodily strain.

Predicted Four-hour Sweat Rate

If one single response is to be taken as an index of the stresses due to heat and work the rate of sweating has much to commend it for that purpose. It is easily measured experimentally and relatively small changes in stress cause substantial changes in the loss of sweat. The rate of sweating was chosen by McArdle, Dunham, Holling, Ladell, Scott, Thomson and Weiner (1947) as the basis of a scale for the assessment of stress. This scale, the predicted 4-hour sweat rate (P4SR) was designed to enable one to predict from a knowledge of the thermal environment, the rate of working, and the clothing worn, the amount of sweat that would be produced over a period of 4 hours by a fit, acclimatized young man at dry-bulb or

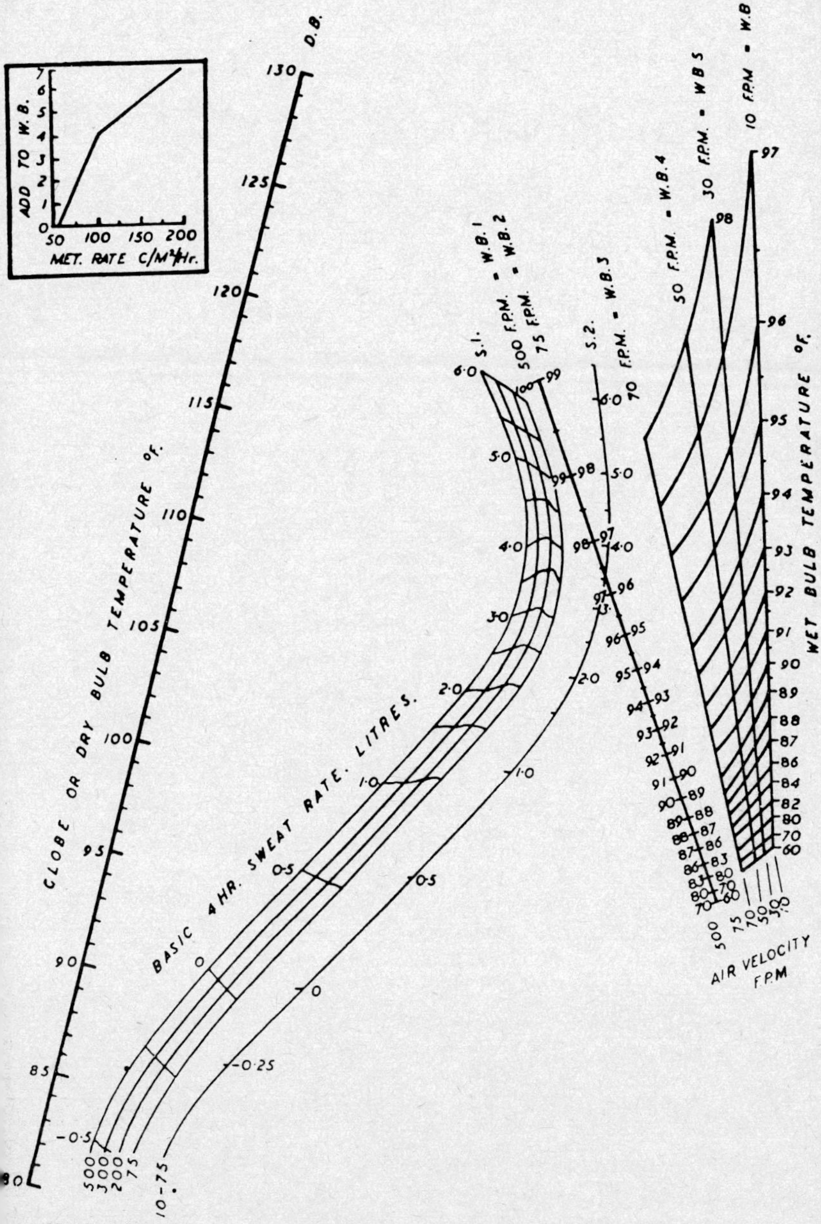

Fig. 115. Nomogram for the prediction of the 4-hour sweat rate. (M.R.C. Memo. No. 29, H.M.S.O.)

globe-thermometer temperatures within the range of 80° to 130°F with wet-bulb temperatures of 80° to 97°F, and air speeds of 10 to 500 ft/min, with metabolic rates ranging from 54 to 200 kcal/m^2 of body surface/hour. When the clothing consisted only of shorts, or when overalls were worn over shorts, or in a few experiments when bush shirts and drill trousers were worn, the predicted sweat rate was said to agree closely with the observed rates of sweating in a series of more than 1,000 individual experiments when the metabolic rates were from 54 to 157 kcal/m^2/hour. No formula is given for this scale and the predicted sweat rate is read from a nomogram (McArdle *et al.*, 1947; Smith, F. E., 1955).

The method of using the chart is briefly as follows:

> The nomogram consists essentially of three scales. The left-hand scale corresponds to the globe-thermometer temperature or, if the surrounding surfaces are at air temperature, to the dry-bulb temperature. The straight lines on the right (scales W.B. 1 to 6) represent the wet-bulb temperature, the line for the appropriate air speed being used as indicated. The curves in the centre of the nomogram, S. 1 and S. 2, are the scales from which the *basic 4-hour sweat rate* is determined. S. 1 and S. 2 consist of a series of lines, each of which is a scale corresponding to the air speed indicated at the lower end of the curve. If the predicted 4-hour sweat rate is required for men sitting in shorts when the globe-thermometer temperature is the same as the dry-bulb temperature, the *basic* 4-hour sweat rate is equal to the *predicted* 4-hour sweat rate. If the men are active, *or* if the globe-thermometer temperature differs from the dry-bulb temperature, *or* if more clothing than shorts is worn, then the predicted 4-hour sweat rate is determined in three stages.
>
> 1. In the first stage the wet-bulb temperature may need correction, depending on the amount of radiation, on the metabolic rate, or on the amount of clothing, as follows:
>
> (i) If the globe thermometer (G.T.) differs from the dry-bulb temperature (D.B.), add to the wet-bulb temperature 0.4 (G.T.–D.B.)°F.
>
> (ii) If the metabolic rate exceeds 54 kcal/m^2/hour add to the wet-bulb temperature an amount (°F) according to the metabolic rate indicated by the small inset. If the metabolic rate is not known add the amount according to the grading of the activity as follows:

Sedentary	55-75	kcal/sq. m/hour	add 1.9°F
Light	75-125	,,	add 4.0°F
Moderate	125-200	,,	add 6.0°F
Heavy	200-250	,,	add 7.5°F

(iii) If the clothing is as heavy as overalls worn with shorts (1,140 g) add 1.8°F to the wet-bulb temperature. A proportional adjustment is made for heavier or lighter clothing, *i.e.* 1°F for every 300 g of clothing above the basic weight of 600 g for which no adjustment is needed.

2. Using the corrected wet-bulb temperature, and using the globe-thermometer temperature if this differs from the dry-bulb temperature, determine the basic 4-hour sweat rate from the nomogram.
3. Having obtained the basic 4-hour sweat rate, the predicted 4-hour sweat rate is determined by adding to the former an amount which depends upon the metabolism and the amount of clothing:

(i) *Men sitting in shorts.* The predicted 4-hour sweat rate is equal to the basic 4-hour sweat rate.
(ii) *Men working in shorts.* The predicted 4-hour sweat rate is equal to the basic 4-hour sweat rate plus 0.014 (M – 54), where M is the metabolic rate (kcal/m^2/hour).
(iii) *Men sitting in overalls worn over shorts.* The predicted 4-hour sweat rate is equal to the basic 4-hour sweat rate plus 0.25 litre.
(iv) *Men working in overalls worn over shorts.* The predicted 4-hour sweat rate is equal to the basic 4-hour sweat rate plus 0.25 plus 0.02 (M – 54), where M is the metabolic rate.

It is remarked by McArdle and his associates that when the amount of sweat lost was less than about 5 litres there was a reasonably close correlation between sweat rate and rectal temperature, but that in more severe hot dry conditions there was a tendency for the sweat rate to be disproportionately high compared with the rectal temperature. This effect was to some extent allowed for in constructing the nomogram, so that in these circumstances the P4SR tends slightly to underestimate the sweat rate, and if the rise in rectal temperature is in turn estimated from the P4SR it is slightly overestimated. It is also remarked that when the P4SR exceeds 5 litres the predicted values tend to be greater than the actual rates of sweating. It is suggested that the main cause of this is that under the conditions represented by these high predicted sweat rates the subjects become progressively less able to sweat so profusely.

While in such conditions the P4SR may not give a very accurate indication of the true rate of sweating, it may nevertheless be a very good index of the overall stress imposed on the subject. Thus, although the nomogram is expressly intended for the prediction of the rate of sweating, there is an implicit admission that in some circumstances the rate of sweating of itself may be inadequate as an index of stress and that rectal temperature should be taken into account.

The results of a large number of experiments carried out at the Royal Naval Tropical Research Unit at Singapore showed that in general the P4SR index predicts with remarkable accuracy the sweat losses of highly acclimatized, healthy young men. The sweat loss was predicted as well in conditions in which the mean radiant temperature was above air temperature as in conditions in which the air temperature and the mean radiant temperature were the same (Macpherson, 1960).

Index for Evaluating Heat Stress

This is an index proposed by Belding and Hatch (1955) on the basis of earlier work by Haines and Hatch (1952).

As a method of assessing thermal stresses in industry, Haines and Hatch applied the heat balance equation to ascertain the rate of evaporation that is required to maintain heat balance without a significant rise in body temperature, and they compared that rate with the maximum evaporative capacity. By thus comparing evaporative needs with evaporative capacity one can evaluate the heat exposure in terms of the excess of one over the other. Haines and Hatch pointed out that the method is essentially an engineering procedure, and that literally it implies a precision of relationships between the environmental stress and the physiological response of man which may not be entirely acceptable to a student of thermal physiology. The justification for offering the method was one of practical engineering. It is assumed that if thermal balance can be maintained with a mean skin temperature of 95°F the accompanying stress is not of serious consequence, and in the calculations of heat loss the skin temperature is therefore taken as 95°F.

In applying this method the rate of change in the heat content of the body and the rate of heat exchange by respiration are considered to be of little practical significance, and they are ignored. Then, when the rates of loss by radiation and convection are subtracted from the rate of heat production, the difference represents the rate of evaporation required to maintain heat balance. The heat-loss equations used are based on those of Nelson and others (1947). They refer to unclothed persons, but it is remarked that the rate of heat loss by convection/unit of body surface area from light clothing was found not to differ greatly from that of a nude man, despite the greater area of the clothed surface.

Belding and Hatch (1955) suggested that the evaporative requirement expressed as a percentage of the maximum evaporative capacity of the air should be used as an index for evaluating heat stress. In very dry air, however, the evaporative capacity may exceed the capacity of the body to produce sweat, so an upper limit of 2,400 Btu/hour is placed on the maximum evaporative capacity.

The calculations are made with reference to a body surface area of 20 sq. ft, which represents a man about 5 ft 8 in tall and weighing 154 lb.

The rate of evaporation required to maintain heat balance (E_{req}) is given by

$$E_{\mathrm{req}} = M + R + C$$

where M is the rate of heat production within the body;
R is the rate at which the body gains heat by radiation; and
C is the rate at which the body gains heat by convection.

When these quantities are expressed in Btu/hour the values of R and C are given by

$$R = 22(t_w - 95)$$

$$C = 2\sqrt{V}(t_a - 95)$$

where t_w is the mean radiant temperature, in °F;
t_a is the air temperature, in °F; and
V is the air speed, in ft/min.

The sum of $R + C$ may also be estimated from the globe thermometer temperatures and the air speed.

The maximum evaporative capacity is given by

$$E_{max} = 10.3V^{0.4}(42 - P_a)$$

where P_a is the vapour pressure of the air in mm of mercury. Belding and Hatch provide a series of flow charts by the use of which E_{req} and E_{max} can be rapidly determined.

A heat stress value of 100 is assigned when E_{req} equals E_{max}, or when E_{req} equals 2,400 Btu/hour. The maximal value of 2,400 Btu for a heat stress figure of 100 represents a rate of sweating of 1 litre/hour.

At other stresses, when E_{max} is less than 2,400 the index value is E_{req} expressed as a percentage of E_{max}, and when E_{max} exceeds 2,400 the index value is E_{req} as a percentage of 2,400.

An index figure of 100 is said to represent the maximum strain that will be tolerated daily by fit, acclimatized young men.

Turner (1958) reported estimates of heat stress on foundrymen during 98 typical exposures in non-ferrous foundries. He compared the rates of heat loss by the evaporation of sweat as calculated by the method of Belding and Hatch with the rates he actually observed, and found close agreement, especially in the hoter environments. There was evidence to support the view that men will not willingly work for prolonged periods under conditions for which the value of the heat-stress index exceeds 100.

Caution is necessary in applying the index if appreciable amounts of clothing are worn. An instance in which the index clearly overestimated the stress of the industrial environment is reported by Belding, Hertig and Riedesel (1960). From subsequent laboratory studies it appeared that the clothing that was worn blocked about 40 per cent of the incident radiation.

Comparison of Effective Temperature and Predicted Four-hour Sweat Rate

McArdle and his co-workers (1947; see also Macpherson, 1960) drew attention to certain apparent anomalies in the effective temperature scale. It was said that at temperatures near the upper limit of endurance effective temperature does not give proper weighting to the relative effects of dry-bulb and wet-bulb temperatures, that in certain conditions it does not show the deleterious effects of very low air speeds, and that at very high temperatures it fails to show the increased stress caused by high air speeds.

F. E. Smith (1955) made a statistical comparison of effective temperature and the predicted 4-hour sweat rate, using the records of the experiments on which the P4SR index was based. Since effective temperature makes no allowance for the rate of working, and since there are separate scales for persons stripped to the waist and for those wearing normal clothing, separate comparisons were made for given combinations of clothing and rate of working. The individual rates of sweating observed were correlated (*a*) with the values given by the P4SR scale, and, since the relation of the rate of sweating with effective temperature is curvilinear, (*b*) with a quadratic expression in terms of effective temperature. It was found that predictions of rates of sweating can be made somewhat more accurately from the P4SR index than from effective temperature, but the error in predicting individual rates is large with either index.

Smith also concluded that the results of his analysis bore out the criticisms made of effective temperature.

In experiments carried out at Singapore, Adam and Cague (1952) also found that when each work-clothing combination was taken separately the predictive accuracies of the two scales were about the same, although there was a slight advantage in favour of the P4SR. They found that in hot conditions effective temperature does not make accurate allowance for the effects of changes in air speed. They remark that the P4SR scale can only be applied within the range of conditions which cause people to sweat, and that it cannot replace effective temperature under the conditions for which it was primarily designed and where sweating does not occur.

Corrected Effective Temperature

Earlier reference has been made to the adjustment of the original effective temperature scales to allow for the effects of radiant heat. The adjustment is made by using the temperature indicated by the globe thermometer instead of the air temperature on the dry-bulb temperature scale of the effective temperature chart. There is evidence that the correction is a reasonable one. Humphreys and his colleagues (1946) studied the physiological responses of people exposed to high effective temperatures when the mean radiant temperature was higher than thc air temperature, with effective temperatures of 80° to 95° F the elevation of the mean radiant temperature above the dry-bulb temperature varied between 0° and 40° F, so that the radiation effects were considerable. The air velocity throughout these experiments was between 15 and 25 ft/min.

Fig. 116 gives curves showing various combinations of effective temperature and elevation of mean radiant temperature above dry-bulb temperature which produced the same rectal temperatures in the subjects of the tests. Thus a combination of 87° F effective temperature with 30° F elevation of mean radiant temperature produced a rectal temperature of 99.8° F. The curve for a rectal temperature of 99.8° cuts the abscissa at the point corresponding with an effective temperature of 91° F. Hence, with rectal temperature as the criterion of the physio-

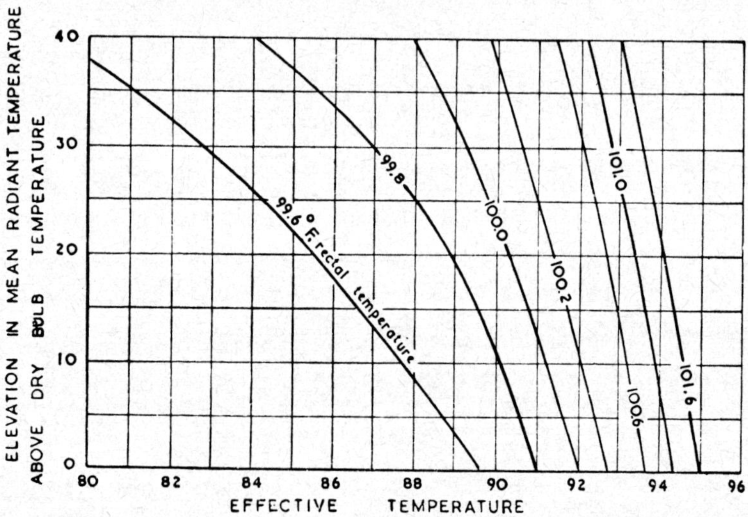

Fig. 116. Combination of effective temperature and elevation of mean radiant temperature producing the same body temperature.

logical effect of the environment an effective temperature of 91° F, when the dry-bulb temperature and the mean radiant temperature were equal, was equivalent to an effective temperature of 87° when the mean radiant temperature excess was 30° F.

These observations can be used to estimate the validity of the method of correcting effective temperature by the use of globe thermometer readings. From each of the curves shown in Fig. 116 we have read various combinations of effective temperature and elevation of mean radiant temperature which produced the same rectal temperature, and have computed the corresponding corrected effective temperatures. If rectal temperature is of itself an adequate index of the physiological effects of heat, the corrected effective temperatures corresponding with any one rectal temperature should all be equal to the uncorrected effective temperature when the elevation of mean radiant temperature is zero. Corrected effective temperatures were calculated for an air velocity of 20 ft/min, the mean of the range used by Humphreys and his colleagues, and for relative humidities of 30, 50 and 80 per cent.

In Fig. 117 the corrected effective temperatures calculated from four of the curves given by Humphreys and his colleagues are shown. Separate curves are given for humidities of 30 per cent and 80 per cent, and for comparison the observed (uncorrected) effective temperatures are shown in broken lines. The curve for conditions causing a rectal temperature of 100° F, when the humidity was low, indicates that the corrected effective temperature was constant at 92° F—the observed effective temperature with zero elevation of mean radiant temperature—even when the excess of mean radiant temperature over the dry-bulb temperature was as much as 40°, while the maximum error at 80 per cent relative humidity was 1° F. In cooler environments the radiation correction was scarcely sufficient, for, as the amount of radiation increased, the corrected effective temperature fell below the value obtained when the elevation of the mean radiant temperature was zero. On the other hand, with hotter conditions the correction was to some extent overdone.

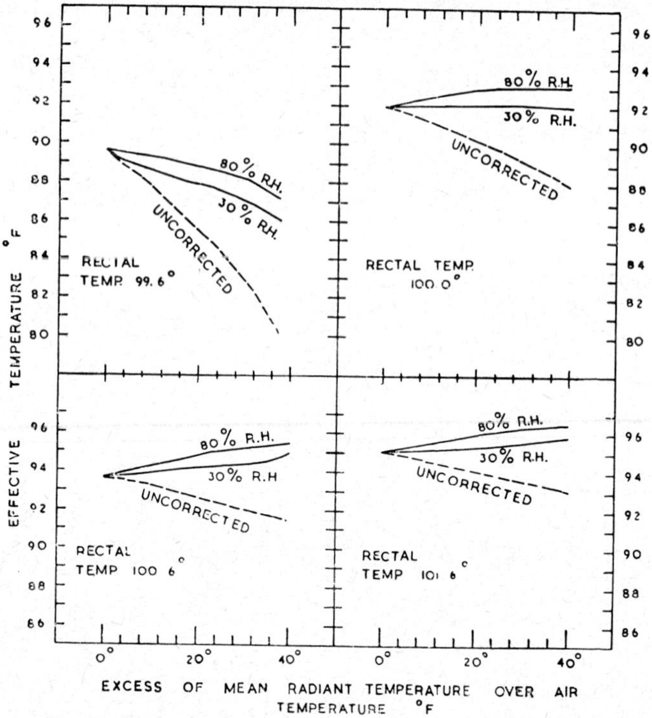

Fig. 117. Corrected effective temperature computed from the data of Humphreys et al. *(Uncorrected effective temperature shown in broken lines.)*

At the Royal Naval Tropical Research Unit, Singapore, the contribution of radiant heat to environmental stress was studied (Macpherson, 1960). Men alternately sat and 'step-climbed' in a rectangular tunnel the floor, ceiling and two sides of which were panels containing heating elements, so that the mean radiant temperature could be raised considerably above the air temperature. In different experiments the men were exposed to mean radiant temperatures ranging up to 32° F above the temperature of the air. The air temperature was 90° or 110° F, the water vapour pressure was 16.5 or 27.0 mm of mercury, and the air speed 150, 200 or 350 ft/min.

The relation between the rate of sweating/kg of body weight and corrected effective temperature was exponential in form. To describe the results for subjects wearing overalls two equations were necessary, one for the higher vapour pressure and one for the lower. For the results on subjects wearing only shorts four equations were necessary, for the two vapour pressures and for different wind speeds at each vapour pressure. The need for more than one equation for each condition of clothing was due to the defects in the effective temperature scales as indices of heat stress—that at high temperatures the weighting for changes in humidity is not perfect, and that the allowance for air speed is at fault. Yet, when the curves for the equations were drawn and the observed sweat losses plotted the points fell closely about the appropriate curves, whether the mean radiant temperature and the air temperature were the same or the mean radiant temperature exceeded the air temperature by more than 30° F. There was also a close relation between the mean skin temperature and the corrected effective temperature.

These findings give strong evidence of the validity of the allowance for radiant heat that is included in the scales of corrected effective temperature.

LIMITING CONDITIONS OF HEAT

The maximum conditions of heat that can be borne without the development of unpleasant symptoms depend on the rate of working, on the amount of clothing worn by the worker, and on his state of acclimatization to heat. The environmental stress depends very largely on the wet-bulb temperature, but air movement and radiant heat are also important factors.

That pioneer of heat physiology, J. S. Haldane (1905; 1907) found that the endurable limit for his subjects, when they were resting stripped to the waist, was a wet-bulb temperature of 88° F in calm air, and of 93° F in an air current of about 150 ft/min. When his subjects climbed steps at the rate of 14 steps/min their endurable limit of wet-bulb temperature was reduced by 8° to 10° F below the resting values.

In the light of later evidence Haldane's limiting temperatures appear to be rather low. Vernon (1923) could tolerate a wet-bulb temperature of 86° F when unclothed and climbing steps at the rate of 13½ steps/min, and coal miners in some British mines produce at wet-bulb temperatures of 86° F or even higher.

Dreosti (1935) and Weiner (1938*b*) found that unclothed workers, accustomed to working in the hot and deep gold mines of South Africa, could do hard work for about 1 hr in air practically saturated at 95° F. Discussing heat sickness amongst Indian miners in the Kolar Gold Field, Caplan (1943) said that heat collapse occurred only rarely when the wet-bulb temperature did not exceed 90° F and the dry-bulb temperature lay between 110° and 120° F. At wet-bulb temperatures of 91° to 93° F Caplan found that heat casualties began to occur, and at wet-bulb temperatures of 94° F or over they were fairly frequent.

Clearly, if one is to specify limiting conditions of heat for any kind of work such limits will depend on the duration of exposure. One may want to know the upper limit of tolerance for sustained work during spells of some hours for several consecutive days, or for much shorter periods during some emergency, as for example in mine-rescue work or fire fighting underground, or one may be concerned with the upper limits of heat at which men can work safely and efficiently day after day and year after year.

Limits for Sustained Work

Robinson and his colleagues (1945) carried out experiments to determine the most severe conditions in which men, clad only in shorts, could maintain thermal equilibrium after the second hour of 6-hour exposures. The subjects of these experiments were all in good physical condition and well acclimatized to working in the heat. When the men were at rest the limiting air temperatures were 96.8° with 98 per cent relative humidity, or 122° F with 34 per cent relative humidity (*i.e.* wet-bulb temperatures of 96° and 93° F respectively). When they walked up an incline so that their metabolic rates were 188 Kcal/m^2/hour, the limiting temperatures were 93.2° F with 91 per cent relative humidity, and 122° F with 21 per cent relative humidity (wet-bulb temperatures of 91° and 84° F). The air velocity in these experiments was 165 ft/min.

Other investigations were made by Eichna and others (1945) with young soldiers as their subjects. The subjects were nude or wore only cotton shorts, and their work consisted in marching at 3 m.p.h. carrying a 20-lb pack. Their energy expenditure was 250 to 300 Kcal/hour, and since their average body surface area was 1.89 m^2, their metabolic rates were of the order of 150 Kcal/m^2/hour. The limiting wet-bulb temperature was 96° F in a saturated atmosphere, and 94° F when the dry-bulb temperature was 120° F. Under these conditions the moderately hard work led rapidly to total disability in most men, with excessive and often disturbing physiological changes. Those men who worked for longer than 1 hour did so inefficiently and with marked physiological disturbances. At wet-bulb temperatures between 91° and 94° F men were capable of prolonged moderately hard work, but they worked inefficiently and with difficulty, lost vigour and alertness, sustained undesirable physiological changes, and at times became heat casualties. With wet-bulb temperatures below 91° F they worked easily and efficiently, and with only mild physiological changes. The air velocity in these experiments was effectively that caused by marching at 3 m.p.h. (260 ft/min). The findings of Eichna and his colleagues are summarized in the Table.

Limiting wet-bulb temperatures (*after Eichna* et al.)

Atmospheric conditions	*Wet-bulb temperatures, °F, when moderately hard work was*		
	Relatively easy	*Difficult*	*Impossible*
Saturated air	92.5	94	96
Air at 120°F, dry-bulb	90	92	94

McArdle and his colleagues (1947) concluded that conditions that correspond to a predicted 4-hour sweat rate of 4.5 litres, which figure they calculated to represent closely the 'relatively easy' conditions of Eichna and his co-workers, should be accepted as the upper limit for fit, acclimatized young men. In hotter environments an increasing number of such men would find the conditions beyond their endurance.

At Singapore naval subjects carried out a routine of step-climbing while wearing naval drill overalls over shorts (Ellis *et al.*, 1953; 1954; Macpherson, 1960). They were exposed to the test conditions for 4 hours on alternate mornings, and on the intervening mornings they carried out other 2-hour experiments unclothed. When the air temperature was 100° F or less the wet-bulb temperature above which some men were incapacitated was 92° F. Above 100° F the limiting wet-bulb temperature was lowered by about 1° F for each increase of 10° F in air temperature up to 130° F. Very small increments in the wet-bulb temperature had a profound effect on the level of stress. All the subjects completed their task with ease when the dry- and wet-bulb combinations were 120/89° F and 100/92° F, but most of them were incapacitated when the wet-bulb temperatures were increased by only 2° F.

The wet-bulb temperatures above which casualties occurred in Singapore were only 1° to 2° F above those represented by a predicted 4-hour sweat rate of 4.5 litres.

Limits for Short Exposures

If men are exposed to an environment so hot that the thermal balance of the body cannot be maintained, or can only be maintained at an inordinately high body temperature, they will only be able to remain in that environment for a limited time.

Roantree (1950–1) reported on investigations made in Mysore, India, on teams engaged in fighting fires underground while wearing Proto breathing apparatus. It was concluded that environmental temperatures of up to 115° F dry-bulb with 104° F wet-bulb could be regarded, tentatively, as permissible for an average, fresh, acclimatized and trained team for a normal shift of 20 min and for not more than 30 min. For the most urgent of reasons work at yet higher temperatures might be permissible for picked men; then temperatures not exceeding

127° F dry-bulb and 110° F wet-bulb might be allowed for not more than 20 min. Roantree gave a chart showing the lengths of exposure to various wet-bulb temperatures that were regarded as permissible for less severe conditions than those mentioned. Thus, wearing Proto apparatus, an exposure of 1 hour to a wet-bulb temperature of 90° F might be allowed.

More recently Lind and his colleagues (1957) have studied mine rescue teams in Great Britain, wearing breathing apparatus and working in both saturated and unsaturated environments. With either of the two types of breathing apparatus the times for which men could be expected to endure severe conditions were distinctly shorter than the corresponding times found by Roantree to be permissible. Thus, in a saturated atmosphere at 100° F the expected tolerance times are only about 19 and 20 min for the two types of apparatus, whereas Roantree gave 20 or even 30 min for an atmosphere with a wet-bulb temperature of 104° F and a distinctly higher dry-bulb temperature.

In comparing their results for saturated and unsaturated environments Lind and his colleagues found that the tolerance times could not be estimated satisfactorily from the effective temperature, but that they could be estimated much better from a weighted mean temperature calculated by adding 0.85 times the wet-bulb temperature to 0.15 times the dry-bulb temperature. This weighted temperature has lately been referred to as the Oxford index.

Provins and others (1962) studied the tolerance to heat of fit young men who were engaged in sedentary work while wearing boiler suits. When the tolerance times for a wide range of conditions were plotted against the weighted mean temperatures obtained as above the points were fitted by a curve with the equation

$$y = b + \frac{c}{x - a}$$

where y is the tolerance time in minutes, x is the weighted mean temperature, and a, b and c are constants.

Later experiments (Bell *et al.*, 1962) were made on young

men, again wearing boiler suits, who either stood and periodically walked a few paces, or carried out a routine of step-climbing. The tolerance times for these two rates of working were also fitted by curves of the same form. The original curves were derived from the mean tolerance times for various weighted mean environmental temperatures, so that at any given temperature some could not endure an exposure for the time given by the equation. Exposure for three-quarters of the mean tolerance times could be endured in about 95 per cent of the trials, and periods representing 75 per cent of the mean times are therefore regarded as the safe tolerance times. The values of the constants in the foregoing equation which give these safe tolerance times in minutes when the weighted mean temperature (0.85 wet bulb + 0.15 dry bulb) is expressed in degrees Centigrade are shown below:

Activity	*Value of constant*		
	a	*b*	*c*
Sitting	36.01	+5.99	135.75
Standing	34.78	−2.35	170.83
Working	32.04	−2.36	175.58

The values of the constant *a* represent the maximum weighted mean temperatures (°C) that will be tolerable for continuous exposures of longer than 4 hours.

Limits for Industrial Work

In the discussion of limiting conditions for sustained work reference has been made to the conditions that can be tolerated by physically fit and highly acclimatized young men for 3 or 4 hours a day. Many industrial workers are no longer young, they do not customarily work dressed only in shorts, and they commonly work eight or more hours a day year after year. On this point observations by Fleisher, Stacey, Houghten and Ferderber (1939) are of interest, for these workers were studying industrial requirements.

From their observations Fleisher and his colleagues concluded that no worker should be expected to carry out his task in an environment so hot that his body temperature will rise by more than $1\frac{1}{2}°$ F during the period of his exposure. With $1\frac{1}{2}°$ F as the maximum rise of body temperature in a susceptible worker, the average temperature-rise in workers suited for work in hot industries will be about $\frac{1}{2}°$ F. The amount of the rise in body temperature depends on the heaviness of the work as well as on the warmth of the environment, and the following Table indicates the effective temperatures at which, according to Fleisher and his colleagues, a maximum rise of $1\frac{1}{2}°$ F in body temperature can be expected with different degrees of activity.

Effective temperatures which may cause a rise of 1.5° F in body temperature

Activity	*Metabolic rate of man of average size (Btu/hour)*	*Effective temperature °F*
Rest	385	92.8
Light bench work (*e.g.* metal-worker)	826	86.5
Heavier work (*e.g.* stone-mason)	1490	72.8

In the interests of general health, safety and efficiency it is essential that the temperature of the industrial environment should be kept well below that at which incapacitation or heat sickness may occur. Investigations carried out in industry over many years have shown that at temperatures distinctly below those that can be tolerated for occasional exposures the thermal environment affects the efficiency and well-being of the workers. The results of some of these investigations are summarized in the following pages.

EFFECTS OF HEAT ON WORKING EFFICIENCY

Observations on Miners

At the Morro Velho mine in Brazil the output was increased by 12 per cent following the introduction of air conditioning by means of which the wet-bulb temperature was reduced from 89° to 80° F (Davies, 1921–2).

In British coal mines observations were made on individual miners (Vernon, Bedford and Warner, 1927). In industrial investigations working efficiency is commonly measured in terms of output, but in coal mines the difficulty of hewing coal

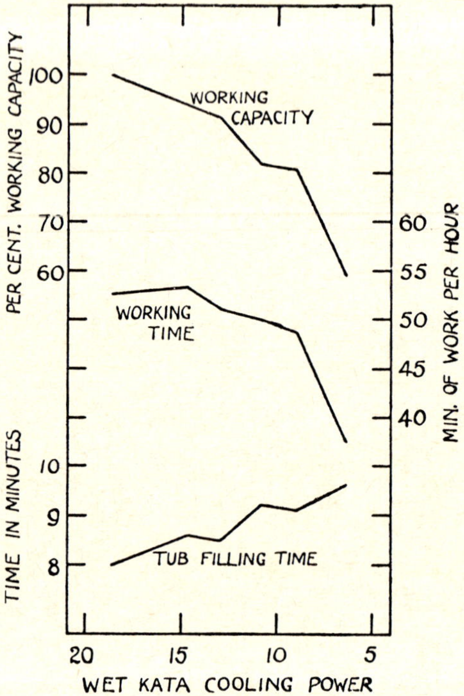

Fig. 118. Working capacity of coal miners in relation to wet kata cooling power.

varies from place to place, even in the same colliery, so that in this study output could not be taken as a measure of the amount of work done. Instead two other tests were used. As an index of the rate of working the time taken to fill a 10-cwt tub or tram with coal was measured; and the amount of time taken in short rests from work was also noted. From these data it was possible to calculate approximately the relative efficiency of men working in different places.

The dry-bulb temperature observed varied from 63° to 89° F

and the wet-bulb temperatures ranged from 59° to 82° F. The wet kata-thermometer cooling power, which makes some allowance for humidity as well as for temperature and air movement, varied from 6 to 19. In the hotter working places the time taken for resting, and the time required to load 10 cwt of coal, both increased, and it was calculated that the working efficiency of the men in the hottest places was only 59 per cent of that of the men working in the coolest places.

Moss and Halls (1935) later confirmed that miners working in hot places take a greater amount of time for resting than men working in cooler places.

Wyndham (1961) refers to a study of the effects of the hot and humid conditions in South African gold mines on the productivity of the African mine workers. Tests were carried out at wet-bulb temperatures ranging from 80° to 96° F, with high and low air speeds. At an air speed of 100 ft/min an increase in the wet-bulb temperature from 80° to 85° F resulted in a decline of 4 per cent in performance. At a wet-bulb temperature of 89° F the performance fell to 84 per cent of the 80° F value, at 92° F it fell to 62 per cent, and at 96° F it was only 20 per cent of the 80° F value. With the high air speed of 800 ft/min the decline in performance with rise in wet-bulb temperature was much less. At 90° F the performance was 94 per cent, at 92° F it was 88 per cent, and at 96° F it was 59 per cent of the 80° F value for that air speed.

Seasonal Variations of Output in Heavy Industries

Further evidence of the adverse effects of excessive warmth on industrial efficiency can be obtained from an examination of seasonal fluctuations of output. Vernon (1919; 1920) showed that in heavy work, especially when it involves exposure to high temperatures, output undergoes a consistent seasonal variation. It is generally greatest in winter, least in summer, and intermediate between the two in spring and autumn. Similar variations were noted by Farmer, Brooke and Chambers (1923) in glass manufacture. The variations in output noted by these observers are shown in the Table.

Seasonal changes of output in heavy work

Occupation	*Relative output during* Winter (December-March)	Spring and Autumn	Summer (June September)
Men charging blast furnaces	105.0	106.5	88.5
Open hearth steel melters at Works *C*	107.7	102.0	92.2
Open hearth steel melters at Works *D*	85.7	85.1	84.8
Open hearth steel melters at Works *F*	102.5	100.5	97.2
Open hearth steel melters at Works *J*	102.5	104.6	102.8
Rolling mill men at Works III	107.0	106.5	107.0
Rolling mill men at Works V	124.0	118.0	111.0
Rolling mill men at Works VI	105.2	98.4	96.5
Men puddling wrought iron	104.7	102.2	96.0
Tinplate mill men	103.3	99.9	96.9
Glass bottle making	101.7	99.6	98.7

Vernon (1919) showed the close relationship between the external temperature and the output from tinplate rolling mills. When the relative rate of output was plotted on the same diagram as the mean monthly temperature, the latter being shown on a reversed scale, the two curves were found to follow practically the same course (Fig. 119). The output in August was about 10 per cent less than that in December and January.

In factories devoted to hot industries, such as tinplate manufacture, ventilation is of much importance. Copious supplies of cool air reduce the temperature, and the beneficial effect of this lowered temperature is enhanced by the additional air movement. The effect of good ventilation was shown by Vernon in a comparison of the seasonal variations of output in different tinplate works (Fig. 120). In the best-ventilated factory the output in summer was only 3 per cent less than in winter, while in the worst-ventilated factory it was over 13 per cent less. Vernon estimated that in the worst-ventilated factory the output on the hottest days was probably 30 per cent less than in the coldest weather.

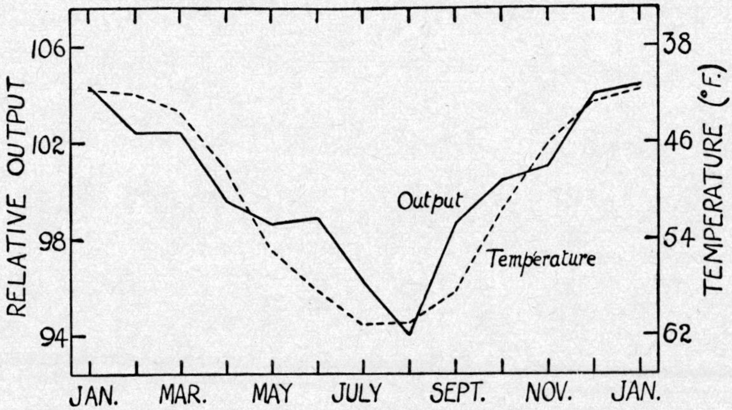

Fig. 119. Seasonal variation of output and of temperature (tinplate manufacture).

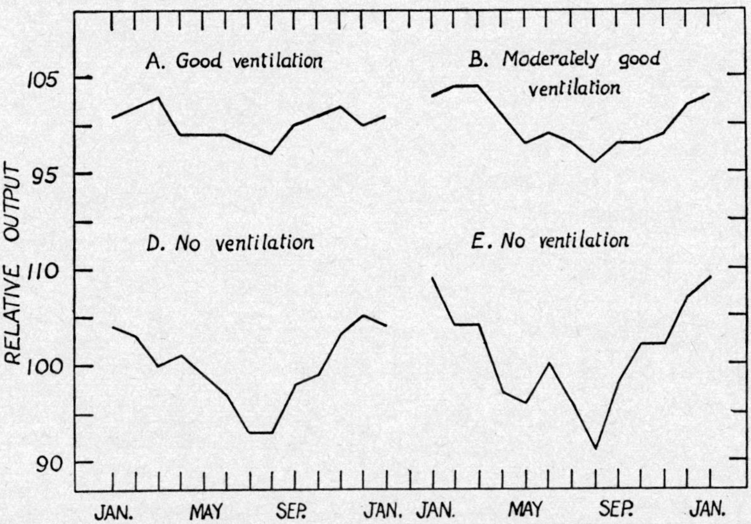

Fig. 120. Seasonal variations of output in individual tinplate factories.

Efficiency in Weaving

The materials used in some industries require special temperature conditions for their manufacture. In some instances it is possible to fulfil these requirements by local methods without much influencing the general room temperature, but in others the whole work-room is kept at the required temperature.

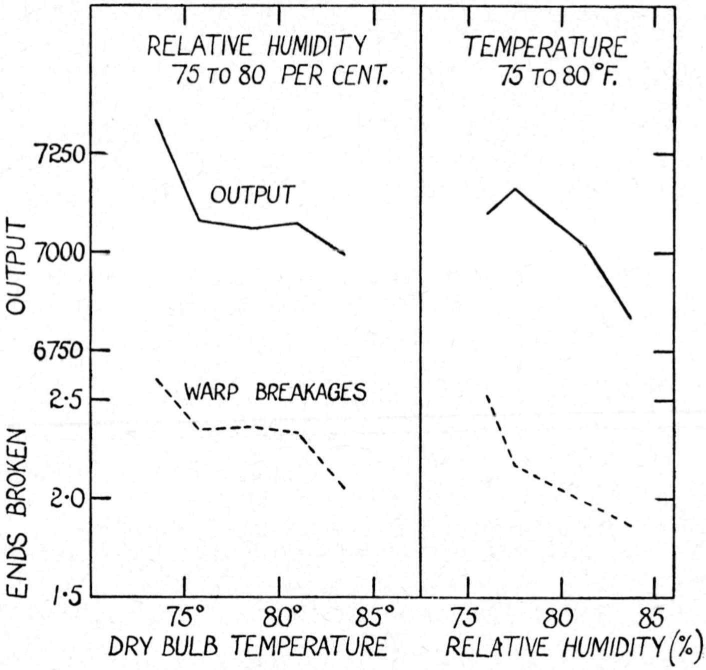

Fig. 121. Output and warp breakages in cotton weaving.

In the weaving of certain qualities of cotton and linen cloths it is the general custom to carry on the work in artificially humidified air at a high temperature. In hot and moist air the warp breakages are less frequent, and production is greater, than in cooler or drier atmospheres. This relation holds up to a certain point, but at still higher temperatures production falls. During an investigation of linen weaving Weston (1922) found that production began to fall off when the wet-bulb temperature exceeded about 73° F. At such temperatures the adverse physiological effects on the weavers more than counterbalanced the favourable physical conditions for weaving. A similar effect was noted by Wyatt and others (1926) in cotton weaving. With high humidity production fell off when the air temperature exceeded 75° F, even though the number of warp breakages, and consequently the number of stoppages, continued to decrease.

These observations were made on British textile workers

acclimatized to the conditions prevailing in Great Britain and Northern Ireland. Workers in factories in the tropics can be expected to maintain efficiency at wet-bulb temperatures rather higher than 73° F, but there can be little doubt that the conditions which exist in many Indian factories are inimical to efficiency. I visited many Indian textile factories (Bedford, 1946*b*). Early in March the temperatures in some spinning rooms reached 90° F, and I learned that the temperature had recently been 105° F when the external temperature was slightly over 100° F. I was told at some mills that a temperature of not less than 90° F was required in spinning rooms to prevent breakages. Opposed to that view is a recommendation of a Sub-Committee of the Government of India Post-War Planning Committee (Textiles). The Sub-Committee stated that it was generally recognized that the ideal temperature for mill interior was 75° to 80° F, but in view of the climate of India, and from considerations of economy, they recommended temperatures of between 80° F and 85° F as the target to be aimed at. It seems very likely that even with Indian workers there will be a decline in efficiency if the wet-bulb temperature is much higher than 80° F.

Efficiency in Mental Work

It is a matter of common experience that one's inclination for mental effort may diminish in very hot environments, but the effects of heat on the capacity for mental work have not formerly been demonstrated by experiment. That is probably due to the experimental arrangement employed in some of the earlier investigations, for in those studies different types of tests were used during any particular work spell, so that at intervals of 20 or 30 min there were changes of activity which introduced the powerful stimulus of variety. Many tasks are monotonous rather than variable, and to assess the influence of temperature on the performance of such tasks protracted experiments are necessary. Such experiments were carried out by Mackworth (1946) with the result that a diminished performance in hot environments was clearly shown.

Mackworth studied the performance of 11 physically fit wireless telegraphy operators. Before the main tests the men were acclimatized by exposure to effective temperatures of from 87.5° to 97° F for 3 hours a day, and 5 or 6 days a week, over a period of 7 to 11 weeks. On the test days work was done for 3 hours, during which spell 9 messages each of 250 groups were transmitted at a speed of 22 words/min.

The air velocity was 100 ft/min throughout the tests, and the dry-bulb temperature ranged between 85° and 105° F, with the wet-bulb temperature in each instance 10° F below the dry-bulb. These conditions represented effective temperatures of from 79° to 97° F. The men were dressed only in shorts and gymnasium shoes.

Hot and moist atmospheres seriously impaired the accuracy with which these trained men could record the morse messages heard over the telephone. The average numbers of errors/subject/hour are shown in the Table.

Average number of mistakes/hour

Condition	Dry bulb, °F	85	90	95	100	105
	Wet bulb, °F	75	80	85	90	95
	Effective temperature, °F	79	83	87.5	92	97
Average number of mistakes/subject/hour		12.0	11.5	15.3	17.3	94.7

The reduction in accuracy when the effective temperature reached 87.5° F was statistically significant.

The data given above are the averages for all subjects. It is noteworthy that the exceptionally competent members of the group were less affected by the difficult atmospheric conditions than men regarded as very good operators, while men of ordinary ability showed the greatest deterioration of all.

The exceptionally skilled men showed no deterioration until the effective temperature exceeded 92° F, while in the competent group the errors at 87.5° F effective temperature were one-third more numerous than at 83° F. This is an interesting point, for other studies of industrial fatigue have shown that

Average number of mistakes/hour for various groups

Efficiency category	*Average number of errors/hour at effective temperature (°F) of* 79	83	87.5	92	97
Exceptionally skilled	2.9	2.7	2.5	1.9	6.2
Very good	6.2	6.6	7.9	12.9	99.8
Competent	30.9	28.7	40.7	41.8	174.9

under difficult conditions the more competent workers show less signs of fatigue than their fellows.

There were no signs of cumulative effects from day to day but during a 3-hour spell the number of errors/hour rose progressively.

Average number of mistakes/hour during tests

Period of test	*Average number of errors/subject/hour, when effective temperature (°F) was* 79	83	87.5	92	97
1st hour	5.2	5.0	6.6	7.9	68.5
2nd hour	11.2	11.4	13.0	15.2	32.9
3rd hour	19.7	18.3	26.4	29.1	182.4

On the whole, errors were twice as numerous in the second hour as in the first, and three or four times as numerous in the third hour.

The rectal temperatures of these men responded to changes in environmental temperature, but there were persistent individual differences in these body temperatures. Partly because of these differences and partly on account of variations in individual ability in wireless reception, there was no significant correlation between rectal temperature and accuracy of reception at constant room temperature.

Bell (1967) has reviewed the extensive literature on the effects of hot environments and performance and notes that the most important characteristic of the data provided by studies in the laboratory is the extent to which the performances of people vary in 'normal' as well as abnormal conditions.

He points out that though decrements in performance have been demonstrated there are no references to the physiological basis for such decrements which may arise from discomfort, physiological changes associated with hyperthermia or both. Moreover he concludes that, whilst most workers in this field accept that productivity and control of the thermal environment are related, they have not yet provided a handbook or guide to hot environments and performance for the industrialist.

For further information on this subject, the reader is referred to the books written by Leithead and Lind (1964), and McK. Kerslake (1972).

CHAPTER XXII

EFFECTS OF ENVIRONMENTAL CONDITIONS ON ACCIDENT FREQUENCY AND SICKNESS

WE have seen that an excessively hot or cold environment, besides being uncomfortable, may have a serious effect on the capacity for doing work. But the effect of such an environment is not limited to the causation of discomfort and the hindrance of production. Adverse temperature conditions also have their effect on health and on the risk of accidents.

ACCIDENTS IN RELATION TO TEMPERATURE

Accidents in Munition Factories

The first recorded observations on the influence of environmental conditions on accident frequency appear to be those made by Vernon (1918) during the war of 1914–18. At a large munition factory the average temperature and the number of cuts and other minor accidents during each work spell was recorded. From these records it appeared that accidents were least frequent when the temperature was between 65° and 69° F. Further observations by Osborne and Vernon (1922) in other factories confirmed this conclusion (Fig. 122). As the temperature fell below 65° F, or rose above 69° F, accidents increased in frequency. Compared with the rate at temperatures of 65° to 69° F, the mean accident frequency for men and women was 23 per cent greater when the temperature rose above 75° F, and with temperatures below 55° F the accident frequency rose by 32 to 35 per cent.

Accidents to Miners

The introduction of a refrigeration plant at the Morro Velho gold mine in Brazil resulted in the cooling of the intake air from

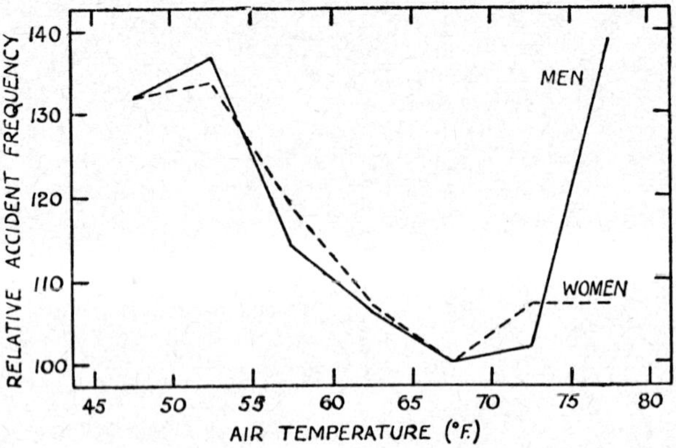

Fig. 122. Accident frequency in relation to shop temperature (munition workers).

about 67° down to 43° F, and the wet-bulb temperature in the deepest workings, some 6,000 ft below the surface, was reduced from 89° to 80° F. The reduced underground temperatures were accompanied by a decline in the number of fatal accidents. In the sixteen months previous to the installation of the cooling plant there were twenty fatal accidents, and in the sixteen months after its introduction only six.

From a preliminary study made in two British coal mines, Vernon, Bedford and Warner (1927) concluded that miners working in the hottest seams were more liable to accidents than were those exposed to less trying conditions. The investigation was extended, and records of all the accidents incurred by 18,000 underground workers employed at 10 collieries over a two-year period which entailed absences from work of three days, and for which compensation was paid, were examined. Later, the records for a further two years were analysed. A survey of the environmental conditions in each seam at each colliery was made, and it was thus possible to ascertain the average conditions to which the men in each group were exposed (Vernon, Bedford and Warner, 1928 and 1931).

When the accidents were classified according to the temperatures of the places in which they were incurred, it was found that the temperature of the working place had a distinct bearing on the liability to accident.

Men working at the coal face are about twice as liable to incur accidents as those working elsewhere underground. In the two periods 1924–5 and 1927–8 the accident frequency rates for coal face workers employed in seams where the temperature was 80° F or more were 77 per cent and 30 per cent higher than the corresponding rates for men working at temperatures below 70° F, while for the men working at temperatures of 70° to 79° F the rates were intermediate. Amongst the underground workers other than those employed at the coal face the effect of temperature was more pronounced. At temperatures of 70° to 74° F the accident frequency was 42 to 50 per cent more than at temperatures below 70° F, and at temperatures of 75° F or more the excess was 61 to 72 per cent.

Accident rates of coal miners in relation to dry-bulb temperature

Temperature range °F	*Mean temperature °F*	*Accident frequency/ million hours worked*		*Accident severity (days lost/1,000 hours worked)*		*Days lost/ accident*	
		1924–25	1927–28	1924–25	1927–28	1924–25	1927–28
Coal-face workers:							
Under 70°	63.9	104	133	3.5	4.4	33	33
70° to 79°	77.3	144	144	3.6	3.8	25	26
80° or more	82.8	184	173	4.5	4.8	24	28
Others underground:							
Under 70°	62.0	58	57	1.84	1.79	32	31
70° to 74°	72.7	87	81	2.41	2.45	28	30
75° or more	78.5	100	92	2.29	2.63	23	29

If instead of considering the number of accidents we measure the accident severity in terms of the number of days lost on account of accidents/1,000 hours worked, we find that on the whole the accident severity was greatest in the hottest pits. But the increase in accident severity at the higher temperatures was not as large in proportion as the increase in accident frequency, nor was it as consistent. On the average, men took less time off work after accidents that occurred in hot places than after those that were incurred in the cooler places.

The differences in the severity of accidents suffered at various temperatures were further examined by subdividing the accidents according as they caused absences of under 10 days, 10 to 19 days, or of periods longer than 19 days. In Fig. 123 the relative frequencies of accidents causing various lengths of absence are plotted against the temperature of the workings.

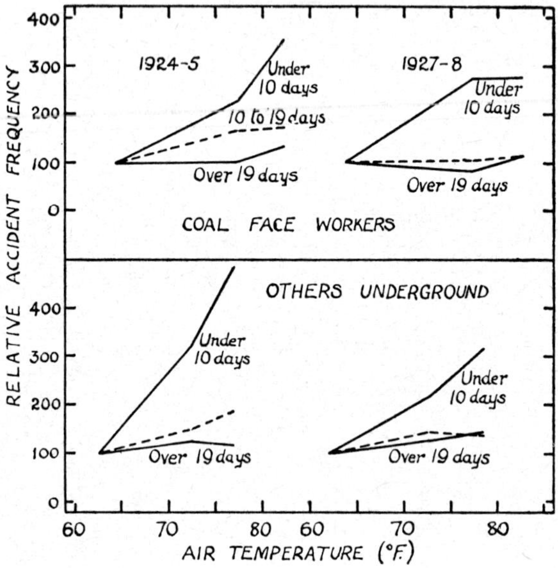

Fig. 123. Accident disability of coal miners in relation to temperature. (I.H.R.B. Rept. No. 62, H.M.S.O.)

For both coal-face workers and other underground workers, in both statistical periods, the frequency of accidents causing absences of 20 days or more was practically the same at all temperatures, but the frequency of minor accidents varied with temperature. Accidents causing absences of less than 10 days were three times as numerous at the higher temperatures as at the lower ones.

In considering these data it must be borne in mind that the records did not include all accidents, but only those for which compensation was paid. No compensation was payable for accidents followed by absences of three days or less, and such accidents were not recorded.

If we assume that the liability to accidents is measured by the frequency of the more serious accidents, it appears that over the range of temperature covered by these observations the accident risk did not vary a great deal. There seems to be no obvious reason why the ratio between the frequency of major and minor accidents should be influenced by temperature, and one might have expected the number of minor accidents also to be about the same at the different temperatures.

When a man suffers an accident sufficiently serious to warrant his absence from work for 20 days or more he has not much option about staying away. If instead he receives only a minor injury he may not stay away at all, or he may be away only for a day or two. The decision whether he stays away a few days longer and draws compensation rests largely with the miner, and this decision may be influenced by considerations other than the plain fact of recovery from injury. Economic factors and working conditions both influence this decision. Uncomfortable conditions of any kind can be a factor in determining absenteeism on account of sickness or of accidents, and it is reasonable to suppose that the apparent excess of minor accidents in the hotter mines was due, at any rate in part, to a greater tendency amongst the men working there to gain respite from unpleasant conditions by staying away from work when they incurred such accidents.

Age and Liability to Accidents

This study of coal miners showed that the temperature of the working places also affects the relation between age and liability to accidents. Various investigators, and notably Newbold (1926), have shown that there is a definite association between age and the frequency of industrial accidents. Young people are more liable to incur accidents than older persons. That is perhaps what one would expect, for the older workers are generally more experienced than the younger ones, and they are also probably more cautious. In her study of accidents in a number of industrial occupations, Newbold distinguished between the effects of experience and of age. She found that accidents certainly diminished with increased experience, but when allowance was made for the correlation between age and experience there was no independent association between experience and accident liability. It appeared from her results that age is a more important factor than experience.

In the study of coal miners it was found that the relationship between accident rate and age differed considerably from that observed in other workers, for, taking the figures as a whole, the frequency and severity of accidents were both at a minium in men aged between 30 and 40 years. The age effect can be seen in the records of both coal-face workers and other underground workers, but as the coal-face workers form a more homogeneous group it is best to confine an examination of age effects to this group.

Accidents amongst miners (coal-face workers) in relation to age

Age group	*Accident frequency*		*Average number of days lost/accident*	
	1924-5	1927-8	1924-5	1927-8
Under 20 years	288	309	21	23
20 to 29 years	258	271	22	23
30 to 39 years	230	232	23	27
40 to 49 years	272	244	28	29
50 or more years	296	349	31	38

The increased accident frequency amongst the older men was not due to a disproportionate number of minor accidents, for the proportion of accidents causing absences of less than 20 days decreased steadily with increasing age, while the proportion of more serious accidents increased. On the average an accident to men of 50 or over entailed about 40 per cent longer absence than that caused by an accident to a man of between 30 and 40.

That the relation between age and accident frequency is much influenced by the temperature of the working place is shown in Fig. 124. Men working at temperatures below 70° F had the lowest accident rates, and those at temperatures of 80° F or over had the highest. In the coolest mines the accident rates declined with increasing age, reaching a minimum frequency for the 40 to 49 age group, and then increased slightly in the older men. In the remaining two warmer groups of pits

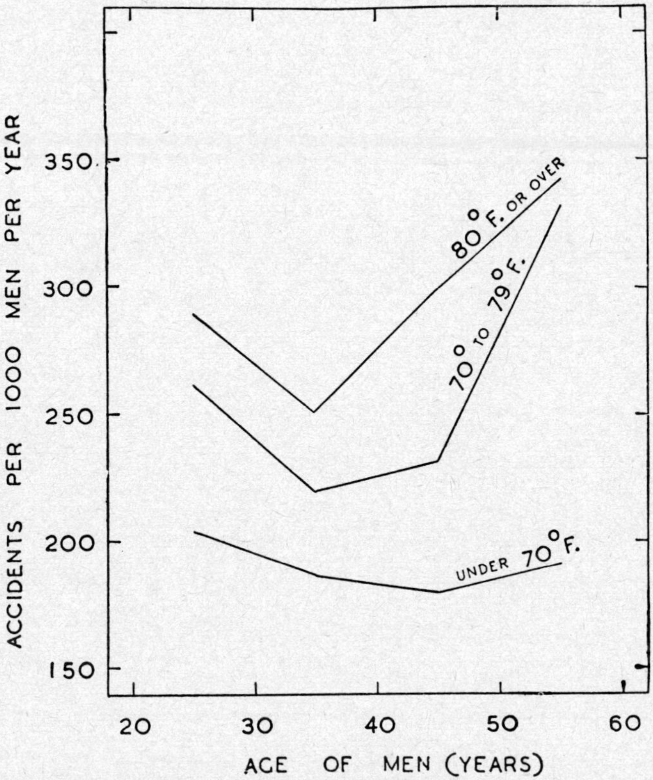

Fig. 124. Accident frequency rates of coal miners in relation to age and to temperature of working place. (M.R.C. War Memo. No. 17, H.M.S.O.)

the minimum frequency was reached at ages of 30 to 39; but whereas in the pits at 70 to 79° F the men aged 40 to 49 had only slightly more accidents than those aged 30 to 39, in the pits at 80° F or higher temperatures there was a sharp increase in accident frequency from the 30 to 39 group to the 40 to 49 group. Fatigue can be an important factor in accident causation, and it seems likely that the higher accident rates amongst the older miners were due to the severity of their work, for older men are more easily fatigued than younger ones. Hard work is more trying when done at high temperatures, and it can readily be imagined that the older men would be specially fatigued by work under such conditions.

SICKNESS IN RELATION TO VENTILATION

Sickness Amongst Cotton Operatives

It is generally accepted that overcrowding and bad ventilation are injurious to health, and the effects of lack of ventilation on industrial sickness have long been recognized.

One hundred and fifty years ago, Dr. Percival and his colleagues of the Manchester Board of Health found that those who worked in cotton factories were peculiarly disposed to be affected by fevers, and that even where no particular disease prevailed those employed in the factories had their constitutions undermined by the close confinement in hot and impure atmospheres. They found also that in well-regulated factories those evils could be in a considerable degree avoided; and they pressed for legislation for the regulation of factories.

In 1802 the Health and Morals of Apprentices Act provided that cotton mills where poor-law apprentices were employed should at all times be properly ventilated.

Since that time much attention has been given to the effects of impure atmospheres on health. Numerous diseases of occupation are known to be caused by breathing certain kinds of dust, and the dusty air of some departments of cotton mills has been the source of much respiratory disease amongst the operatives. Bradford Hill (1930) found that respiratory disease was far more prevalent amongst the workers in these dusty operations than amongst other cotton operatives. Figure 125 shows that amongst cotton strippers and cotton grinders there was about three times as much respiratory disease as amongst ring-room and warehouse workers.

As regards the effects of hot and humid atmospheres the

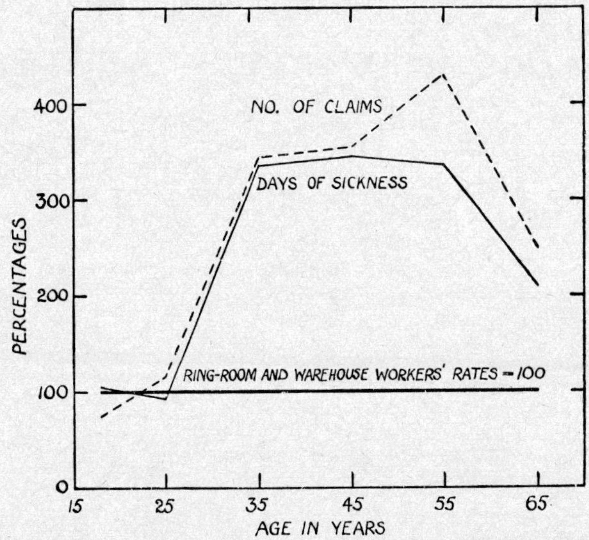

Fig. 125. Rates of sickness from respiratory causes for cotton strippers and grinders (males) as percentages of the rates for males in other departments.

position is less clear. Life in the open air makes for more vigorous health, and it is known that exposure to the open air increases the metabolism even in resting people.

For many years it was asserted that the hot and moist conditions which prevail in a large number of cotton sheds were the cause of ill-health, and 35 years ago a Home Office Committee reported that injury to health might be caused by continuous work for long periods in such atmospheres; but there was no statistical evidence of greater sickness in the hotter mills. Later, Bradford Hill (1927) made a statistical study of the effects of humid atmospheres on the sickness rates of cotton weavers. The sickness rates for weaving sheds where the air was artificially humidified were compared with the rates for sheds which were not humidified. No material difference could be found between the humid and the non-humid sheds, either in the amount or in the nature of the sickness suffered. When the sheds were grouped according to their 'characteristic' temperatures and relative humidities, no increase was shown in the amount of sickness as the temperature rose or as the humidity became greater; so that over the range of conditions met with in those sheds neither temperature nor humidity appeared to have influenced sickness.

Sickness in Iron and Steel Workers

Men who perform heavy work at high temperatures, and who are liable to be exposed to sudden changes of temperature, experience more sickness than those who work under more equable conditions.

Vernon and Rusher (1920) tabulated the sickness records of 20,000 iron and steel workers over a six-year period. The Table shows the sickness rates of various groups of men expressed in terms of the percentage variation from the average of all workers combined. The rates of sickness from all causes for the groups of men engaged in hot and heavy work were greater than those for men working at ordinary temperatures. The steel melters had a sickness rate 23 per cent greater than that for the whole body of workers, the rate for puddlers was 20 per cent greater, and the rates for tinplate millmen and rolling-millmen were 12 to 8 per cent greater.

Sickness of various groups of steel workers as percentage variation from the average of all workers combined (Vernon and Rusher)

	Percentage variation from average of all workers				
Occupation	*Rheumatism*	*Respiratory diseases*	*Injuries*	*Other causes*	*All causes*
Steel melters, teemers and pitmen	+44	+10	+71	+14	+23
Puddlers	+78	+35	–	−3	+20
Tinplate millmen	–	+5	+14	+21	+12
Rolling-millmen, soakermen, hot-bank men	–	–	+29	+10	+8
Engine-men, crane-men locomotive-men	+33	+10	−43	−3	−8
All other workers (largely labourers)	−11	−5	−29	−7	−9
Average number of days sickness/year, all workers	0.9	2.0	0.7	2.9	6.5

The puddlers and steel melters, who suffered from an excessive amount of rheumatism and respiratory disease, incur greater risk of chills, because their work necessitates occasional bouts of very heavy work followed by intervals of rest or lighter work. The tinplate millmen work almost continuously throughout their shift, only resting for brief intervals. These short rests would hardly be expected to allow time for chilling, and the millmen did not lose any more time from rheumatism than was lost on the average by all workers.

Vernon and Rusher also compared the mortality of steel workers with that of all occupied and retired civilian males. The mortality of the steel workers from all causes was 5 per cent below that of all occupied and retired civilian males, but their mortality from respiratory disease was twice as great.

Yaglou (1937*a*) quotes the experience of the Metropolitan Life Insurance Company. The mortality figures for white male industrial policy holders, of ages from 16 to 64 years, engaged in various hot industries are compared with the figures for all occupations. Iron foundry workers had a mortality rate from pneumonia 120 per cent greater than that for all occupations, that for iron and steel mill workers was 42 per cent greater, while the mortality of coal miners showed an excess of 50 per cent.

Sickness Amongst Miners

Further evidence of the association of high sickness rates with high temperatures was obtained by Vernon, Bedford and

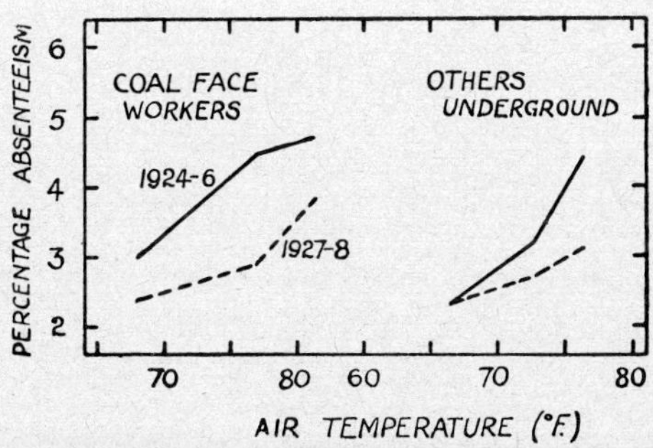

Fig. 126. Sickness of coal miners in relation to temperature.

Warner (1928, 1931) in their study of the absenteeism of coal miners. Records of absenteeism were collected from a number of collieries during two statistical periods, and in both periods there was a distinct relation between temperature and the amount of time lost on account of sickness.

Both groups of workers considered, the coal-face workers and other underground workers, experienced more sickness in the hotter pits than in the cooler ones.

Sickness in Other Industries

Some observations made by Vernon, Bedford and Warner (1926*a*) suggest that relatively small differences in atmospheric conditions may exert an influence on the amount of sickness experienced by factory workers.

At a canister factory where 800 women and girls were employed a large number of temperature observations were made at different seasons of the year, and records of the time lost on account of sickness over a two-year period were obtained. These data are summarized in the Table.

In workrooms (*a*) to (*d*) the sickness rates over the two periods were very similar and averaged 1.55 per cent of possible working time. In workroom (*e*) the plenum heating installation was inadequate, with the result that the temperature in winter averaged only about 54° F. From the columns to the right of the table it will be noticed that the sickness rate in room (*e*) was higher than that for the whole factory throughout the year, but the excess was greatest in the coldest months, smaller in spring and autumn, and only very slight in the summer. Workroom (*f*) consistently had a temperature some degrees higher than those found in the remaining rooms, so that in the summer months, when the average temperature over the whole factory was barely 70° F, that in workroom (*f*) was 78° F. Throughout the year the sickness in this room was above the average for the factory, but in cold weather when its temperature, though higher than the average, was only 67° F, the sickness rate was only 2.00 per cent as compared with 1.90 for the whole factory. In room (*c*) the winter temperature was identical with

Time lost by sickness at a canister factory

Workroom	No. of women employed	Average age, in years	Mean workroom temperature, °F		Time lost by sickness (% of possible time) over 2-year period	Differences of time lost from mean for all 6 workrooms during		
			Winter	Summer		Dec. to March	April, May, Oct. and Nov.	June to Sept.
a	149	20.7	59.8	66.2	1.47	−0.24	−0.35	−0.06
b	147	19.7	63.0	67.2	1.51	−0.15	−0.30	−0.10
c	170	21.4	61.3	71.7	1.60	−0.18	−0.31	+0.21
d	102	19.7	62.5	68.7	1.64	+0.14	+0.12	−0.42
e	94	21.4	54.3	65.9	1.88	+0.32	+0.14	+0.09
f	147	18.8	67.1	77.9	2.05	+0.10	+0.70	+0.29
Mean	(809)	20.3	61.3	69.6	1.69	(1.90)	(1.64)	(1.54)

the average for the whole factory, and with that for the most comfortably warmed rooms, but in summer this room was warmer than any of the other workrooms with the exception of room (*f*). It is interesting to note, therefore, that while in winter and in the intermediate seasons the sickness rate in room (*c*) was below the average for the factory, it was above the average in summer.

The women concerned in this study were drawn from the same social class, lived under similar conditions, and were of approximately the same age. When they were recruited their allocation to one or another of the workrooms depended on the staff requirements at the time, and it can be assumed that they were essentially drawn from the same population. While it cannot be asserted that the higher sickness rates discussed above were caused by differences in warmth conditions, the figures seem to suggest that relatively small differences in temperature may have an effect on health.

Further evidence obtained by Vernon, Bedford and Warner at another factory seemed to point to ventilation as the chief factor responsible for some differences in sickness rates. In two of the rooms (*a*) and (*b*) girls were employed on the same task of tying up small packages. Over a two-year period the average time lost in room (*a*) was 2.44 per cent of possible time, while in room (*b*) it was 3.73 per cent, or 53 per cent more. The temperatures in the two rooms were almost identical, but the air movement was distinctly less in room (*b*) than in room (*a*). The average air velocity in room (*a*) was 34 ft/min in winter and 40 ft/min in summer, while the corresponding velocities for room (*b*) were only 15 and 27 ft/min. There was no moving machinery in these rooms, and in each of them the air velocity was created by the plenum ventilation plant and open windows. The lower velocity in room (*b*) was an indication of poor ventilation.

The poor ventilation of room (*b*) was not due to lack of windows, for in relation to the size of the room the window area available for ventilation was three times that of room (*a*). The windows were not utilized as fully as they might have been, for when the outside temperature was 65° F only 27 per cent of

Sickness and ventilation at factory D

		Workroom (a)	Workroom (b)
No. of workers		81	54
Floor area (sq. ft)		5,940	2,950
Window area (sq. ft), available for ventilation, per 100 sq. ft of floor		3.5	10.3
Per cent of window area open at outside temperature of	35°	11	10
	45°	21	11
	55°	46	26
	65°	61	27
Plenum air supply, air changes/hour		4	1
Mean air velocity, (ft/min)	winter	34	15
	summer	40	27
Time lost through sickness over a 2-year period, as percentage of possible working time		2.44	3.73

the available window area was open. Further, the room was shut in by other buildings, and the windows were thus screened from the effects of wind, whereas room (*a*) was in an exposed block. Again, the plenum plant only supplied 1 air change/hour to room (*b*), while room (*a*) had 4 air changes/hour.

Sickness Amongst School Children

In a report issued by the New York State Commission on Ventilation in 1923, the conclusion was reached that the most suitable temperature for schools in New York was 68° F, and that variations of 2° or 3° F from that temperature had an appreciable effect on the amount of sickness. Later studies undertaken by that Commission (1931) failed to confirm this finding, however, for from the later work it appeared that such slight differences in room temperature exerted no measurable influence on the incidence of respiratory diseases. Those diseases were more closely related to outside weather conditions than to conditions within the school room.

Vernon, Bedford and Warner (1930) studied the absenteeism records for about 5,000 children in 20 elementary schools in

Derbyshire over a period of three years, 1926 to 1929. At the time it was not practicable to measure the air turnover in the school rooms, but an attempt was made to grade the rooms according to the facilities for cross-ventilation. Rooms were marked 1.0 if they had windows on one side only, 2.0 if there were windows on opposite sides affording a large opening area, and between 1.0 and 2.0 if windows were on adjacent sides. Some of the rooms had windows on three sides and they were graded 2.8. The average absenteeism during the two years 1927 to 1929 in the various groups of schools is shown in the Table.

Ventilation and absenteeism of school children

Average grading of ventilation	*No. of schools*	*No. of children*	*Average age of children (years)*	*Per cent absenteeism in 1927-9*
2.2	6	1,560	11.7	7.7
1.5	8	1,934	11.2	9.2
1.2	6	1,466	10.9	10.7
Semi-open-air	6	1,352	11.6	8.9

The Table shows that schools with an average ventilation grading of 2.2 had an absenteeism rate of 7.7 per cent, while in those graded 1.5 the rate was 9.2 per cent, and in the group with the poor ventilation it was 10.7 per cent. The semi-open-air schools which were more freely ventilated than the ordinary schools had an average absenteeism rate of 8.9 per cent, a figure nearly equal to the average for the other schools but higher than that for the schools with a ventilation grading of 2.2. This relatively high figure for the semi-open-air schools was apparently due not to lack of ventilation, but to low indoor temperatures in cold weather. Over all the schools included in the investigation, there was a clear relationship between absenteeism and the temperature in the class rooms. When the indoor temperatures were low absenteeism increased, and in the semi-open-air schools the winter temperatures averaged 2° to 3° F lower than those for ordinary schools.

Sickness in the Royal Navy

Further evidence of the influence of climate on health is given by the results of a prolonged investigation carried out in the Royal Navy. F. E. Smith (1958) examined monthly returns of sickness in H.M. Ships, large and small, and stationed in different parts of the world. The returns gave the numbers of men on the attending list, *i.e.* the numbers who were attending the sick bay for treatment of minor ailments; and they also reported measurements of temperature made on the upper deck at noon. When the external temperature at noon was about 50° F some $3\frac{1}{2}$ per cent of the men were on the attending list, and the proportion increased as the temperature rose above that level. At 80° F the proportion on the attending list was about 5 per cent, and at temperatures above 90° F it approached 9 per cent. In temperate waters about 30 per cent of those attending did so on account of skin diseases, while in very hot climates skin diseases accounted for nearly half. At temperatures below 50° F about 23 per cent of those attending suffered from upper respiratory infections, while in hot climates only about 7 per cent of those attending did so for that cause.

It is estimated that when the external air temperature is about 80° F the effective temperature on the mess decks will reach 80° F, which is the figure earlier recommended as being the effective temperature that should not be exceeded in H.M. Ships. The rapid increase in minor sickness at higher temperature supports that recommendation. Smith remarks that on most foreign stations climatic temperatures so often exceed 80° F that ventilation alone cannot be relied upon to keep the thermal conditions in ships down to recommended standards, and that this can only be done by the use of air conditioning.

Air Conditioning and Sickness

In 1932 the Metropolitan Life Insurance Company of New York opened a new office building. This huge structure, 28 storeys in height, and with a total floor area of 23 acres, is air conditioned. Alongside the new building is the company's older office which is not air conditioned. Here the ventilation depends on the opening of windows, and there is no artificial humidification in the heating season, nor is there cooling in summer. The clerical staffs in the two buildings are comparable in sex- and age-distribution, are of similar economic status, and are under the same supervision.

When the new building was opened it was decided to check whether the sickness experienced in the old and the new buildings reflected any beneficial influence of air conditioning (McConnell and others, 1937). In the new building temperatures ranged between 68° and 75° F, and relative humidities were between 30 and 40 per cent during the heating season. During summer the temperature indoors varied according to the outdoor temperature, and a maximum temperature of 79° F was maintained when the outdoor temperature was above 90° F, and the range of relative humidity during summer was 43 to 53 per cent.

There was no doubt that the air-conditioned building was distinctly the more comfortable, and it was thought that the controlled temperatures probably increased efficiency. Yet for neither respiratory diseases nor for non-respiratory illness was there any significant difference between the incidence of absence or the amount of time lost from sickness in the two buildings.

Another example of the effects of air conditioning on sickness is provided by experience in the Royal Navy (Smith, 1958). The living spaces in two sloops were air conditioned. Sickness returns were obtained from these ships before the air conditioning was installed, and afterwards. The effect of the air conditioning was striking. The proportion of the ships' companies attending the sick bay for minor illness declined from 11 per cent to about 7 per cent, while the proportion attending on account of skin diseases was reduced by almost one-half.

BIBLIOGRAPHY

Adam, J. M. and Cague, W. S. (1952): Metabolic cost of resting and step-climbing in Singapore. *Med. Res. Coun., Lond., R. N. P. Rep.* No. 52/731.

Aitken, J. (1884): Thermometer screens. *Proc. Roy. Soc. Edinburgh,* **12**, 661.

Aitken, J. (1887): Addition to thermometer screens. *Ibid.,* **14**, 428.

Aldrich, L. B. (1928): A study of body radiation. *Smithsonian Misc. Coll.,* **81**, Publ. No. 2980.

Allen, J. R., and Walker, J. H. (1931): *Heating and ventilation.* New York: McGraw-Hill Book Co.

Ambler, H. R. (1955): Notes on climate of Nigeria with reference to personnel. *J. Trop. Med.,* **58**, 99.

American Congress of Governmental Industrial Hygienists (1961): Threshold limit values for 1961. *Arch. Environmental Hlth.,* **3**, 489.

American Society of Heating and Ventilating Engineers (1942): Measurement of the physical properties of the thermal environment: Report of the A.S.H.V.E. Research Technical Advisory Committee on Instruments. *Heat. Pip. and Air Condit.,* **14**, 382.

American Society of Heating and Ventilating Engineers (1945): *Heating, ventilating, air-conditioning guide.* New York: Amer. Soc. Heat. Vent. Engrs.

Andrew, L. W. and Wills, D. R. (1961): Heat services by gas in houses and flats in Great Britain. *8th Internat. Gas Conference, Stockholm.*

Andrewes, C. H. (1940): Control of air-borne infection in air-raid shelters and elsewhere. *The Lancet,* II, p. 770, Dec. 21.

Angus, T. C. (1923): The ventilation of English factories and workshops in hot weather. *J. Industr. Hyg.,* **4**, 479.

Angus, T. C. (1936): The kata-thermometer and its uses. *J. Instn. Heat. Vent. Engrs., Lond.,* **4**, 50.

Angus, T. C. (1938): Air jet ventilation in hot weather. *Heat. and Vent. Engr.,* **12**, 2.

Arbuthnot, J. (1733): *An essay concerning the effects of air on human bodies.* London: Tonson and Draper.

Artenstein, M. S. and Millet, W. S. (1966): Air sampling for respiratory disease agents in army recruits. *Bacteriol. Rev.,* **30**, 571-572.

Baetjer, A. M. (1924): Threshold air currents in ventilation. *Amer. J. Hyg.,* **4**, 650.

Bailey, G. B. (1962): The valance system and its development. *J. Instn. Heat. Vent. Engrs., Lond.,* **29**, 406.

Bailey, N. P. (1932): The measurement of surface temperatures. *Mech. Engr.,* **54**, 553.

Baines, F. (1931): Memorandum regarding the ventilation of the Chamber of the House of Commons. (Cmd. 3871). London: H.M. Stationery Office.

Barker, A. H. (1931): The principles of calculation of low-temperature radiant heating. *Proc. Instn. Heat. Vent. Engrs., Lond.*, **30**, 212.

Barracks and Hospitals Commission (1861): General Report of the Commission appointed for improving the sanitary condition of Barracks and Hospitals.

Bazett, H. C. (1949): The regulation of body temperatures. In *Physiology of heat regulation and the science of clothing* (L. H. Newburgh, Ed.), Philadelphia: W. B. Saunders Co.

Bazett, H. C. and McGlone, B. (1927): Temperature gradients in the tissues in man. *Amer. J. Physiol.*, **82**, 415.

Bazett, H. C. and McGlone, B. (1932*a*): Studies in sensation. II. The mode of stimulation of cutaneous sensations of cold and warmth. *Arch. Neur. and Psychiatr.*, **27**, 1031.

Bazett, H. C. and McGlone, B. (1932*b*): Studies in sensation. III. The chemical factor in the stimulation of end-organ giving temperature sensations. *Ibid.*, **28**, 71.

Bazett, H. C., McGlone, B., Williams, R. G. and Lufkin, H. M. (1932): Studies in sensation. I. Depth, distribution and probable identification in the prepuce of sensory end-organs concerned in sensations of temperature and touch; thermometric conductivity. *Ibid.*, **27**, 489.

Bedford, T. (1935): The effective radiating surface of the human body. *J. Hyg., Camb.*, **35**, 303.

Bedford, T. (1936): The warmth factor in comfort at work. *Rep. Industr. Hlth. Res. Bd., Lond.*, No. 76.

Bedford, T. (1946*a*): Environmental warmth and its measurement. *Med. Res. Coun., Lond., War Memo.* No. 17.

Bedford, T. (1946*b*): The health of the industrial worker in India. *Rept. to Government of India* (*Dept. of Health*).

Bedford, T. and Warner, C. G. (1931): The reduction of mine air temperatures. *J. Industr. Hyg.*, **13**, 135.

Bedford, T. and Warner, C. G. (1933): The influence of radiant heat and air movement on the cooling of the kata-thermometer. *J. Hyg., Camb.*, **33**, 330.

Bedford, T. and Warner, C. G. (1934*a*): On methods of measuring skin temperature. *Ibid.*, **34**, 81.

Bedford, T. and Warner, C. G. (1934*b*): The globe thermometer in studies of heating and ventilation. *Ibid.*, **34**, 458.

Bedford, T. and Warner, C. G. (1939): Subjective impressions of freshness in relation to environmental conditions. *Ibid.*, **39**, 498.

Bedford, T., Warner, C. G. and Chrenko, F. A. (1943): Observations on the natural ventilation of dwellings. *J. Roy. Inst. Brit. Arch.*, **51**, 3rd series, 7.

Belding, H. S. and Hatch, T. F. (1955): Index for evaluating heat stress in terms of resulting physiological strains. *Heat., Pip. & Air Condit.*, **27**, No. 8, 129.

Belding, H. S., Hatch, T. F., Hertig, B. A. and Riedesel, M. L. (1960). Laboratory simulation of a hot industrial job to find effective heat stress and resulting physiologic strain. *Amer. Indust. Hyg. Assoc. J.*, **21**, 25.

Bell, C. R. (1967): Hot environments and performance, in The effects of abnormal physical conditions at work, pp. 60-73. E. and S. Livingstone Ltd., Edinburgh and London.

Bell, C. R., Hellon, R. F., Hiorns, R. W. and Provins, K. A. (1962): Tolerance to extreme heat at two metabolic rates. *Med. Res. Coun., Lond., W. V. P. Rep.* No. 82.

Benedict, F. G. and Root, H. F. (1926): Insensible perspiration: its relation to human physiology and pathology. *Arch. Int. Med.*, **38**, 1.

Bernan, W. (1845): *On the history and art of warming and ventilating rooms and buildings, etc.* London: G. Bell.

Berry, E. W. and Taylor, J. (1960): The design and performance of convector gas fires. *London & Southern Jun. Gas. Assoc.*

Billings, J. S. (1923): Cit. from New York State Commission on Ventilation. p. 209.

Billings, J. S., Mitchell, S. W. and Bergey, D. H. (1895): The composition of expired air and its effects upon animal life. *Smithsonian Contr. to Knowledge*, 989, Washington.

Blowers, R. and Crew, B. (1960): Ventilation of operating theatres. *J. Hyg., Camb.*, **58**, 427.

Blowers, R., Lidwell, O. M. and Williams, R. E. O. (1962): Infection in operating theatres in relation to air conditioning equipment. *J. Instn. Heat. Vent. Engrs., Lond.*, **30**, 244.

Bohnenkamp, H. and Ernst, H. W. (1931): Untersuchungen zu den Grundlagen des Energie- und Stoffwechsels. I. Über die Strahlungsverluste des Menschen. Das energetische Oberflächengesetz. *Pflüg. Arch.*, **228**, 40.

Bohnenkamp, H. and Pasquay, W. (1931): III. Ein neuer Weg zur Bestimmung der für die Wärmestrahlung massgebenden Oberfläche des Menschen. Die 'mittlere Strahlungstemperatur' des Menschen und seiner Kleideroberfläche. *Ibid.*, **228**, 79.

Bourdillon, R. B., Lidwell, O. M. and van den Ende, M. (1948): Air disinfection by ultraviolet radiation. *Med. Res. Coun., Lond., Spec. Rep. Ser.*, No. 262, Pt. III.

Bourdillon, R. B., Lidwell, O. M. and Thomas, J. C. (1941): A slit sampler for collecting and counting air-borne bacteria. *J. Hyg., Camb.*, **41**, 197.

Bowditch, M., Drinker, C. K., Drinker, P., Haggard, H. H. and Hamilton, A. (1940): Code for safe concentrations of certain common toxic substances used in industry. *J. Industr. Hyg. & Toxicol.*, **22**, 251.

Brandt, A. D. (1933): The influence of atmospheric ionization upon the human organism. *J. Industr. Hyg.*, **15**, 354.

British Standards Institution (1949): Heating and thermal insulation of buildings. *Code of Practice*, 3.

Brown-Séquard, C. E. and D'Arsonval, A. (1887): Démonstration de la puissance toxique des exhalaisons pulmonaires provenant de l'homme et du chien. *Compt. Rend. Soc. de Biol.*, **39**, 814.

Brown-Séquard, C. E. and D'Arsonval, A. (1888*a*): Nouvelles recherches démontrant que les poumons sécrètent un poison extrêmement violent qui en sorte avec l'air expiré. *Ibid.*, **40**, 33.

Brown-Séquard, C. E. and D'Arsonval, A. (1888*b*): Nouvelles remarques à l'égard du poison pulmonaire. *Ibid.*, **40**, 54.

Brown-Séquard, C. E. and D'Arsonval, A. (1888*c*): Remarques sur la valeur des faits qui nous ont servi à démontrer la toxicité de l'air expiré. *Ibid.*, **40**, 99.

Brown-Séquard, C. E. and D'Arsonval, A. (1888*d*): Recherches démontrant que l'air expiré par l'homme et les mammifères, à l'état de santé, contient un agent toxique très puissant. *Comptes rend. Acad. Sci.*, **106**, 106.

Brown-Séquard, C. E. and D'Arsonval, A. (1889): Nouvelles recherches démontrant que la toxicité de l'air expiré ne dépend pas de l'acide carbonique. *Ibid.*, **108**, 267.

Bruce, H. H. (1936): Panel warming calculations: the thermometer and comfort. *J. Instn. Heat. Vent. Engrs., Lond.*, **4**, 427.

Buchbinder, L., Solowey, M. and Solotorovsky, M. (1945): Comparative quantitative studies of bacteria in air of enclosed places. *Heat. Pip. & Air Condit.*, **17**, 389.

Building Research Board (1931): Report for the year 1930.

Building Research Board (1932): Report for the year 1931.

Building Research Board (1934): Report for the year 1933.

Building Research Board (1935): Report for the year 1934.

Building Research Board (1938): Report for the year 1937.

Building Research Board (1939*a*): Report for the year 1938.

Building Research Board (1939*b*): A study of the ventilation of a warmed room, carried out at the Building Research Station. *J. Instn. Heat. Vent. Engrs., Lond.*, **7**, 328.

Burfoot, A. H. (1966): Experience in newly constructed hospitals symposium, hospital engineering services–present and future, conducted by the Instn. Heat. Vent. Engrs. Lond. 12 January 1966.

Burton, A. C. (1935): Human calorimetry. II. The average temperature of the tissues of the body. *J. Nutrition*, **9**, 261.

Burton, A. C. and Edholm, O. G. (1955): Man in a cold environment. Edward Arnold (Publishers) Ltd. London.

Büttner, K. (1932): Physikalisches zum Warmehaushalt des Menschen insbesondere über die Wärmeabgabe durch Leitung. *Klin. Wochenschr.*, 2, 1508.

Calver, G. W. (1930): Ventilation problem of the Halls of the Senate and House of Representatives. *U.S. Nav. Med. Bull.*, **28**, 177.

Campbell, J. F. (1857): General Board of Health (1857), Appendix.

Caplan, A. (1943): A critical analysis of collapse in underground workers on the Kolar Gold Field. *Bull. Instn. Min. & Metall.*, No. 463.

Carpenter, T. M. (1937): The constancy of the atmosphere with respect to carbon dioxide and oxygen content. *J. Amer. Chem. Soc.*, **59**, 358.

Carson, R. (1965): Silent spring. Penguin Books in association with Hamish Hamilton.

Chamberlain, A. C. (1967): Deposition of particles to natural surfaces, pp. 138-164 in Airborne Microbes, the 17th Symposium of the Society for General Microbiology, edited by P. M. Gregory and J. L. Monteith. Cambridge University Press.

Chance, P. R. and Dance, E. W. G. (1961): Laboratory and field experience in warm air heating. *J. Instn. Heat. Vent. Engrs., Lond.*, **29**, 288.

De Chaumont, F. S. B. F. (1875): *Lectures on State Medicine.* London: Smith, Elder and Co.

Chrenko, F. A. (1953*a*): Heated ceilings and comfort. *J. Instn. Heat. Vent. Engrs., Lond.*, **20**, 375 and **21**, 145.

Chrenko, F. A. (1953*b*): Probit analysis of subjective reactions to thermal stimuli–a study of radiant panel heating in buildings. *Br. J. Psychol.*, (General Section) **XLIV**, 248-256.

Chrenko, F. A. (1955): The assessment of subjective reactions in heating and ventilation research. *J. Instn Heat. Vent. Engrs.*, **23**, 281.

Chrenko, F. A. (1956): Heated floors and comfort. *Ibid.*, **23**, 385.

Chrenko, F. A. (1957): The effects of the temperatures of the floor surface and of the air on thermal sensations and the skin temperature of the feet. *Brit. J. Indust. Med.*, **14**, 13.

Chrenko, F. A. (1962); Human factors in heating (part 2). Design, **166**, pp. 50-53.

Chrenko, F. A. (1964): Threshold intensities of thermal radiation evoking sensations of warmth. *J. Physiol.*, **73**, pp. 1-12.

Chrenko, F. A. (1967): Heating and ventilation for a human environment. *Proc. Instn. Mech. Engrs.*, **182**, part 3E, pp. 1-7.

Clifford, P. and Beecher, J. L. (1964): Nasopharyngeal cancer in Kenya. *Brit. J. Cancer.*, **18**, 25.

Colburn, A. P. and Hougen, O. A. (1930): Studies in heat transmission. I. Measurements of fluid and surface temperatures. *Ind. & Eng. Chem.*, **22**, 522.

Cole, R., Kimball, D. D., Lee, F. S., Palmer, G. T., Phelps, E. B., Thorndike, E. L. and Winslow, C.-E. A. (1931): A study of rural school ventilation in Cattaraugus County, N.Y. *Amer. J. Hyg.*, **14**, 49.

Collier, H. E. (1940): *Outlines of industrial medical practice.* London: Edward Arnold and Co.

Collins, B. G. and Smith, D. B. (1955): The measurement of ventilation-rate using a radioactive tracer. *J. Instn. Heat. Vent. Engrs., Lond.*, **23**, 270.

Copping, G. L. (1944): Unit heaters. *J. Instn. Heat. Vent. Engrs., Lond.*, **12**, 206.

Crowden, G. P. (1934*a*): Metallic insulation. *Ibid.*, **2**, 422.

Crowden, G. P. (1934*b*): Insulation against heat and cold for human comfort: the uses of bright metallic surfaces. *The Lancet,* **I**, 37.

Crowden, G. P., Hetherington, M. and Luxton, W. R. (1933): A comparative study of the physiological effects on children of school heating by hot-water radiators and by radiant heat from electric ceiling and wall panels. *J. Hyg., Camb.*, **33**, 151.

Cruikshank, R. and Muir, C. (1940): Air-borne streptococcal infection following influenza. *The Lancet,* **II**, 1155.

DallaValle, J. M. and Hollaender, A. (1940): A study of the role of ventilating systems in the transmission of bacteria. *Publ. Hlth. Repts., Washington,* **55**, 1268.

Davy, J. (1814): An account of some experiments on arterial heat. *Phil. Trans. Roy. Soc., Lond.,* **104**, 590.
Daynes, H. A. (1920): The theory of the katharometer. *Proc. Roy. Soc. Lond. A,* **97**, 273.
Deighton, T. (1933): Physical factors in body temperature maintenance and heat elimination. *Physiol. Reviews,* **13**, 427.
Departments of the Army, Navy and Air Force, Washington (1957): TB MED 175, NAVMED P-5052-5, AFP 160-4-1A.
Dessauer, F. (1931): *Zehn Jahre Forschung auf dem Physikalische-Medizinischen Grenzegebeit.* Leipzig: Georg Thieme.
Dick, J. B. (1949): Experimental studies in natural ventilation of house. *J. Instn. Heat. Vent. Engrs., Lond.,* **17**, 420.
Dimitroff, J. M., Kuppenheim, H. F., Graham, I. C. and McKeehan, C. W. (1956): Spectral reflectance of the skin of rats, rabbits and hairless mice in the regions 243-700 mμ and 0.707-2.660 μ. *J. Appl. Physiol.,* **8**, 532.
Douglas, C. G. (1938): Memorandum, in Building Research Board (1938): Appendix I.
Dreosti, A. O. (1935): The results of some investigations into the medical aspect of deep mining on the Witwatersrand. *J. Chem. Metall. Min. Soc., So. Africa,* **36**, 102.
Drysdale, J. W. (1950): Climate and the design of buildings. *Commonwealth Exptl. Bldg. Stn., Dupl. Document* No. 32.
Dubois, D. and Dubois, E. F. (1916): Clinical calorimetry: X. A formula to estimate approximate surface area if height and weight be known. *Arch. Int. Med.,* **17**, 863.
Dubois, E. F. (1937): *Lane Medical Lectures: The mechanism of heat loss and temperature regulation.* Stanford University Press.
Dufton, A. F. (1927): The warming of walls. *Phil. Mag.,* 7th ser., **4**, 888.
Dufton, A. F. (1929): The eupatheostat. *J. Sci. Instr.,* **6**, 249.
Dufton, A. F. (1931*a*): The warming of a room. *Phil. Mag.,* 7th ser., **11**, 1233.
Dufton, A. F. (1931*b*): Radiant heat. *Rep. Instn. Vent. Engrs., Lond.,* **31**, 230.
Dufton, A. F. (1931*c*): A coke fire. *Gas J.,* **196**, 434.
Dufton, A. F. (1932): The equivalent temperature of a room and its measurement. *Bldg. Res. Tech. Pap.,* No. 13.
Dufton, A. F. (1933*a*): The use of kata-thermometers for the measurement of equivalent temperature. *J. Hyg., Camb.,* **33**, 349.
Dufton, A. F. (1933*b*): The measurement of equivalent temperature. *Ibid.,* **33**, 474.
Dufton, (1933*c*): The heat requirements of a house. *J. Instn. Heat. Vent. Engrs., Lond.,* **1**, 99, 156, and 209.
Dufton, A. F. (1934): Metallic wall-paper. *Ibid.,* **2**, 218.
Dufton, A. F. (1936): The equivalent temperature of a warmed room. *Ibid.,* **4**, 227.
Dufton, A. F. (1939): Freshness in relation to environmental conditions. *Ibid.,* **7**, 397.
Dufton, A. F. (1942): Improvement of the open fire. *J. Roy. Inst. Brit. Arch.,* **49**, 3rd ser., 82 (March).

Dufton, A. F. and Bedford, T. (1933): Nose-opening rays. *J. Hyg., Camb.*, **33**, 476.

Dufton, A. F. and Marley, W. G. (1934): Measurement of the rate of air change. *J. Instn. Heat. Vent. Engrs., Lond.*, **1**, 645.

Dufton, A. F. and Marley, W. G. (1942): The comparison of heating systems. *Ibid.*, **10**, 17.

Edward, D. G. ff. (1944): Haemolytic streptococci in the dust of hospital wards, and their relationship to infection. *J. Hyg., Camb.*, **43**, 256.

Edward, D. G., Lush, D. and Bourdillon, R. B. (1943): Studies on air-borne virus infections: II. The killing of virus aerosols by ultra-violet radiation. *Ibid.*, **43**, 11.

Egerton, A. C. G. (1945): Trends in the design of heating and installations for domestic purposes. *J. Instn. Heat. Vent. Engrs., Lond.*, **13**, 147.

Eichna, L. W., Ashe, W. F., Bean, W. B. and Shelley, W. B. (1945): The upper limits of environmental heat and humidity tolerated by acclimatized men working in hot environments. *J. Industr. Hyg. & Toxicol.*, **27**, 59.

Elford, W. J. and van den Ende, J. (1942): An investigation of the merits of ozone as an aerial disinfectant. *J. Hyg. Camb.*, **42**, 240.

Elford, W. J. and van den Ende, J. (1945): Studies on the disinfecting action of hypochlorous acid gas and sprayed solution of hypochlorite against bacterial aerosols. *Ibid.*, **44**, 1.

Electricity Council (1968): Packaged air conditioning–a guide. The Electricity Council Marketing Department. London.

Ellis, F. P. (1952): Thermal comfort in warm, humid atmospheres: observations in a warship in the tropics. *J. Hyg., Camb.*, **50**, 415.

Ellis, F. P. (1953): Thermal comfort in warm and humid atmospheres: observations on groups and individuals in Singapore. *Ibid.*, **51**, 386.

Ellis, F. P., Ferres, H. M., Lind, A. R. and Newling, P. S. B. (1953): The upper tolerable levels of warmth for acclimatized European men working in the tropics. *Med. Res. Coun., Lond., R. N. P. Rep.* No. 53/759.

Emswiller, J. E. and Randall, W. C. (1928): The weathertightness of rolled section steel windows. *J. Amer. Soc. Heat. Vent. Engrs.*, **34**, 477.

Evans, G. V. and Webb, J. W. (1971): An examination of radioisotope techniques for the measurement of ventilation rate. Nuclear Physics Division, Atomic Energy Research Establishment, Harwell, Berkshire. AERE-R 6709. H.M.S.O. London.

Faber, O. (1944): Report from the Select Committee on House of Commons (Rebuilding). Appendix 4.

Faber, O. (1946): Factory heating and fuel savings. *Heat. & Vent. Engr.*, **20**, 204.

Faber, O. and Kell, J. R. (1943): *Heating and air conditioning of buildings.* (2nd Edn.). London. Architectural Press.

Faber, O. and Kell, J. R. (1951): The heating and air conditioning of the House of Commons. *J. Instn. Heat. Vent. Engrs., Lond.*, **19**, 375.

Falconer, W. (1781): *Remarks on the influence of climate, etc., on the disposition, etc., of mankind.* London: C. Dilly.

Fanger, P. O. (1967): *Calculation of thermal comfort:* Introduction of a basic equation. ASHRAE, Trans. 73, II.

Fanger, P. O. (1970): *Thermal comfort* (in English). Danish Technical Press. Copenhagen.

Faraday, M. (1835): Cit. from Tomlinson (1850), p. 18.

Farmer, E., Brooke, R. St. C. and Chambers, E. G. (1923): A comparison of different shift systems in the glass trade. *Rep. Industr. Fatigue Res. Bd.*, No. 24.

Fechner, G. T. (1860): Elemente der Psychophysik. Leipzig: Breitkopf and Härtel.

Finney, D. J. (1947): *Probit analysis.* Cambridge: The University Press.

Fishenden, M. (1925): *House heating.* London: H. F. & G. Witherby.

Fishenden, M. and Dufton, A. F. (1929): Heat transmission. *Build. Res. Bd. Spec. Rep.* No. 11.

Fishenden, M. and Saunders, O. A. (1932): *The calculation of heat transmission.* London: H.M. Stationery Office.

Fishenden, M. and Willgress, R. E. (1925): The heating of rooms. *Fuel Res. Bd. Tech. Paper* No. 12.

Fleisher, W. L., Stacey, A. E., Houghten, F. C. and Ferderber, M. B. (1939): Air conditioning in industry: physiological reactions of individual workers to high effective temperatures. *Trans. Amer. Soc. Heat. Vent. Engrs.*, **45**, 59, and *J. Instn. Heat. Vent. Engrs., Lond.*, **7**, 48.

Flügge, C. (1905): Ueber Luftverunreinigung, Wärmestauung und Lüftung in geschlossenen Räumen. *Zeitschr. f. Hyg.*, **49**, 363.

Gagge, A. P. (1936): The linearity criterion as applied to partitional calorimetry. *Amer. J. Physiol.*, **116**, 656.

Gagge, A. P. (1937): A new physiological variable associated with sensible and insensible perspiration. *Ibid.*, **120**, 277.

Gagge, A. P., Herrington, L. P. and Winslow, C.-E. A. (1937): Thermal interchanges between the human body and its atmospheric environment. *Amer. J. Hyg.*, **26**, 84.

Gagge, A. P. and Moriyama, I. M. (1935): The annual and diurnal variation of ions in an urban community. *Terrest. Mag. & Atmosph. Electr.*, **40**, 295.

Gagge, A. P., Winslow, C.-E. A. and Herrington, L. P. (1938): The influence of clothing on the physiological reactions of the human body to varying environmental temperatures. *Amer. J. Physiol.*, **124**, 30.

Gagge, A. P., Burton, A. C. and Bazett, H. C. (1941): A practical system of units for the description of the heat exchange of man with his environment. *Science*, **94**, 245-248.

Galton, D. (1880): *Healthy dwellings.* Oxford: Clarendon Press.

Gauger, N. (1714): *La mechanique du feu.* Amsterdam. H. Schelte.

General Board of Health (1857): Report by the Commissioners appointed to inquire into the warming and ventilation of dwellings.

Glickman, N., Inouye, T., Keeton, R. W. and Fahnestock, M. K. (1950): Physiological examination of the effective temperature index. *Heat., Pip. & Air Condit.*, **22**, 157.

Griffiths, E. (1921): Some modified forms of hygrometers. *Physical Soc., Lond. Discussion on Hygrometry.* London: The Fleetway Press, Ltd.

Griffiths, E. (1941): Some properties of insulating materials. *J. Instn. Heat. Vent. Engrs., Lond.*, **9**, 177.

Griffiths, E. and Davis, A. H. (1931): The transmission of heat by radiation and convection. *Food Investigation Spec. Rept.* No. 9.

Haines, G. F., Jr. and Hatch, T. (1952): Industrial heat exposures—evaluation and control. *Heating & Ventilating,* **49,** No. 11, 93.

Haldane, J. S. (1899): *Report of the Departmental Committee appointed to enquire into the manufacture and use of water gas and other gases containing a large proportion of carbonic oxide.* Appendix I., C. 9164.

Haldane, J. S. (1905): The influence of high air temperatures. *J. Hyg., Camb.,* **5,** 494.

Haldane, J. S. (1907): Work under pressure and in great heat. *Science Progress,* N.S., **2,** 378.

Haldane, J. S. (1929): Discussion on paper by D. Penman. *Trans. Instn. Min. Engrs., Lond.,* **76,** 353.

Haldane, J. S. and Priestley, J. G. (1905): The regulation of the lung ventilation. *J. Physiol.,* **32,** 225.

Haldane, J. S. and Priestley, J. G. (1935): *Respiration.* Oxford: Clarendon Press.

Haldane, J. S. and Smith, J. L. (1892-3*a*): The physiological effects of air vitiated by respiration. *J. Path. & Bact.,* **1,** 168.

Haldane, J. S. and Smith, J. L. (1892-3*b*): The toxic action of expired air. *Ibid.,* **1,** 318.

Haldane, T. G. N. (1930): The operation of the heat pump and its possible application to heating problems, particularly swimming bath heating. *Proc. Instn. Heat. Vent. Engrs., Lond.,* **29,** 141.

Hale, A. L. (1946): Characteristics of modern unit heaters. *J. Instn. Heat. & Vent. Engrs., Lond.,* **14,** 153.

Hallett, C. G. H. (1931): Notes on the theory of radiant heating. *Proc. Instn. Heat. Vent. Engrs., Lond.,* **29,** 206.

Hancock, W. (1927*a*): Local air-conditioning underground. *Trans. Instn. Min. Engrs., Lond.,* **72,** 342.

Hancock, W. (1927*b*): Local air-conditioning underground: Efficiencies of small auxiliary ventilating units. *Ibid.,* **73,** 313.

Harcourt, A. V. (1915): On the measurement of the efficiency of domestic fires, and on a simple and smokeless grate. *J. Roy. Soc. Arts,* **63,** 570.

Harding, L. A. and Willard, A. C. (1929): *Mechanical equipment of buildings. Vol. I. Heating and ventilation.* 2nd Edn. New York: John Wiley & Sons.

Hardy, J. D. and Du Bois, E. F. (1937): Regulation of heat loss from the human body. *Proc. Nat. Acad. Sci.,* **23,** 624.

Hardy, J. D. and Du Bois, E. F. (1938): The technic of measuring radiation and convection. *J. Nutrition,* **15,** 461.

Hardy, J. D., Du Bois, E. F. and Soderstrom, G. F. (1938): Basal metabolism, radiation, convection and vaporization at temperatures of 22 to 35° C. *Ibid.,* **15,** 477.

Hardy, J. D. and Muschenheim, C. (1934): The radiation of heat from the human body. IV. The emission, reflection and transmission of infra-red radiation by the human skin. *J. Clin. Invest.,* **13,** 817.

Hardy, J. D. and Muschenheim, C. (1936): Radiation of heat from the human body. V. The transmission of infra-red radiation through skin. *Ibid.,* **15,** 1.

Hardy, J. D. and Oppel, T. W. (1937): Studies in temperature sensation. III. The sensitivity of the body to heat and the spatial summation of the end-organ responses. *Ibid.*, **16**, 533.

Hardy, J. D. and Oppel, T. W. (1938): Studies in temperature sensation. IV. The stimulation of cold sensation by radiation. *Ibid.*, **17**, 771.

Hartley, H. (1934): Radiant heat and comfort. *J. Instn. Heat. Vent. Engrs., Lond.*, **2**, 273.

Health of Munition Workers Committee (1916): Ventilation and lighting of munition factories and workshops. *Memo.* No. 9, Cd. 8215.

Health of Munition Workers Committee (1917): Industrial efficiency and fatigue. *Interim Report*, Cd. 8511.

Health of Munition Workers Committee (1918): Industrial health and efficiency. *Final Report*, Cd. 9065.

Heath, W. R. and Criqui, A. E. (1944): The axial flow fan and its place in ventilation. *Heat., Pip. & Air Condit.*, **16**, 107; and *Trans. Amer. Soc. Heat. Vent. Engrs.*, **50**, 197.

Heberden, W. (1826): An account of the heat of July, 1825; together with some remarks upon sensible cold. *Phil. Trans. Roy. Soc. Lond.*, Pt. II, 69.

Hermans, J. T. F. (1883): Über die vermeintliche Ausathmung organischer Substanzen durch den Menschen. *Arch. f. Hyg.*, **1**, 1.

Herrington, L. P. (1935): The influence of ionized air upon normal subjects. *J. Clin. Invest.*, **14**, 70.

Herrington, L. P. and Kuh, C. (1938): The reaction of hypertensive patients to atmospheres containing high concentrations of heavy ions. *J. Industr. Hyg. & Toxicol.*, **20**, 179.

Herrington, L. P. and Lorenzi, R. J. (1949): Effect of panel location on skin and clothing surface temperatures. *Heat., Pip. & Air Condit.*, **21**, 107.

Herrington, L. P. and Smith, K. L. (1935): The effect of high concentrations of light negative atmospheric ions in the growth and activity of the albino rat. *J. Industr. Hyg.*, **17**, 283.

Herrington, L. P., Winslow, C.-E. A. and Gagge, A. P. (1937): The relative influence of radiation and convection upon vasomotor temperature regulation. *Amer. J. Physiol.*, **120**, 133.

Hertz, H. (1896): *Miscellaneous papers.* London: Macmillan & Co.

Hickish, D. E. (1955): Thermal sensations of workers in light industry in summer. A field study in Southern England. *J. Hyg., Camb.*, **53**, 112.

Hill, A. Bradford (1927): Artificial humidification in the cotton weaving industry. Its effect upon the sickness rates of weaving operatives. *Rep. Industr. Fatigue Res. Bd., Lond.*, No. 48.

Hill, A. Bradford (1930): Sickness among operatives in Lancashire cotton spinning mills. *Rep. Industr. Hlth. Res. Bd., Lond.* No. 59.

Hill, L. (1914): Report on ventilation and the effect of open air and wind on the respiratory metabolism. *Rep. Loc. Govt. Bd. Publ. Hlth., N.S.*, No. 100.

Hill, L. (1919): The science of ventilation and open air treatment. Part I. *Sp. Rep. Ser. Med. Res. Coun., Lond.*, No. 32.

Hill, L. (1920): The science of ventilation and open air treatment. Part II. *Ibid.*, No. 52.

Hill, L. (1931): The influence of infra-red radiation in relation to ventilation and heating. *J. State Med.*, **39**, 683.

Hill, L. (1932): Infra-red rays and ventilation. *Proc. Physiol. Soc., J. Physiol.*, **74**, 1P.

Hill, L. (1934): Infra-red rays and comfort. *J. Instn. Heat. Vent. Engrs., Lond.*, **2**, 20.

Hill, L. (1935): 'Nose-opening' rays. *J. Hyg., Camb.*, **35**, 75.

Hill, L. (1939): Recent work on the quality of heat rays. *Brit. Med. J.*, May 13, 977.

Hill, L., Flack, M., McIntosh, J., Rowlands, R. A. and Walker, H. B. (1913): The influence of the atmosphere on our health and comfort in confined and crowded places. *Smithsonian Miscell. Coll.*, **60**, No. 23, Publn. No. 2170.

Ho, P. Y. (1952): Correlations of equatorial climatic factors with comfort. *J. Instn. Heat. Vent. Engrs., Lond.*, **20**, 196.

Home Office (1907): *Second Report of the Departmental Committee appointed to enquire into the ventilation of factories and workshops.* London: H.M. Stationery Office.

Home Office (1909): *Report of the Departmental Committee on humidity and ventilation in cotton weaving sheds.* London: H.M. Stationery Office.

Houghten, F. C. and Ingels, M. (1927): Infiltration through plastered and unplastered brick walls. *J. Amer. Soc. Heat. Vent. Engrs.*, **33**, 249.

Houghten, F. C. and McDermott, P. (1933): Cold walls and their relation to the feeling of warmth. *Heat., Pip. & Air Condit.*, 53; and *Trans. Amer. Soc. Heat. Vent. Engrs.*, **39**, 83.

Houghten, F. C., Teague, W. W., Miller, W. E. and Yant, W. P. (1929): Thermal exchanges between the human body and its atmospheric environment. *Amer. J. Physiol.*, **88**, 386.

Houghten, F. C., Teague, W. W., Miller, W. E. and Yant, W. P. (1931*a*): Heat and moisture losses from men at work and application to air conditioning problems. *Trans. Amer. Soc. Heat. Vent. Engrs.*, **37**, 541.

Houghten, F. C., Teague, W. W., Miller, W. E. and Yant, W. P. (1931*b*): Thermal exchanges between the bodies of men working and the atmospheric environment. *Amer. J. Hyg.*, **13**, 415.

Houghten, F. C. and Yagloglou, C. P. (1923): Determining lines of equal comfort. *Trans. Amer. Soc. Heat. Vent. Engrs.*, **29**, 163.

Houghten, F. C. and Yagloglou, C. P. (1924): Cooling effect on human beings produced by various air velocities. *Ibid.*, **30**, 193.

Humphreys, C. M., Imalis, O. and Gutberlet, C. (1946): Physiological response of subjects exposed to high effective temperatures and elevated mean radiant temperatures. *Heat., Pip. & Air Condit.*, **18**, 101.

Humphreys, C. M. and Jennings, B. H. (1962): Atmospheric ions in relation to comfort and other responses of normal individuals. *A.S.H.R.A.E. Journal*, Sept., **4**, 55.

Huntington, E. (1915): *Civilization and climate.* New Haven: Yale University Press.

Hutchinson, F. W. (1945): A single-equation design procedure for radiant panel systems. *Heat., Pip. & Air Condit.*, **17**, 637.

Government of India Post-War Planning Committee (Textiles) (1946): *Report of the Sub-Committee on minimum standards for buildings, etc., affecting the health and comfort of operatives in the post-war period.* Bombay: Office of the Textile Commissioner. (Reprinted from *The Indian Textile Journal,* Bombay, Jan. 1946.)

Institution of Heating and Ventilating Engineers (1942): *The computation of heat requirements for buildings.* London: Instn. Heat. Vent. Engrs.

Inter-Departmental Committee on Terminology, etc., in Heating and Ventilation Research (1932): Summary of recommendations. *Building Res. Bd.* (1932), Appendix I.

Jacquez, J. A., Huss, J., McKeehan, W., Dimitroff, J. M. and Kuppenheim, H. F. (1955): Spectral reflectance of human skin in the region 0.7-2.6 μ. *J. Appl. Physiol.,* **8**, 297.

Jennings, B. H., Bigg, E. and Olson, F. C. W. (1944): The use of glycol vapours for air sterilization and the control of air-borne infection. *Heat., Pip. & Air Condit.,* **16**, 538.

Joselin, E. L. (1934): *Ventilation.* London: Edward Arnold & Co.

Joseph, E. L. (1938): Air purification. *J. Instn. Heat. Vent. Engrs., Lond.,* **6**, 170.

Kennedy, D. M. (1970): Environmental control and human factors. *J. Inst. Gas Engrs., Lond.* Sept. pp. 645-662.

Kerslake, McK. D. (1972): The stress of hot environments. Cambridge University Press.

Koch, W. (1962): Atmospheric ions and comfort. *J. Hyg., Camb.,* **60**, 183.

Koch, W., Jennings, B. H. and Humphreys, C. M. (1960): Sensation responses to temperature and humidity under still-air conditions in the comfort range. *A.S.H.R.A.E. Journal,* April, **2**, 63.

Kornblueh, I. H., Piersol, G. M. and Speicher, F. P. (1958): Relief from pollinosis in negatively ionized rooms. *Amer. J. Phys. Med.,* **37**, 18.

Krueger, A. P. and Smith, R. F. (1957): Effects of air ions on isolated rabbit trachea. *Proc. Soc. Exp. Biol. & Med.,* **96**, 807.

Krueger, A. P. (1962): Air ions and physiological function. *J. Gen. Physiol.,* **45**, (4), pp. 233-241.

Küster, E. and Janitzky, A. (1932): Versuche über Raumbelüftung. *Zeitschr. f.d. ges. Physikalische Therap.,* **43**, 12.

Ladell, W. S. S. (1951): Assessment of group acclimatization to heat and humidity. *J. Physiol.,* **115**, 296.

Langdon, F. J. and Loudon, A. G. (1970): Discomfort in schools from overheating in summer. *J. Instn. Heat. Vent. Engrs., Lond.,* **37**, 265-274.

Larson, G. L., Braatz, C. and Nelson, D. W. (1929): Air infiltration through various types of brick wall construction. *J. Amer. Soc. Heat. Engrs.,* **35**, 125.

League of Nations Health Organization (1937): The hygiene of housing: Report by the Housing Commission. *Bull. Health Organization of the League of Nations,* **6**, Extract No. 12.

Leblanc, F. (1842): Recherches sur la composition de l'air confiné. *Ann. de chimie et de physique,* **5**, 223.

Lehmann, G. (1939): Der L. Hillsche Nasenreflex und seine Bedeutung für die Kenntnis der thermoregulatorischen Hautreflexe. *Arbeitsphysiol.*, **10**, 418.

Leithead, C. S. and Lind, A. R. (1964): *Heat stress and heat disorders.* Cassell. London.

Leopold, C. S. (1947): Conditions for comfort. *Trans. Amer. Soc. Heat. Vent. Engrs.*, **53**, 295.

Leslie, J. (1804): *An experimental inquiry into the nature and propagation of heat.* London: J. Mawman.

Lewis, Thomas (1930): Observations upon the reactions of the vessels of the human skin to cold. *Heart*, **15**, 177.

Lewis, Thornton (1926): Observing warm air in circulation. *J. Amer. Soc. Heat. Vent. Engrs.*, **32**, 451.

Lidwell, O. M. (1946): Bactericidal effects of the partial irradiation of a room with ultra-violet light. *J. Hyg., Camb.*, **44**, 333.

Lidwell, O. M. and Lovelock, J. E. (1946): Some methods of measuring ventilation. *Ibid.*, **44**, 326.

Lidwell, O. M. and Williams, R. E. O. (1960): The ventilation of operating theatres. *J. Hyg., Camb.*, **58**, 449.

Lidwell, O. M. (1966): The ventilation of operating suites. Symposium on hospital engineering services—present and future. *Instn. Heat. Vent. Engrs., Lond.*, pp. 84-90.

Lind, A. R., Weiner, J. S., Hellon, R. F. and Jones, R. M. (1957): Reactions of mines-rescue personnel to work in hot environments. *Nat. Coal Bd., Med. Res. Mem.* No. 1.

Lorenzi, R. J., Herrington, L. P. and Winslow, C.-E. A. (1946): The influence of an interior coating of aluminium paper on internal thermal conditions and on heat economy. *Heat., Pip. & Air Condit.*, **18**, 109.

Lusk, G. (1928): *The science of nutrition.* 4th Edn. Philadelphia: W. B. Saunders Co. p. 19.

Macfarlane, D. (1938): Axial flow screw fans. *Paper read before Belfast Assoc. of Engineers*, Feb. 10th.

Mackworth, N. H. (1946): Effects of heat on wireless operators. *Brit. J. Industr. Med.*, **3**, 143.

Macmillan, R. G. (1967): Air conditioning of operating theatre suites and intensive care units. Report for Royal School of Military Engineers.

Macpherson, R. M. (1960): Physiological responses to hot environments. *Med. Res. Coun., Lond., Spec. Rep. Ser.* No. 298.

Macpherson, R. M. (1962): The assessment of the thermal environment. *Brit. J. Industr. Med.*, **19**, 151.

Margolis, A. (1935): Experiences with district heating in Europe and the U.S.A. and its further development. *J. Instn. Heat. Vent. Engrs., Lond.*, **3**, 361.

Markham, S. F. (1937): Discussion on the effects of water-vapour content in the atmosphere of interest to heating and ventilating engineers. *Ibid.*, **4**, 570.

Markham, S. F. (1942): *Climate and the energy of nations.* Oxford University Press.

Marley, W. G. (1935): The measurement of the rate of air change. *J. Instn. Heat. Vent. Engrs., Lond.*, **2**, 499.

Marley, W. G. (1937): The first year's work in the heating laboratory, *Ibid.*, **5**, 283.

Marley, W. G. (1938): Experiments with extended surface heating in the heating laboratory. *Ibid.*, **6**, 202.

Mavrogordato, A. and Pirow, H. (1927): Deep level mining and high temperatures. *J. So. African Instn. Engrs.*, **25**, 101.

May, K. R. (1967): Physical aspects of sampling airborne microbes, in *Airborne Microbes.* Cambridge University Press.

McAdams, W. H. (1933): *Heat transmission.* New York: McGraw-Hill Book Co.

McArdle, B., Dunham, W., Holling, H. E., Ladell, W. S. S., Scott, J. W., Thomson, M. L. and Weiner, J. S. (1947): The prediction of the physiological effects of warm and hot environments. *Med. Res. Coun., Lond., R.N.P. Rep.* No. 47/391.

McConnell, W. J., Fellows, H. H. and Stephens, M. G. (1927): Some observations on the health aspects of air conditioning. *14th Ann. Conference, Life Office Management Assoc., Chicago.*

McCoy, G. W. (1934): Psittacosis among the personnel of the hygiene laboratory. *J. Infect. Dis.*, **55**, 156.

Medical Research Council (1941): The prevention of 'hospital infection' of wounds. *Med. Res. Coun., Lond., War Memo.* No. 6.

Medical Research Council (1942): Report of the Committee on Tuberculosis in War-Time. *Sp. Rep. Ser. Med. Res. Coun., Lond.*, No. 246.

Medical Research Council (1944): The control of cross infection in hospitals. *Med. Res. Coun., Lond., War Memo.* No. 11.

Medical Research Council (1954): Air disinfection with ultra-violet irradiation: its effect on illness among school-children. *Spec. Rep. Ser. Med. Res. Coun., Lond.*, No. 283.

Mills, C. A. and Ogle, C. (1938): Control of body heat loss through radiant means. *Trans. Amer. Soc. Heat. Vent. Engrs.*, **44**, 167.

Ministry of Fuel and Power (1943): Thermal insulation of buildings. *Fuel Efficiency Bulletin* No. 12.

Ministry of Health (1937): The acute general hospital. *First Report of the Departmental Committee on the Cost of Hospitals and other Public Buildings.* London: H.M.S.O.

Ministry of Health (1945): Technical note on some types of permanent prefabricated houses. *Circular* 182/45, Oct. 16th.

Ministry of Health (1957): Operating theatre suites. *Hospital Building Bulletin,* No. 1.

Ministry of Health (1966): Annual Report of the Chief Medical Officer. London: H.M.S.O.

Ministry of Housing and Local Government (1961): Homes for today and tomorrow. London: H.M.S.O. (The 'Parker Morris' Report.)

Ministry of Labour and National Service (1940): Ventilation of factories. (4th Edn.) *Welfare Pamphlet* No. 5.

Ministry of Works (1944*a*): Solid fuel installations. *Post-War Building Studies,* No. 10.

Ministry of Works (1944*b*): Walls, floors and roofs. *Ibid.*, No. 15.

Ministry of Works (1945): Heating and ventilation of dwellings. *Ibid.*, No. 19.

Missenard, A. (1933): *Étude physiologique et technique de la ventilation.* Paris: Librairie de l'Enseignement Technique.

Missenard, A. (1935): Théorie simplifée du thermomètre résultant-thermostat résultant. *Chauffage et Ventilation,* **12**, 347.

Missenard, A. (1944): Influence des conditions thermiques ambiantes sur la capacité de travail, la morbidité et la mortalité des ouvriers et la construction des lieux de travail. *Com. d'Organisation du Bâtiment et des Travaux Publics. Circular,* series L. No. 8, Feb. 5th.

Miura, U. (1931): The effect of variations in relative humidity upon skin temperature and sense of comfort. *Amer. J. Hyg.*, **13**, 432.

Molloy, E. (1945): *Principles and practice of heating and ventilating.* London: George Newnes, Ltd.

Moon, P. (1940): Solar radiation curves. *J. Franklin Inst.*, **230**, p. 604.

Morin, A. (1863): *Études sur la ventilation.* Paris: Librairie de L. Hachette et Cie.

Moss, K. N. (1924): Some effects of high air temperatures upon the miner. *Trans. Instn. Min. Engrs., Lond.*, **66**, 284.

Moss, K. N. and Halls, J. L. (1935): A time study of coal face workers. *Ibid.*, **90**, 62.

Mudd, S., Goldman, A. and Grant, S. B. (1921): Reactions of the nasal cavity and postnasal space to chilling of the body surface. I. Vasomotor reactions. *J. Exp. Med.*, **34**, 11.

Munro, A. F. and Chrenko, F. A. (1949): The effect of radiation from the surroundings on subjective impressions of freshness. *J. Hyg., Camb.*, **47**, 288.

Nelson, N., Eichna, L. W., Horvath, S. M., Shelley, W. B. and Hatch, T. F. (1947): Thermal exchanges of man at high temperatures. *Amer. J. Physiol.*, **151**, 626.

Newbold, E. M. (1926): A contribution to the study of the human factor in the causation of accidents. *Rep. Industr. Fatigue Res. Bd., Lond.*, No. 34.

Newcombe, S. F. (1944): Heat insulation in domestic buildings. *J. Roy. Inst. Brit. Arch.*, 3rd Ser., **51**, 253.

Newling, P. S. B. (1954): The measurement of mean radiant temperature and the determination of the amount of radiant heat gained by a man. *Med. Res. Coun., Lond., R.N.P. Rep.* No. 54/784.

Newton, A. B., Houghten, F. C., Gutberlet, C. and Qualley, R. W. (1938*a*): Summer cooling requirements of 275 workers in an air-conditioned office. *Trans. Amer. Soc. Heat. Vent. Engrs.*, **44**, 337.

Newton, A. B., Houghten, F. C., Gutberlet, C. and Qualley, R. W. (1938*b*): Shock experiences of 275 workers after entering and leaving cooled and air-conditioned offices. *Ibid.*, **44**, 571.

New York Commission on Ventilation (1931): *Principles and practices in school ventilation.* New York: Bureau of Publications, Teachers' College, Columbia University.

New York Commission on Ventilation. *Report of the New York State Commission on Ventilation* (1923): New York: E. P. Dutton & Co.

Noble, W. C. (1967): Sampling airborne microbes: handling the catch, in *Airborne Microbes.* Cambridge University Press.

Onslow, D. V. (1945): The heat pump for space heating. *Brit Elect. Engrs Res. Assoc., Tech. Pap.*, ref. Y/T7.

Oppel, T. W. and Hardy, J. D. (1937*a*): Studies in temperature sensation. I. A comparison of the sensation produced by infra-red and visible radiation. *J. Clin. Invest.*, **16**, 517.

Oppel, T. W. and Hardy, J. D. (1937*b*): Studies in temperature sensation. II. The temperature changes responsible for the stimulation of the heat end-organs. *Ibid.*, **16**, 525.

Osborne, E. E., Vernon, H. M. and Muscio, B. (1922): Two contributions to the study of accident causation. *Rep. Industr. Fatigue Res. Bd., Lond.*, No. 19.

Othmer, D. F. and Coats, H. B. (1928): Measurement of surface temperature. *Ind. & Eng. Chem.*, **20**, 124.

Owens, J. S. (1927): *Proc. Instn. Heat. Vent. Engrs. Lond.*, **26**, 126, cited by Vernon, H. M. (1934), *The principles of heating and ventilation.* Edward Arnold & Co. London.

Pallot, A. C. (1943): Thermal insulation at medium temperatures. *J. Roy. Soc. Arts*, **91**, 122.

Partridge, R. C. and Maclean, D. L. (1935): Determination of the comfort zone for school children. *J. Industr. Hyg.*, **17**, 66.

Pearse, R. J. (1961): Observations on panel heating in hospital wards. *Heating*, **23**, 415.

Pettenkofer, M. (1862-3): Über die Respiration. *Ann. d. Chem. u. Pharm., Supplement*, **2**, 1.

Phelps, E. B. and Vold, A. (1934): Studies in ventilation. I. Skin temperature as related to atmospheric temperature and humidity. *Amer. J. Publ. Hlth.*, **24**, 959.

Pickering, G. W. (1932): The vasomotor regulation of heat-loss from the human skin in relation to external temperatures. *Heart*, **16**, 115.

Powell, R. W. (1940): Further experiments on the evaporation of water from saturated surfaces. *Trans. Instn. Chem. Engrs.*, **18**, 36.

Powell, R. W. and Griffiths, E. (1935): The evaporation of water from plane and cylindrical surfaces. *Ibid.*, **13**, 175.

Provins, K. A., Hellon, R. F., Bell, C. R. and Hiorns, R. W. (1962): Tolerance to heat of subjects engaged in sedentary work. *Ergonomics*, **5**, 93.

Provins, K. A. and Bell, C. R. (1962): The effects of heat on human performance. *Biometerorology*. Ed. S. W. Tromp. Pergamon Press.

Pulvertaft, R. J. V. (1944): The use of antiseptic sprays for air sterilization. *J. Hyg., Camb.*, **43**, 352.

Pulvertaft, R. J. V., Lemon, G. C. and Walker, J. W. (1939): Atmospheric and surface sterilization by aerosols. *The Lancet*, **I**, 443.

Raber, B. F. and Hutchinson, F. W. (1943): Optimum surface distribution in panel heating and cooling systems. *Heat., Pip. & Air Condit.*, **15**, 602; and *Trans. Amer. Soc. Heat. Vent. Engrs.*, **50**, 231.

Raber, B. F. and Hutchinson, F. W. (1945): *Refrigeration and air conditioning engineering.* New York: John Wiley & Sons.

Raber, B. F. and Hutchinson, F. W. (1947): *Panel heating and cooling analysis.* New York: J. Wiley & Sons, Inc.

Reichenbach, H. and Heymann, B. (1907): Untersuchungen über die Wirkungen Klimatischer Faktoren auf den Menschen. I. Beziehung zwischen Haut- und Lufttemperatur. *Zeitsch. Hyg. Infektkr.*, **57**, 1.

Reid, D. B. (1844): *Illustrations of the theory and practice of ventilation.* London: Longman, Brown, Green and Longmans.

Reid, D. D. (1964). *Proc. Roy. Soc. Med.*, **57**, 965, cited by J. R. Goldsmith in *Air Pollution* (1968) vol. I. Ed. A. C. Stern: Academic Press.

Roantree, W. B. (1950-1): Work in high air temperatures in a fire in Mysore Mine, Kolar Gold Field. *Trans. Instn. Min. & Metall., Lond.*, **60**, 513.

Robinson, S., Turrell, E. S., Belding, H. S. and Horvath, S. M. (1943): Rapid acclimatization to work in hot climates. *Amer. J. Physiol.*, **140**, 168.

Robinson, S., Turrell, E. S. and Gerking, S. D. (1945): Physiologically equivalent conditions of air temperature and humidity. *Ibid.*, **143**, 21.

Roscoe, H. E. (1857): Report on the chemical relations of ventilation. *Appendix to General Board of Health* (1857).

Rowley, F. B., Jordan, R. C. and Snyder, W. E. (1947): Comfort reactions of 275 workers during occupancy of air conditioned offices. *Heat., Pip. & Air Condit.*, **19**, 113.

Royal College of Physicians of London (1936): *Domestic heating by gas considered from the point of view of health and comfort.* London: Harrison & Sons, Ltd.

Royal Commission on Environmental Pollution (1971): First report. London: H.M.S.O.

Rubner, M. (1896): Zur Bilanz unserer Wärmeökonomie. *Arch. f. Hyg.*, **27**, 69.

Rubner, M. (1897): Experimentelle Untersuchungen über die modernen Bekleidungssysteme. II. Hygienische Gesichtspunkte zur Beurteilung einer Kleidung. *Ibid.*, **31**, 142.

Rubner, M. and Lewaschew, –(1897): Ueber den Einfluss der Feuchtigkeitsschwankungen unbewegeter Luft auf den Menschen während körperlicher Ruhe. *Ibid.*, **29**, 1.

Rusk, D. O., Cherry, V. H. and Boelter, L. (1932): Air infiltration through steel-framed windows. *Heat., Pip. & Air Condit.*, **4**, 696; and *Trans. Amer. Soc. Heat. Engrs.*, **39**, 169.

Rydberg, J. and Norbäak, P. (1949): Air distribution and draft. *Trans. Amer. Soc. Heat. Vent. Engrs.*, **55**, 225.

Shaw, W. N. (1892): Warming and ventilation. In Stevenson and Murphy (1892).

Shaw, W. N. (1907): *Air currents and the laws of ventilation.* Cambridge University Press.

Shaw, W. N. (1933): The drama of weather. Cambridge University Press.

Smith, D. V. H. (1945): District heating and the smokeless city. *J. Roy. San. Inst.*, **65**, 28.

Smith, F. E. (1955): Indices of heat stress. *Med. Res. Coun., Lond., Mem.* No. 29.

Smith, F. E. (1958): Climate and health in the Royal Navy. *Brit. J. Industr. Med.*, **15**, 197.

Smith, J. H. (1930): Cit. from Vernon (1934).

Soderstrom, G. F. and Dubois, E. F. (1917): The water elimination through the skin and respiratory passages in health and disease. *Arch. Int. Med.*, **19**, 931.

Spanswick, R. (1972): Total control at the new London Stock Exchange. *J. Instn. Heat. Vent. Engrs., Lond.*, **39**, A29.

Stamp D. L. (1964): Some aspects of medical geography. Oxford University Press.

Stead, K. and Rose, B. T. (1961): Planning of mechanical services for hospitals. *J. Instn. Heat. Vent. Engrs., Lond.*, **29**, 157.

Stevenson, T. and Murphy, S. F. (1892): *A treatise on hygiene and public health.* Vol. I. London: J. and A. Churchill.

Teale, T. P. (1883): *Economy of coal in house fires.* London.

Tomlinson, C. (1850): *A rudimentary treatise on warming and ventilation.* London: J. Weale.

Tredgold, T. (1824): *Principles of warming and ventilating public buildings, dwelling-houses, etc.* London: Josiah Taylor.

Tredre, B. E. (1960): The physiological assessment of radiant heat in indoor environments. *Univ. of London: Ph.D. Thesis.*

Turner, D. (1955): Radiative and convective stimuli of threshold intensity. *J. Hyg., Camb.*, **53**, 32.

Turner, D. (1958): Heat stress in non-ferrous foundries. *Brit. J. Industr. Med.*, **15**, 38.

Tuve, G. L. and Priester, G. B. (1944): Control of air streams in large spaces. *Heat., Pip. & Air Condit.*, **16**, 39; and *Trans. Amer. Soc. Heat. Vent. Engrs.*, **50**, 153.

Twort, C. C., Baker, A. H., Finn, S. R. and Powell, E. O. (1940): The disinfection of closed atmospheres with germicidal aerosols. *J. Hyg., Camb.*, **40**, 253.

van Dishoeck, H. A. E. (1935): Infra-red radiation and nasal obstruction. *J. Hyg., Camb.*, **35**, 185.

van den Ende, M., Edwards, D. G. ff. and Lush, D. (1941): Reduction of dust-borne infection by treatment of bedclothes. *The Lancet*, **I**, 716.

van den Ende, M., Lush, D. and Edwards, D. G. ff. (1940): Reduction of dust-borne bacteria by treating floors. *Ibid.*, **II**, 133.

van den Ende, M. and Thomas, J. C. (1941): Treatment of bedclothes with dust laying oils: use of oil-in-water emulsions. *Ibid.*, **II**, 755.

Vernon, H. M. (1918): An investigation of the factors concerned in the causation of industrial accidents. *Health of Munition Workers Committee Memo.* No. 21, Cd. 9046.

Vernon, H. M. (1919): The influence of hours of work and of ventilation on output in tinplate manufacture. *Rep. Industr. Fatigue Res. Bd., Lond.*, No. 1.

Vernon, H. M. (1920): Fatigue and efficiency in the iron and steel industry. *Ibid.*, No. 5.

Vernon, H. M. (1923): The index of comfort at high atmospheric temperatures. *Sp. Rep. Ser., Med. Res. Coun., Lond.*, No. 73, p. 116.

Vernon, H. M. (1930): The measurement of radiant heat in relation to human comfort. *J. Physiol.*, **70**, 15P.

Vernon, H. M. (1932*a*): The measurement, in relation to human comfort, of the radiation produced by various heating systems. *Proc. Instn. Heat. Vent. Engrs., Lond.*, **31**, 160.

Vernon, H. M. (1932*b*): The measurement of radiant heat in relation to human comfort. *J. Industr. Hyg.*, **14**, 95.

Vernon, H. M. (1934): *The principles of heating and ventilation.* London: Edward Arnold & Co.

Vernon, H. M., Bedford, T. and Warner, C. G. (1926*a*): A physiological study of the ventilation and heating in certain factories. *Rep. Industr. Fatigue Res. Bd., Lond.*, No. 35.

Vernon, H. M., Bedford, T. and Warner, C. G. (1926*b*): The influence of cooling power and of variability of air currents on sensations of air movement. *Sp. Rep. Ser., Med. Res. Coun., Lond.*, No. 100, Pt. III.

Vernon, H. M., Bedford, T. and Warner, C. G. (1927): The relation of atmospheric conditions to the working capacity and the accident rate of miners. *Rep. Industr. Fatigue Res. Bd., Lond.*, No. 39.

Vernon, H. M., Bedford, T. and Warner, C. G. (1928): A study of absenteeism in a group of ten collieries. *Ibid.*, No. 51.

Vernon, H. M., Bedford, T. and Warner, C. G. (1930): A study of heating and ventilation in schools. *Rep. Industr. Hlth. Res. Bd., Lond.*, No. 58.

Vernon, H. M., Bedford, T. and Warner, C. G. (1931): Two studies of absenteeism in coal mines. *Ibid.*, No. 62.

Vernon, H. M. and Manley, J. J. (1926): The measurement of variations in the velocity and temperature of air currents. *Sp. Rep. Ser. Med. Res. Coun., Lond.*, No. 100, Pt. I.

Vernon, H. M. and Rusher, E. A. (1920): Fatigue and efficiency in the iron and steel industry: The health of the workers. *Rep. Industr. Fatigue Res. Bd., Lond.*, No. 5, Pt. V.

Vernon, H. M., Vernon, M. D. and Lorrain-Smith, I. (1928): A physiological investigation of the radiant heating in various buildings. *Ibid.*, No. 46.

Vernon, H. M. and Warner, C. G. (1932): The influence of the humidity of the air on capacity for work at high temperatures. *J. Hyg., Camb.*, **32**, 431.

Walker, J. H. (1929): District heating service in Detroit, U.S.A. *Publication of Detroit Edison Co., Detroit, Mich.*

Ward, E. F. (1930): The measurement of skin temperature in its relation to the sensation of comfort. *Am. J. Hyg.*, **12**, 130.

Warner, C. G. (1940): Measurements of the ventilation of dwellings. *J. Hyg., Camb.*, **40**, 125.

Webb, C. G. (1952): On some observations of indoor climate in Malaya. *J. Instn. Heat. Vent. Engrs., Lond.*, **20**, 189.

Webb, C. G. (1959): An analysis of some observations of thermal comfort in an equatorial climate. *Brit. J. Industr. Med.*, **16**, 297.

Webb, C. G. (1960): Thermal discomfort in an equatorial climate. *J. Instn. Heat. Vent. Engrs., Lond.*, **27**, 297.

Weiner, J. S. (1938*a*): Discussion on paper by E. L. Joseph (1938). *J. Instn Heat. Vent. Engrs. Lond.*, **6**, 170.

Weiner, J. S. (1938*b*): An experimental study of heat collapse. *J. Industr. Hyg. & Toxicol.*, **20**, 389.

Wellner, P. (1932): Untersuchungen über die Grösse des natürlichen Lüftwechsels in Wohnungen alterer und neuer Bauweise. *Dissertation, Sachsischen Technischen Hochschule, Dresden.*

Wells, W. F. (1933): Apparatus for study of the bacterial behaviour of air. *Amer. J. Publ. Hlth.*, **23**, 58.

Wells, W. F. (1935): Air-borne infection and sanitary control. *J. Industr. Hyg.*, **17**, 253.

Wells, W. F. and Brown, H. W. (1936): Recovery of influenza virus suspended in air. *Science*, **84**, 68.

Wells, W. F. and Fair, G. M. (1935): Viability of *B. coli.* exposed to ultra-violet radiation in air. *Science*, **82**, 280.

Wells, W. F. and Wells, M. W. (1936): Air-borne infection. *J. Amer. Med. Assoc.*, **107**, 1698 and 1805.

Weston, H. C. (1922): A study of efficiency in fine linen weaving. *Rep. Industr. Fatigue Res. Bd., Lond.*, No. 20.

Whitehouse, A. G. R., Hancock, W. and Haldane, J. S. (1932): The osmotic passage of water and gases through the human skin. *Proc. Roy. Soc. B*, **111**, 412.

Widdows, G. H. (1921): School design. *The Builder*, **121**, 706.

Willard, A. C. and Fahnestock, M. K. (1930): The effect of two types of cast-iron steam radiators on air temperatures in room heating. *Heat, Pip. & Air Condit.*, **2**, 185.

Willard, A. C., Kratz, A. P. and Fahnestock, M. K. (1933): Application of eupatheoscope for measuring performance of direct radiators and convectors in terms of equivalent temperature. *Ibid.*, **5**, 369; and *Trans. Amer. Soc. Heat. Vent. Engrs.*, **39**, 303.

Willard, A. C., Kratz, A. P., Fahnestock, M. K. and Konzo, S. (1929): Investigation of heating rooms with direct steam radiators equipped with enclosures and shields. *Trans. Amer. Soc Heat. Vent. Engrs.* **35**, 77.

Williams, R. E. O. (1967): Spread of airborne bacteria pathogenic for man. *Airborne Microbes.* Cambridge University Press.

Winslow, C.-E. A., Gagge, A. P. and Herrington, L. P. (1939): The influence of air movement upon heat losses from the clothed human body. *Amer. J. Physiol.*, **127**, 505.

Winslow, C.-E. A., Greenburg, L. and Herrington, L. P. (1934): The influence of heat and light upon nasal obstruction. *Amer. J. Hyg.*, **20**, 195.

Winslow, C.-E. A. and Herrington, L. P. (1935): Subjective reactions of human beings to certain outdoor atmospheric conditions. *Heat., Pip. & Air Condit.*, **7**, 551.

Winslow, C.-E. A. and Herrington, L. P. (1936): The influence of odour upon appetite. *Amer. J. Hyg* , **23**, 143.

Winslow, C.-E. A., Herrington, L. P. and Gagge, A. P. (1936*a*): A new method of partitional calorimetry. *Amer. J. Physiol.*, **116**, 641.

Winslow, C.-E. A., Herrington, L. P. and Gagge, A. P. (1936*b*): The determination of radiation and convection exchanges by partitional calorimetry. *Ibid.*, **116**, 669.

Winslow, C.-E. A., Herrington, L. P. and Gagge, A. P. (1937*a*): Relations between atmospheric conditions, physiological reactions and sensations of pleasantness. *Amer. J. Hyg.*, **26**, 103.

Winslow, C.-E. A., Herrington, L. P. and Gagge, A. P. (1937*b*): Physiological reactions of the human body to varying environmental temperatures. *Amer. J. Physiol.*, **120**, 1.

Winslow, C.-E. A., Herrington, L. P. and Gagge, A. P. (1937*c*): Physiological reactions of the human body to various atmospheric humidities. *Ibid.*, **120**, 288.

Winslow, C.-E. A., Herrington, L. P. and Gagge A. P. (1938*a*): The reactions of the clothed human body to variations in atmospheric humidity. *Ibid.*, **124**, 692.

Winslow, C.-E. A., Herrington, L. P. and Gagge, A. P. (1938*b*): Physiological reactions and sensations of pleasantness under varying atmospheric conditions. *Trans. Amer. Soc. Heat. Vent. Engrs.*, **44**, 179.

Winslow, C.-E. A., Herrington, L. P. and Gagge, A. P. (1938*c*): The relative influence of radiation and convection upon the temperature regulation of the clothed body. *Amer. J. Physiol.*, **124**, 51.

Winslow, C.-E. A., Herrington, L. P. and Nelbach, J. H. (1942): The influence of atmospheric temperature and humidity upon the dryness of the oral mucosa. *Amer. J. Hyg.*, **35**, 27.

Winslow, C.-E. A. and Palmer, G. T. (1915): The effect upon appetite of the chemical constituents of the air of occupied rooms. *Proc. Soc. Exp. Biol. & Med.*, **12**, 141.

Witheridge, W. N. (1944): Control of industrial atmospheres, dusts, fumes, mists, vapours and gases. *Heat., Pip. & Air Condit.*, **16**, 712; and *Trans. Amer. Soc. Heat. Vent. Engrs.*, **51**, 227.

Witheridge, W. N. and Yaglou, C. P. (1939): Ozone in ventilation–its possibilities and limitations. *Heat., Pip. & Air Condit.*, **11**, 648; and *Trans. Amer. Soc. Heat. Vent. Engrs.*, **45**, 509.

Wolpert, H. (1905): Über den Einfluss der landhausmässigen Bebauung auf die natürliche Ventilation der Wohnräume. *Arch. f. Hyg.*, **52**, 46.

Wyatt, S., Fraser, J. A. and Stock, F. G. L. (1926): Fan ventilation in a humid weaving shed. *Rep. Industr. Fatigue Res. Bd., Lond.*, No. 37.

Wyndham, C. H. (1961): Applied physiological research in the South African gold mining industry. *Proc. 7th Commonwealth Mining & Metall. Congress.*

Yaglou, C. P. (1926): The thermal index of atmospheric conditions and its application to sedentary and to industrial life. *J. Industr. Hyg.*, **8**, 5.

Yaglou, C. P. (1927): Temperature, humidity, and air movement in industries: the effective temperature index. *Ibid.*, **9**, 297.

Yaglou, C. P. (1935): Observations of physiological efficiency in ionized air. *Ibid.*, **17**, 280.

Yaglou, C. P. (1937*a*): Abnormal air conditions in industry; their effects on workers, and methods of control. *Ibid.*, **19**, 12.

Yaglou, C. P. (1937*b*): Physical and physiologic principles of air conditioning. *J. Amer. Med. Assoc.*, **108**, 1708, and **109**, 945.

Yaglou, C. P. (1938): The heated thermometer anemometer. *J. Industr. Hyg.*, **20**, 497.

Yaglou, C. P. (1941): Present trends and recent developments in air conditioning. *Amer. Publ. Hlth. Assoc., 11th Ann. Year Book*, 1940-41. *Suppl. to Amer. J. Publ. Hlth*, **31**, 116.

Yaglou, C. P., Benjamin, C. L. and Choate, S. P. (1932): Changes in ionic content of air in occupied rooms ventilated by natural and by mechanical methods. *Trans. Amer. Soc. Heat. Vent. Engrs.* **38**, 191.

Yaglou, C. P., Brandt, A. D. and Benjamin, C. L. (1933): Observations on a group of subjects before, during and after exposure to ionized air. *J. Industr. Hyg.*, **15**, 341.

Yaglou, C. P. and Drinker, P. (1928): The summer comfort zone: climate and clothing. *Ibid.*, **10**, 350.

Yaglou, C. P. and Messer, A. (1941): The importance of clothing in air conditioning. *J. Amer. Med. Assoc.*, **117**, 1261.

Yaglou, C. P. and Miller, W. E. (1925): Effective temperature with clothing. *Trans. Amer. Soc. Heat. Vent. Engrs.*, **31**, 89.

Yaglou, C. P. and Minard, D. (1957): Control of heat casualties at military training centers. *A.M.A. Arch. Industr. Hlth.*, **16**, 302.

Yaglou, C. P., Riley, E. C. and Coggins, D. I. (1936): Ventilation requirements. *Heat., Pip. & Air Condit.*, **8**, 65; and *Trans. Amer. Soc. Heat. Vent. Engrs.*, **42**.

Yaglou, C. P. and Witheridge, W. N. (1937): Ventilation requirements (Part 2). *Heat., Pip. & Air Condit.*, **9**, 447; and *Trans. Amer. Soc. Heat. Vent. Engrs.*, **43**, 423.

Yarnold, K. W. and Fitzpatrick, H. D. N. (1940): Radiation and 'stuffiness'. Letter to *Nature*, **146**, 62.

APPENDIX A

THE INTERNATIONAL SYSTEM (S.I.) OF UNITS

The present is a time of transition for the Heating and Ventilating and Construction Industries, in which the 'Système International d'Unités'— internationally known as the S.I. units, has begun to replace older systems.

The programme for change in the Construction Industry given in PD 6030 (Feb. 1967) is that designers and quantity surveyors should change to the production of drawings and documents in S.I. terms for *all* new contracts beginning in 1969, and that main and sub-contractors will follow this lead. The change will be complete by the end of 1972, the only exception being the continuation of large contracts started in the old system and some works of maintenance and repair.

The International System (S.I.) is a coherent system of units, that is, the product or quotient of any two unit quantities in the system being the unit of the resultant quantity, as shown on page 508.

The multiples and sub-multiples of the basic S.I. units are:

Symbol	*Prefix*	*Multiplication factor*
T	tera	10^{12}
G	giga	10^{9}
M	mega	10^{6}
k	kilo	10^{3}
h	hecto	10^{2}
da	deca	10
d	deci	10^{-1}
c	centi	10^{-2}
m	milli	10^{-3}
μ	micro	10^{-6}
n	nano	10^{-9}
p	pico	10^{-12}

Reference can also be made to the following:

BS 3763 : 64. The International System (S.I.) Units.

Changing to the Metric System. Anderton, P. and Bigg, P. H. H.M.S.O.

Thermodynamic and Transport Properties of Fluids in S.I. Units. Mayhew, Y. R. and Rogers, G. F. C. Blackwell.

Quantity	S.I. unit	Recommended multiples of S.I. units	Other recommended units	Conversion factors Unit × Factor = Unit			Remarks
Basic units							
Length	m (metre)	km, mm, μm		ft	3.048×10^{-1}	m	
Mass	kg (kilogramme)	t, g, mg, μg		lb	4.536×10^{-1}	kg	1 t (tonne) = 10^3 kg
Time	s (second)			–		–	Although the minute hour and day are in common use, convenient sized derived units using the second arise in most cases, and multiples of these should be used.
Electric current	A (ampere)	mA, μA		–		–	
Thermodynamic temperature	°K (degree kelvin)			°R	5.556×10^{-1}	°K	
Luminous intensity	cd (candela)			candle	9.810×10^{-1}	cd	
Derived units							
Area	m^2	km^2, mm^2		ft^2	9.290×10^{-2}	m^2	
Volume	m^3	1 (litre) ml		ft^3	2.832×10^{-2}	m^3	1l = 10^{-3} m^3 = $1dm^3$
Velocity	m/s			ft/min	5.080×10^{-3}	m/s	
Acceleration	m/s^2			ft/s^2	3.048×10^{-1}	m/s^2	
Rotational frequency	s^{-1}		rev/s				
Volume flow rate	m^3/s	1/s		ft^3/min	4.720×10^{-4}	m^3/s	
Density	kg/m^3			lb/ft^3	1.602×10^{-1}	kg/m^3 = g/1	
Specific volume	m^3/kg			ft^3/lb	6.243×10^{-2}	m^3/kg	
Mass flow rate	kg/s			lb/h	1.260×10^{-4}	kg/s	
Pressure	N/m^2	bar		lbf/in^2	6.895×10^3	N/m^2	1 bar = 10^5 N/m^2
		mb (millibar)		in water	2.492	mb	1 mb = 10^2 N/m^2
Dynamic viscosity	Ns/m^2	cP (centipoise)		1bfs ft^2	4.789×10^1	Ns/m^2	1 cP = 10^{-3} Ns/m^2

Kinematic viscosity	m^2/s	cSt (centistokes)		ft^2/s	9.290×10^{-2}	m^2/s	1 cSt = 10^{-6} m^2/s
Energy, work, quantity of heat	J (joule)	GJ, MJ, kJ		Btu	1055×10^{3}	J	1J = 1 Nm = 1Ws
				(kilocalorie)	4.187×10^{3}	J	
Power (heat flow rate)	W (watt)	kW		(Btu/h	2.931×10^{-1}	W = J/s = Nm/s	
				hp	7.457×10^{2}	W	
Customary temperature			°C				
Temperature interval	deg		deg C	deg F	5.556×10^{-1}	deg C	During the transition period from Fahrenheit to Celsius units, the letter 'C' will be inserted after 'deg'.
Thermal conductivity	*						
Thermal resistivity	*						
Thermal conductance	*						
Thermal resistance	*						
Transmittance coefficient	*						
Heat capacity	J/deg		J/deg C	Btu/deg F	1.899×10^{3}	J/deg C	
Specific enthalpy	J/kg	kJ/kg		Btu/lb	2.326×10^{3}	J/Kg	
Specific heat capacity	J/kg deg	kJ/kg deg	kJ/kg deg C	Btu/lb deg F	4.187×10^{3}	J/kg deg C	
Latent heat	J/kg	kJ/kg		Btu/lb	2.326×10^{3}	J/kg	
Mass calorific value	J/kg	MJ/kg		Btu/lb	2.326×10^{3}	J/kg	
Volumetric calorific value	J/m^3	MJ/m^3		Btu/ft^3	3.726×10^{4}	J/m^3	
Force	N (newton)			lbf	4.448	N	

* See page 509.

Examples in the use of S.I. units

Comparison of existing and S.I. units for basic engineering equations

Force = mass x acceleration
1 pdl = 1 lb x 1 ft/s^2
1 lbf = 1 slug x 1 ft/s^2
1 N = 1 kg x 1 m/s^2
(1 lbf = 4.448 N)

Pressure = force ÷ area
1 lbf/ft^2 = 1 lbf ÷ 1 ft^2
1 N/m^2 = 1 N ÷ 1 m^2
(1 lbf/ft^2 = 47.88 N/m^2) and (1 lbf/in^2 = 6895 N/m^2)

Work = force x distance
1 ft-lbf = 1 lbf x 1 ft
1 J = 1 N x 1 m
(1 ft-lbf = 1.356 J) (Note: 1 J = 1 Nm = 1 Ws)

Power = work ÷ time
1 hp = 33000 ft-lbf ÷ 1 min
1 W = 1 J ÷ 1 s
(1 hp = 745.7 W)

Torque (moment of force) = force x radius
1 lbf ft = 1 lbf x 1 ft
1 Nm = 1 N x 1 m
(1 lbf ft = 1.356 Nm) (Note: A torque of 1 Nm = 1 J/rad)

Power = torque x angle ÷ time
1 ft-lbf/s = 1 lbf ft x 1 rad ÷ 1 s
1 hp = 33000 lbf ft x 2π rev ÷ 1 min
1 W = 1 Nm x 1 rad ÷ 1 s
(1 ft-lbf/s = 1.356 Nm) and (1 hp = 745.7 W)

Terms, symbols and units used in heat transmission *

Conductivity†

$$k \quad \frac{\text{Btu in}}{\text{ft}^2\text{h deg F}} \quad \frac{\text{Wm}}{\text{m}^2\,\text{deg C}} = \frac{\text{W}}{\text{m deg C}} \quad \left(1\ \frac{\text{Btu in}}{\text{ft}^2\text{h deg F}} = 0.144\ \frac{\text{W}}{\text{m deg C}}\right)$$

Thickness‡

$$L \ \text{in} \qquad \text{m} \ \ (1 \ \text{in} = 0.0254 \ \text{m})$$

Resistivity §

$$r\left(=\frac{1}{k}\right) \frac{\text{ft}^2\text{h deg F}}{\text{Btu in}} \quad \frac{\text{m}^2\,\text{deg C}}{\text{Wm}} = \frac{\text{m deg C}}{\text{W}} \quad \left(1\ \frac{\text{ft}^2\text{h deg F}}{\text{Btu in}} = 6.93\ \frac{\text{m deg C}}{\text{W}}\right)$$

Conductance

$$C\left(=\frac{k}{L}\right) \frac{\text{Btu}}{\text{ft}^2\text{h deg F}} \quad \frac{\text{W}}{\text{m}^2\,\text{deg C}} \quad \left(1\ \frac{\text{Btu}}{\text{ft}^2\text{h deg F}} = 5.678\ \frac{\text{W}}{\text{m}^2\,\text{deg C}}\right)$$

Resistance

$$R\left(=\frac{L}{k}\right) \frac{\text{ft}^2\text{h deg F}}{\text{Btu}} \quad \frac{\text{m}^2\,\text{deg C}}{\text{W}} \quad \left(1\ \frac{\text{ft}^2\text{h deg F}}{\text{Btu}} = 0.1761\ \frac{\text{m}^2\,\text{deg C}}{\text{W}}\right)$$

Transmittance coefficient

$$u \quad \frac{\text{Btu}}{\text{ft}^2\text{h deg F}} \quad \frac{\text{W}}{\text{m}^2\,\text{deg C}} \quad \left(1\ \frac{\text{Btu}}{\text{ft}^2\text{h deg F}} = 5.678\ \frac{\text{W}}{\text{m}^2\,\text{deg C}}\right)$$

Rate of heat flow/unit area

$$\frac{q}{A} \quad \frac{\text{Btu}}{\text{h ft}^2} \quad \frac{\text{W}}{\text{m}^2} \quad \left(1\ \frac{\text{Btu}}{\text{h ft}^2} = 3.1546\ \frac{\text{W}}{\text{m}^2}\right)$$

Rate of heat flow

$$q \quad \frac{\text{Btu}}{\text{h}} \quad \text{W} \quad \left(1\ \frac{\text{Btu}}{\text{h}} = 0.29307\ \text{W}\right)$$

Notes

* In the above S.I. units, time as a separate division disappears, the watt being a *rate* of heat flow of 1 Joule/sec. W = J/S.

† In the S.I. system, by using consistent units for thickness and area, the thickness over area, m/m^2, cancels to 1/m.

‡ In many cases the thickness of materials may be given in mm or cm and these must be converted to metres.

§ Note † applies, m^2/m cancels to m.

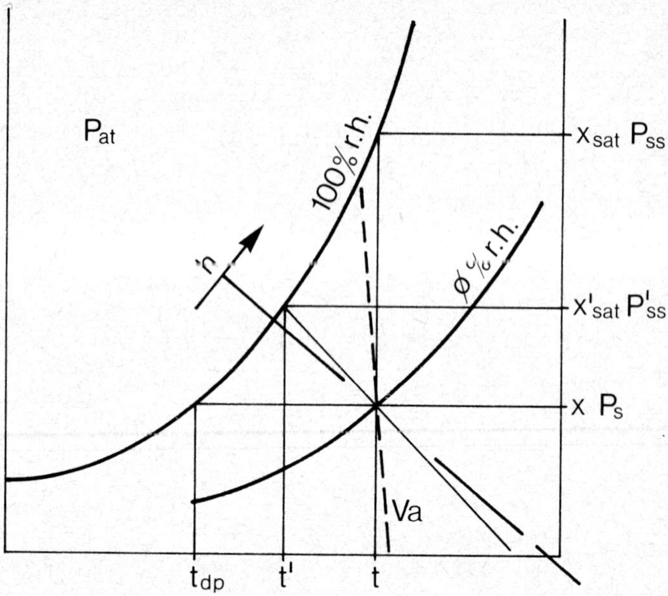

Fig. 127. Psychrometric chart

Terms used in psychrometry

t	dry-bulb temperature	°F d.b.	°C d.b.
t'	wet-bulb temperature	°F w.b.	°C w.b.
p_s	actual vapour pressure		
	$p_s = p'_{ss} - Ap_{at}(t-t')$	mb	N/m^2
p_{ss}	saturation vapour pressure at d.b. temp.	in Hg	mb
p'_{ss}	saturation vapour pressure at w.b. temp.	lbf/ft^2	
p_{at}	barometric pressure		
A	psychrometric constant. For w.b. temp. above 32°F $A = 3.7 \times 10^{-4}$	°F^{-1}	
	For w.b. temp. above 0°C $A = 6.67 \times 10^{-4}$		°C^{-1}
ϕ	relative humidity		
	$\phi = \dfrac{p_s}{p_{ss}} \times 100$	%r.h.	% r.h.

x	humidity ratio $x = \frac{M_s}{M_a} \times \frac{p_s}{p_{at} - p_s}$		
	$= 0.622 \frac{p_s}{p_{at} - p_s}$	lb/lb	kg/kg
	moisture content = 7,000 x	grains/lb	
	moisture content = 1,000 x		g/kg
V_a	humid volume $\frac{V_a}{m_a} = \frac{R_a T_a}{p_{at} - p_s}$	ft^3/lb	m^3/kg
h	enthalpy. Datum: dry air and water at 32°F, 0°C		
	$h = 1 \times 0.24 \times (t - 32) + x \times 0.45 \times (t - 32) + 1{,}075\, x$	Btu/lb	
	$= 1 \times 1.01 \times t + x \times 1.89 \times t + 2{,}500\, x$		kJ/kg
$t_{d.p.}$	dew point temperature	°F d.p.	°C d.p.
ρ	humid density $\rho = \frac{1 + x}{V_a}$	lb/ft^3	kg/m^3

APPENDIX B

The predicted mean comfort vote (PMV) for various levels of activity, expressed in kcal/m² /hour, of human subjects wearing clothing of different insulation values, expressed in 'clo' units, at various environmental temperatures (°C) and at various relative air velocities (m/s).

(From P. O. Fanger, THERMAL COMFORT, Danish Technical Press, Copenhagen, 1970. Republished by McGraw-Hill Book Company, New York, 1973.)

Activity level 50 kcal/m²/hour

Clothing clo	Ambient temp. °C	Relative velocity (m/s)								
		<0.10	0.10	0.15	0.20	0.30	0.40	0.50	1.00	1.50
0	26.	−1.62	−1.62	−1.96	−2.34					
	27.	−1.00	−1.00	−1.36	−1.69					
	28.	−0.39	−0.42	−0.76	−1.05					
	29.	0.21	0.13	−0.15	−0.39					
	30.	0.80	0.68	0.45	0.26					
	31.	1.39	1.25	1.08	0.94					
	32.	1.96	1.83	1.71	1.61					
	33.	2.50	2.41	2.34	2.29					
0.25	24.	−1.52	−1.52	−1.80	−2.06	−2.47				
	25.	−1.05	−1.05	−1.33	−1.57	−1.94	−2.24	−2.48		
	26.	−0.58	−0.61	−0.87	−1.08	−1.41	−1.67	−1.89	−2.66	
	27.	−0.12	−0.17	−0.40	−0.58	−0.87	−1.10	−1.29	−1.97	−2.41
	28.	0.34	0.27	0.07	−0.09	−0.34	−0.53	−0.70	−1.28	−1.66
	29.	0.80	0.71	0.54	0.41	0.20	0.04	−0.10	−0.58	−0.90
	30.	1.25	1.15	1.02	0.91	0.74	0.61	0.50	0.11	−0.14
	31.	1.71	1.61	1.51	1.43	1.30	1.20	1.12	0.83	0.63
0.50	23.	−1.10	−1.10	−1.33	−1.51	−1.78	−1.99	−2.16		
	24.	−0.72	−0.74	−0.95	−1.11	−1.36	−1.55	−1.70	−2.22	
	25.	−0.34	−0.38	−0.56	−0.71	−0.94	−1.11	−1.25	−1.71	−1.99
	26.	0.04	−0.01	−0.18	−0.31	−0.51	−0.66	−0.79	−1.19	−1.44
	27.	0.42	0.35	0.20	0.09	−0.08	−0.22	−0.33	−0.68	−0.90
	28.	0.80	0.72	0.59	0.49	0.34	0.23	0.14	−0.17	−0.36
	29.	1.17	1.08	0.98	0.90	0.77	0.68	0.60	0.34	0.19
	30.	1.54	1.45	1.37	1.30	1.20	1.13	1.06	0.86	0.73
0.75	21.	−1.11	−1.11	−1.30	−1.44	−1.66	−1.82	−1.95	−2.36	−2.60
	22.	−0.79	−0.81	−0.98	−1.11	−1.31	−1.46	−1.58	−1.95	−2.17
	23.	−0.47	−0.50	−0.66	−0.78	−0.96	−1.09	−1.20	−1.55	−1.75
	24.	−0.15	−0.19	−0.33	−0.44	−0.61	−0.73	−0.83	−1.14	−1.33
	25.	0.17	0.12	−0.01	−0.11	−0.26	−0.37	−0.46	−0.74	−0.90
	26.	0.49	0.43	0.31	0.23	0.09	0.00	−0.08	−0.33	−0.48
	27.	0.81	0.74	0.64	0.56	0.45	0.36	0.29	0.08	−0.05
	28.	1.12	1.05	0.96	0.90	0.80	0.73	0.67	0.48	0.37
1.00	20.	−0.85	−0.87	−1.02	−1.13	−1.29	−1.41	−1.51	−1.81	−1.98
	21.	−0.57	−0.60	−0.74	−0.84	−0.99	−1.11	−1.19	−1.47	−1.63
	22.	−0.30	−0.33	−0.46	−0.55	−0.69	−0.80	−0.88	−1.13	−1.28
	23.	−0.02	−0.07	−0.18	−0.27	−0.39	−0.49	−0.56	−0.79	−0.93
	24.	0.26	0.20	0.10	0.02	−0.09	−0.18	−0.25	−0.46	−0.58
	25.	0.53	0.48	0.38	0.31	0.21	0.13	0.07	−0.12	−0.23
	26.	0.81	0.75	0.66	0.60	0.51	0.44	0.39	0.22	0.13
	27.	1.08	1.02	0.95	0.89	0.81	0.75	0.71	0.56	0.48
1.25	16.	−1.37	−1.37	−1.51	−1.62	−1.78	−1.89	−1.98	−2.26	−2.41
	18.	−0.89	−0.91	−1.04	−1.14	−1.28	−1.38	−1.46	−1.70	−1.84
	20.	−0.42	−0.46	−0.57	−0.65	−0.77	−0.86	−0.93	−1.14	−1.26
	22.	0.07	0.02	−0.07	−0.14	−0.25	−0.32	−0.38	−0.56	−0.66
	24.	0.56	0.50	0.43	0.37	0.28	0.22	0.17	0.02	−0.06
	26.	1.04	0.99	0.93	0.88	0.81	0.76	0.72	0.61	0.54
	28.	1.53	1.48	1.43	1.40	1.34	1.31	1.28	1.19	1.14
	30.	2.01	1.97	1.93	1.91	1.88	1.85	1.83	1.77	1.74
1.50	14.	−1.36	−1.36	−1.49	−1.58	−1.72	−1.82	−1.89	−2.12	−2.25
	16.	−0.94	−0.95	−1.07	−1.15	−1.27	−1.36	−1.43	−1.63	−1.75
	18.	−0.52	−0.54	−0.64	−0.72	−0.82	−0.90	−0.96	−1.14	−1.24
	20.	−0.09	−0.13	−0.22	−0.28	−0.37	−0.44	−0.49	−0.65	−0.74
	22.	0.35	0.30	0.23	0.18	0.10	0.04	0.00	−0.14	−0.21
	24.	0.79	0.74	0.68	0.63	0.57	0.52	0.49	0.37	0.31
	26.	1.23	1.18	1.13	1.09	1.04	1.01	0.98	0.89	0.84
	28.	1.67	1.62	1.58	1.56	1.52	1.49	1.47	1.40	1.37

Activity level 60 kcal/m²/hour

Clothing	Ambient temp.	Relative velocity (m/s)								
clo	°C	<0.10	0.10	0.15	0.20	0.30	0.40	0.50	1.00	1.50
0	25.	−1.33	−1.33	−1.59	−1.92					
	26.	−0.83	−0.83	−1.11	−1.40					
	27.	−0.33	−0.33	−0.63	−0.88					
	28.	0.15	0.12	−0.14	−0.36					
	29.	0.63	0.56	0.35	0.17					
	30.	1.10	1.01	0.84	0.69					
	31.	1.57	1.47	1.34	1.24					
	32.	2.03	1.93	1.85	1.78					
0.25	23.	−1.18	−1.18	−1.39	−1.61	−1.97	−2.25			
	24.	−0.79	−0.79	−1.02	−1.22	−1.54	−1.80	−2.01		
	25.	−0.42	−0.42	−0.64	−0.83	−1.11	−1.34	−1.54	−2.21	
	26.	−0.04	−0.07	−0.27	−0.43	−0.68	−0.89	−1.06	−1.65	−2.04
	27.	0.33	0.29	0.11	−0.03	−0.25	−0.43	−0.58	−1.09	−1.43
	28.	0.71	0.64	0.49	0.37	0.18	0.03	−0.10	−0.54	−0.82
	29.	1.07	0.99	0.87	0.77	0.61	0.49	0.39	0.02	−0.22
	30.	1.43	1.35	1.25	1.17	1.05	0.95	0.87	0.58	0.39
0.50	18.	−2.01	−2.01	−2.17	−2.38	−2.70				
	20.	−1.41	−1.41	−1.58	−1.76	−2.04	−2.25	−2.42		
	22.	−0.79	−0.79	−0.97	−1.13	−1.36	−1.54	−1.69	−2.17	−2.46
	24.	−0.17	−0.20	−0.36	−0.48	−0.68	−0.83	−0.95	−1.35	−1.59
	26.	0.44	0.39	0.26	0.16	0.01	−0.11	−0.21	−0.52	−0.71
	28.	1.05	0.98	0.88	0.81	0.70	0.61	0.54	0.31	0.16
	30.	1.64	1.57	1.51	1.46	1.39	1.33	1.29	1.14	1.04
	32.	2.25	2.20	2.17	2.15	2.11	2.09	2.07	1.99	1.95
0.75	16.	−1.77	−1.77	−1.91	−2.07	−2.31	−2.49			
	18.	−1.27	−1.27	−1.42	−1.56	−1.77	−1.93	−2.05	−2.45	
	20.	−0.77	−0.77	−0.92	−1.04	−1.23	−1.36	−1.47	−1.82	−2.02
	22.	−0.25	−0.27	−0.40	−0.51	−0.66	−0.78	−0.87	−1.17	−1.34
	24.	0.27	0.23	0.12	0.03	−0.10	−0.19	−0.27	−0.51	−0.65
	26.	0.78	0.73	0.64	0.57	0.47	0.40	0.34	0.14	0.03
	28.	1.29	1.23	1.17	1.12	1.04	0.99	0.94	0.80	0.72
	30.	1.80	1.74	1.70	1.67	1.62	1.58	1.55	1.46	1.41
1.00	16.	−1.18	−1.18	−1.31	−1.43	−1.59	−1.72	−1.82	−2.12	−2.29
	18.	−0.75	−0.75	−0.88	−0.98	−1.13	−1.24	−1.33	−1.59	−1.75
	20.	−0.32	−0.33	−0.45	−0.54	−0.67	−0.76	−0.83	−1.07	−1.20
	22.	0.13	0.10	0.00	−0.07	−0.18	−0.26	−0.32	−0.52	−0.64
	24.	0.58	0.54	0.46	0.40	0.31	0.24	0.19	0.02	−0.07
	26.	1.03	0.98	0.91	0.86	0.79	0.74	0.70	0.57	0.50
	28.	1.47	1.42	1.37	1.34	1.28	1.24	1.21	1.12	1.06
	30.	1.91	1.86	1.83	1.81	1.78	1.75	1.73	1.67	1.63
1.25	14.	−1.12	−1.12	−1.24	−1.34	−1.48	−1.58	−1.66	−1.90	−2.04
	16.	−0.74	−0.75	−0.86	−0.95	−1.07	−1.16	−1.23	−1.45	−1.57
	18.	−0.36	−0.38	−0.48	−0.55	−0.66	−0.74	−0.81	−1.00	−1.11
	20.	0.02	−0.01	−0.10	−0.16	−0.26	−0.33	−0.38	−0.55	−0.64
	22.	0.42	0.38	0.31	0.25	0.17	0.11	0.07	−0.08	−0.16
	24.	0.81	0.77	0.71	0.66	0.60	0.55	0.51	0.39	0.33
	26.	1.21	1.16	1.11	1.08	1.03	0.99	0.96	0.87	0.82
	28.	1.60	1.56	1.52	1.50	1.46	1.43	1.41	1.34	1.30
1.50	12.	−1.09	−1.09	−1.19	−1.27	−1.39	−1.48	−1.55	−1.75	−1.86
	14.	−0.75	−0.75	−0.85	−0.93	−1.03	−1.11	−1.17	−1.35	−1.45
	16.	−0.41	−0.42	−0.51	−0.58	−0.67	−0.74	−0.79	−0.96	−1.05
	18.	−0.06	−0.09	−0.17	−0.22	−0.31	−0.37	−0.42	−0.56	−0.64
	20.	0.28	0.25	0.18	0.13	0.05	0.00	−0.04	−0.16	−0.23
	22.	0.63	0.60	0.54	0.50	0.44	0.39	0.36	0.25	0.19
	24.	0.99	0.95	0.91	0.87	0.82	0.78	0.76	0.67	0.62
	26.	1.35	1.31	1.27	1.24	1.20	1.18	1.15	1.08	1.05

Activity level 70 kcal/m²/hour

Clothing clo	Ambient temp. °C	Relative velocity (m/s) <0.10	0.10	0.15	0.20	0.30	0.40	0.50	1.00	1.50
0	24.	−1.14	−1.14	−1.35	−1.65					
	25.	−0.72	−0.72	−0.95	−1.21					
	26.	−0.30	−0.30	−0.54	−0.78					
	27.	0.11	0.11	−0.14	−0.34					
	28.	0.52	0.48	0.27	0.10					
	29.	0.92	0.85	0.69	0.54					
	30.	1.31	1.23	1.10	0.99					
	31.	1.71	1.62	1.52	1.45					
0.25	22.	−0.95	−0.95	−1.12	−1.33	−1.64	−1.90	−2.11		
	23.	−0.63	−0.63	−0.81	−0.99	−1.28	−1.51	−1.71	−2.38	
	24.	−0.31	−0.31	−0.50	−0.66	−0.92	−1.13	−1.31	−1.91	−2.31
	25.	0.01	0.00	−0.18	−0.33	−0.56	−0.75	−0.90	−1.45	−1.80
	26.	0.33	0.30	0.14	0.01	−0.20	−0.36	−0.50	−0.98	−1.29
	27.	0.64	0.59	0.45	0.34	0.16	0.02	−0.10	−0.51	−0.78
	28.	0.95	0.89	0.77	0.68	0.53	0.41	0.31	−0.04	−0.27
	29.	1.26	1.19	1.09	1.02	0.89	0.80	0.72	0.43	0.24
0.50	18.	−1.36	−1.36	−1.49	−1.66	−1.93	−2.12	−2.29		
	20.	−0.85	−0.85	−1.00	−1.14	−1.37	−1.54	−1.68	−2.15	−2.43
	22.	−0.33	−0.33	−0.48	−0.61	−0.80	−0.95	−1.06	−1.46	−1.70
	24.	0.19	0.17	0.04	−0.07	−0.22	−0.34	−0.44	−0.76	−0.96
	26.	0.71	0.66	0.56	0.48	0.35	0.26	0.18	−0.07	−0.23
	28.	1.22	1.16	1.09	1.03	0.94	0.87	0.81	0.63	0.51
	30.	1.72	1.66	1.62	1.58	1.52	1.48	1.44	1.33	1.25
	32.	2.23	2.19	2.17	2.16	2.13	2.11	2.10	2.05	2.02
0.75	16.	−1.17	−1.17	−1.29	−1.42	−1.62	−1.77	−1.88	−2.26	−2.48
	18.	−0.75	−0.75	−0.87	−0.99	−1.16	−1.29	−1.39	−1.72	−1.92
	20.	−0.33	−0.33	−0.45	−0.55	−0.70	−0.82	−0.91	−1.19	−1.36
	22.	0.11	0.09	−0.02	−0.10	−0.23	−0.32	−0.40	−0.64	−0.78
	24.	0.55	0.51	0.42	0.35	0.25	0.17	0.11	−0.09	−0.20
	26.	0.98	0.94	0.87	0.81	0.73	0.67	0.62	0.47	0.37
	28.	1.41	1.36	1.31	1.27	1.21	1.17	1.13	1.02	0.95
	30.	1.84	1.79	1.76	1.73	1.70	1.67	1.65	1.58	1.53
1.00	14.	−1.05	−1.05	−1.16	−1.26	−1.42	−1.53	−1.62	−1.91	−2.07
	16.	−0.69	−0.69	−0.80	−0.89	−1.03	−1.13	−1.21	−1.46	−1.61
	18.	−0.32	−0.32	−0.43	−0.52	−0.64	−0.73	−0.80	−1.02	−1.15
	20.	0.04	0.03	−0.07	−0.14	−0.25	−0.32	−0.38	−0.58	−0.69
	22.	0.42	0.39	0.31	0.25	0.16	0.10	0.05	−0.12	−0.21
	24.	0.80	0.76	0.70	0.65	0.57	0.52	0.48	0.35	0.27
	26.	1.18	1.13	1.08	1.04	0.99	0.95	0.91	0.81	0.75
	28.	1.55	1.51	1.47	1.44	1.40	1.37	1.35	1.27	1.23
1.25	12.	−0.97	−0.97	−1.06	−1.15	−1.28	−1.37	−1.45	−1.67	−1.80
	14.	−0.65	−0.65	−0.75	−0.82	−0.94	−1.02	−1.09	−1.29	−1.40
	16.	−0.33	−0.33	−0.43	−0.50	−0.60	−0.67	−0.73	−0.91	−1.01
	18.	−0.01	−0.02	−0.10	−0.17	−0.26	−0.32	−0.37	−0.53	−0.62
	20.	0.32	0.29	0.22	0.17	0.09	0.03	−0.01	−0.15	−0.22
	22.	0.65	0.62	0.56	0.52	0.45	0.40	0.36	0.25	0.18
	24.	0.99	0.95	0.90	0.87	0.81	0.77	0.74	0.65	0.59
	26.	1.32	1.28	1.25	1.22	1.18	1.14	1.12	1.05	1.00
1.50	10.	−0.91	−0.91	−1.00	−1.08	−1.18	−1.26	−1.32	−1.51	−1.61
	12.	−0.63	−0.63	−0.71	−0.78	−0.88	−0.95	−1.01	−1.17	−1.27
	14.	−0.34	−0.34	−0.43	−0.49	−0.58	−0.64	−0.69	−0.84	−0.92
	16.	−0.05	−0.06	−0.14	−0.19	−0.27	−0.33	−0.37	−0.50	−0.58
	18.	0.24	0.22	0.15	0.11	0.04	−0.01	−0.05	−0.17	−0.23
	20.	0.53	0.50	0.45	0.40	0.34	0.30	0.27	0.17	0.11
	22.	0.83	0.80	0.75	0.72	0.67	0.63	0.60	0.52	0.47
	24.	1.13	1.10	1.06	1.03	0.99	0.96	0.94	0.87	0.83

Activity level 80 kcal/m²/hour

Clothing clo	Ambient temp. °C	Relative velocity (m/s) <0.10	0.10	0.15	0.20	0.30	0.40	0.50	1.00	1.50
0	23.	−1.12	−1.12	−1.29	−1.57					
	24.	−0.74	−0.74	−0.93	−1.18					
	25.	−0.36	−0.36	−0.57	−0.79					
	26.	0.01	0.01	−0.20	−0.40					
	27.	0.38	0.37	0.17	0.00					
	28.	0.75	0.70	0.53	0.39					
	29.	1.11	1.04	0.90	0.79					
	30.	1.46	1.38	1.27	1.19					
0.25	16.	−2.29	−2.29	−2.36	−2.62					
	18.	−1.72	−1.72	−1.83	−2.06	−2.42				
	20.	−1.15	−1.15	−1.29	−1.49	−1.80	−2.05	−2.26		
	22.	−0.58	−0.58	−0.73	−0.90	−1.17	−1.38	−1.55	−2.17	−2.58
	24.	−0.01	−0.01	−0.17	−0.31	−0.53	−0.70	−0.84	−1.35	−1.68
	26.	0.56	0.53	0.39	0.29	0.12	−0.02	−0.13	−0.52	−0.78
	28.	1.12	1.06	0.96	0.89	0.77	0.67	0.59	0.31	0.12
	30.	1.66	1.60	1.54	1.49	1.42	1.36	1.31	1.14	1.02
0.50	14.	−1.85	−1.85	−1.94	−2.12	−2.40				
	16.	−1.40	−1.40	−1.50	−1.67	−1.92	−2.11	−2.26		
	18.	−0.95	−0.95	−1.07	−1.21	−1.43	−1.59	−1.73	−2.18	−2.46
	20.	−0.49	−0.49	−0.62	−0.75	−0.94	−1.08	−1.20	−1.59	−1.82
	22.	−0.03	−0.03	−0.16	−0.27	−0.43	−0.55	−0.65	−0.98	−1.18
	24.	0.43	0.41	0.30	0.21	0.08	−0.02	−0.10	−0.37	−0.53
	26.	0.89	0.85	0.76	0.70	0.60	0.52	0.46	0.25	0.12
	28.	1.34	1.29	1.23	1.18	1.11	1.06	1.01	0.86	0.77
0.75	14.	−1.16	−1.16	−1.26	−1.38	−1.57	−1.71	−1.82	−2.17	−2.38
	16.	−0.79	−0.79	−0.89	−1.00	−1.17	−1.29	−1.39	−1.70	−1.88
	18.	−0.41	−0.41	−0.52	−0.62	−0.76	−0.87	−0.96	−1.23	−1.39
	20.	−0.04	−0.04	−0.15	−0.23	−0.36	−0.45	−0.52	−0.76	−0.90
	22.	0.35	0.33	0.24	0.17	0.07	−0.01	−0.07	−0.27	−0.39
	24.	0.74	0.71	0.63	0.58	0.49	0.43	0.38	0.21	0.12
	26.	1.12	1.08	1.03	0.98	0.92	0.87	0.83	0.70	0.62
	28.	1.51	1.46	1.42	1.39	1.34	1.31	1.28	1.19	1.14
1.00	12.	−1.01	−1.01	−1.10	−1.19	−1.34	−1.45	−1.53	−1.79	−1.94
	14.	−0.68	−0.68	−0.78	−0.87	−1.00	−1.09	−1.17	−1.40	−1.54
	16.	−0.36	−0.36	−0.46	−0.53	−0.65	−0.74	−0.80	−1.01	−1.13
	18.	−0.04	−0.04	−0.13	−0.20	−0.30	−0.38	−0.44	−0.62	−0.73
	20.	0.28	0.27	0.19	0.13	0.04	−0.02	−0.07	−0.23	−0.32
	22.	0.62	0.59	0.53	0.48	0.41	0.35	0.31	0.17	0.10
	24.	0.96	0.92	0.87	0.83	0.77	0.73	0.69	0.58	0.52
	26.	1.29	1.25	1.21	1.18	1.14	1.10	1.07	0.99	0.94
1.25	10.	−0.90	−0.90	−0.98	−1.06	−1.18	−1.27	−1.33	−1.54	−1.66
	12.	−0.62	−0.62	−0.70	−0.77	−0.88	−0.96	−1.02	−1.21	−1.31
	14.	−0.33	−0.33	−0.42	−0.48	−0.58	−0.65	−0.70	−0.87	−0.97
	16.	−0.05	−0.05	−0.13	−0.19	−0.28	−0.34	−0.39	−0.54	−0.62
	18.	0.24	0.22	0.15	0.10	0.03	−0.03	−0.07	−0.20	−0.28
	20.	0.52	0.50	0.44	0.40	0.33	0.29	0.25	0.14	0.07
	22.	0.82	0.79	0.74	0.71	0.65	0.61	0.58	0.49	0.43
	24.	1.12	1.09	1.05	1.02	0.97	0.94	0.92	0.84	0.79
1.50	8.	−0.82	−0.82	−0.89	−0.96	−1.06	−1.13	−1.19	−1.36	−1.45
	10.	−0.57	−0.57	−0.65	−0.71	−0.80	−0.86	−0.92	−1.07	−1.16
	12.	−0.32	−0.32	−0.39	−0.45	−0.53	−0.59	−0.64	−0.78	−0.85
	14.	−0.06	−0.07	−0.14	−0.19	−0.26	−0.31	−0.36	−0.48	−0.55
	16.	0.19	0.18	0.12	0.07	0.01	−0.04	−0.07	−0.19	−0.25
	18.	0.45	0.43	0.38	0.34	0.28	0.24	0.21	0.11	0.05
	20.	0.71	0.68	0.64	0.60	0.55	0.52	0.49	0.41	0.36
	22.	0.97	0.95	0.91	0.88	0.84	0.81	0.79	0.72	0.68

Activity level 90 kcal/m²/hour

Clothing clo	Ambient temp. °C	Relative velocity (m/s) <0.10	0.10	0.15	0.20	0.30	0.40	0.50	1.00	1.50
0	22.	−1.05	−1.05	−1.19	−1.46					
	23.	−0.70	−0.70	−0.86	−1.11					
	24.	−0.36	−0.36	−0.53	−0.75					
	25.	−0.01	−0.01	−0.20	−0.40					
	26.	0.32	0.32	0.13	−0.04					
	27.	0.66	0.63	0.46	0.32					
	28.	0.99	0.94	0.80	0.68					
	29.	1.31	1.25	1.13	1.04					
0.25	16.	−1.79	−1.79	−1.86	−2.09	−2.46				
	18.	−1.28	−1.28	−1.38	−1.58	−1.90	−2.16	−2.37		
	20.	−0.76	−0.76	−0.89	−1.06	−1.34	−1.56	−1.75	−2.39	−2.82
	22.	−0.24	−0.24	−0.38	−0.53	−0.76	−0.95	−1.10	−1.65	−2.01
	24.	0.28	0.28	0.13	0.01	−0.18	−0.33	−0.46	−0.90	−1.19
	26.	0.79	0.76	0.64	0.55	0.40	0.29	0.19	−0.15	−0.38
	28.	1.29	1.24	1.16	1.10	0.99	0.91	0.84	0.60	0.44
	30.	1.79	1.73	1.68	1.65	1.59	1.54	1.50	1.36	1.27
0.50	14.	−1.42	−1.42	−1.50	−1.66	−1.91	−2.10	−2.25		
	16.	−1.01	−1.01	−1.10	−1.25	−1.47	−1.64	−1.77	−2.23	−2.51
	18.	−0.59	−0.59	−0.70	−0.83	−1.02	−1.17	−1.29	−1.69	−1.94
	20.	−0.18	−0.18	−0.30	−0.41	−0.58	−0.71	−0.81	−1.15	−1.36
	22.	0.24	0.23	0.12	0.02	−0.12	−0.22	−0.31	−0.60	−0.78
	24.	0.66	0.63	0.54	0.46	0.35	0.26	0.19	−0.04	−0.19
	26.	1.07	1.03	0.96	0.90	0.82	0.75	0.69	0.51	0.40
	28.	1.48	1.44	1.39	1.35	1.29	1.24	1.20	1.07	1.00
0.75	12.	−1.15	−1.15	−1.23	−1.35	−1.53	−1.67	−1.78	−2.13	−2.33
	14.	−0.81	−0.81	−0.89	−1.00	−1.17	−1.29	−1.39	−1.70	−1.89
	16.	−0.46	−0.46	−0.56	−0.66	−0.80	−0.91	−1.00	−1.28	−1.44
	18.	−0.12	−0.12	−0.22	−0.31	−0.43	−0.53	−0.61	−0.85	−0.99
	20.	0.22	0.21	0.12	0.04	−0.07	−0.15	−0.21	−0.42	−0.55
	22.	0.57	0.55	0.47	0.41	0.32	0.25	0.20	0.02	−0.09
	24.	0.92	0.89	0.83	0.78	0.71	0.65	0.60	0.46	0.38
	26.	1.28	1.24	1.19	1.15	1.09	1.05	1.02	0.91	0.84
1.00	10.	−0.97	−0.97	−1.04	−1.14	−1.28	−1.39	−1.47	−1.73	−1.88
	12.	−0.68	−0.68	−0.76	−0.84	−0.97	−1.07	−1.14	−1.38	−1.51
	14.	−0.38	−0.38	−0.46	−0.54	−0.66	−0.74	−0.81	−1.02	−1.14
	16.	−0.09	−0.09	−0.17	−0.24	−0.35	−0.42	−0.48	−0.67	−0.78
	18.	0.21	0.20	0.12	0.06	−0.03	−0.10	−0.15	−0.31	−0.41
	20.	0.50	0.48	0.42	0.36	0.29	0.23	0.18	0.04	−0.04
	22.	0.81	0.78	0.73	0.68	0.62	0.57	0.53	0.41	0.35
	24.	1.11	1.08	1.04	1.00	0.95	0.91	0.88	0.78	0.73
1.25	8.	−0.84	−0.84	−0.91	−0.99	−1.10	−1.19	−1.25	−1.46	−1.57
	10.	−0.59	−0.59	−0.66	−0.73	−0.84	−0.91	−0.97	−1.16	−1.26
	12.	−0.33	−0.33	−0.40	−0.47	−0.56	−0.63	−0.69	−0.86	−0.95
	14.	−0.07	−0.07	−0.14	−0.20	−0.29	−0.35	−0.40	−0.55	−0.63
	16.	0.19	0.18	0.12	0.06	−0.01	−0.07	−0.11	−0.24	−0.32
	18.	0.45	0.44	0.38	0.33	0.26	0.22	0.18	0.06	0.00
	20.	0.71	0.69	0.64	0.60	0.54	0.50	0.47	0.37	0.31
	22.	0.98	0.96	0.91	0.88	0.83	0.80	0.77	0.69	0.64
1.50	−2.	−1.63	−1.63	−1.68	−1.77	−1.90	−2.00	−2.07	−2.29	−2.41
	2.	−1.19	−1.19	−1.25	−1.33	−1.44	−1.52	−1.58	−1.78	−1.88
	6.	−0.74	−0.74	−0.80	−0.87	−0.97	−1.04	−1.09	−1.26	−1.35
	10.	−0.29	−0.29	−0.36	−0.42	−0.50	−0.56	−0.60	−0.74	−0.82
	14.	0.17	0.17	0.11	0.06	−0.01	−0.05	−0.09	−0.20	−0.26
	18.	0.64	0.62	0.57	0.54	0.49	0.45	0.42	0.34	0.29
	22.	1.12	1.09	1.06	1.03	1.00	0.97	0.95	0.89	0.85
	26.	1.61	1.58	1.56	1.55	1.52	1.51	1.50	1.46	1.44

Activity level 100 kcal/m²/hour

Clothing clo	Ambient temp. °C	Relative velocity (m/s)								
		<0.10	0.10	0.15	0.20	0.30	0.40	0.50	1.00	1.50
0	18.		−2.00	−2.02	−2.35					
	20.		−1.35	−1.43	−1.72					
	22.		−0.69	−0.82	−1.06					
	24.		−0.04	−0.21	−0.41					
	26.		0.59	0.41	0.26					
	28.		1.16	1.03	0.93					
	30.		1.73	1.66	1.60					
	32.		2.33	2.32	2.31					
0.25	16.		−1.41	−1.48	−1.69	−2.02	−2.29	−2.51		
	18.		−0.93	−1.03	−1.21	−1.50	−1.74	−1.93	−2.61	
	20.		−0.45	−0.57	−0.73	−0.98	−1.18	−1.35	−1.93	−2.32
	22.		0.04	−0.09	−0.23	−0.44	−0.61	−0.75	−1.24	−1.56
	24.		0.52	0.38	0.28	0.10	−0.03	−0.14	−0.54	−0.80
	26.		0.97	0.86	0.78	0.65	0.55	0.46	0.16	−0.04
	28.		1.42	1.35	1.29	1.20	1.13	1.07	0.86	0.72
	30.		1.88	1.84	1.81	1.76	1.72	1.68	1.57	1.49
0.50	14.		−1.08	−1.16	−1.31	−1.53	−1.71	−1.85	−2.32	
	16.		−0.69	−0.79	−0.92	−1.12	−1.27	−1.40	−1.82	−2.07
	18.		−0.31	−0.41	−0.53	−0.70	−0.84	−0.95	−1.31	−1.54
	20.		0.07	−0.04	−0.14	−0.29	−0.40	−0.50	−0.81	−1.00
	22.		0.46	0.35	0.27	0.15	0.05	−0.03	−0.29	−0.45
	24.		0.83	0.75	0.68	0.58	0.50	0.44	0.23	0.10
	26.		1.21	1.15	1.10	1.02	0.96	0.91	0.75	0.65
	28.		1.59	1.55	1.51	1.46	1.42	1.38	1.27	1.21
0.75	10.		−1.16	−1.23	−1.35	−1.54	−1.67	−1.78	−2.14	−2.34
	12.		−0.84	−0.92	−1.03	−1.20	−1.32	−1.42	−1.74	−1.93
	14.		−0.52	−0.60	−0.70	−0.85	−0.97	−1.06	−1.34	−1.51
	16.		−0.20	−0.29	−0.38	−0.51	−0.61	−0.69	−0.95	−1.10
	18.		0.12	0.03	−0.05	−0.17	−0.26	−0.32	−0.55	−0.68
	20.		0.43	0.34	0.28	0.18	0.10	0.04	−0.15	−0.26
	22.		0.75	0.68	0.62	0.54	0.48	0.43	0.27	0.17
	24.		1.07	1.01	0.97	0.90	0.85	0.81	0.68	0.61
1.00	8.		−0.95	−1.02	−1.11	−1.26	−1.36	−1.45	−1.71	−1.86
	10.		−0.68	−0.75	−0.84	−0.97	−1.07	−1.15	−1.38	−1.52
	12.		−0.41	−0.48	−0.56	−0.68	−0.77	−0.84	−1.05	−1.18
	14.		−0.13	−0.21	−0.28	−0.39	−0.47	−0.53	−0.72	−0.83
	16.		0.14	0.06	0.00	−0.10	−0.16	−0.22	−0.39	−0.49
	18.		0.41	0.34	0.28	0.20	0.14	0.09	−0.06	−0.14
	20.		0.68	0.61	0.57	0.50	0.44	0.40	0.28	0.20
	22.		0.96	0.91	0.87	0.81	0.76	0.73	0.62	0.56
1.25	−2.		−1.74	−1.77	−1.88	−2.04	−2.15	−2.24	−2.51	−2.66
	2.		−1.27	−1.32	−1.42	−1.55	−1.65	−1.73	−1.97	−2.10
	6.		−0.80	−0.86	−0.94	−1.06	−1.14	−1.21	−1.41	−1.53
	10.		−0.33	−0.40	−0.47	−0.56	−0.64	−0.69	−0.86	−0.96
	14.		0.15	0.08	0.03	−0.05	−0.11	−0.15	−0.29	−0.37
	18.		0.63	0.57	0.53	0.47	0.42	0.39	0.28	0.22
	22.		1.11	1.08	1.05	1.00	0.97	0.95	0.87	0.83
	26.		1.62	1.60	1.58	1.55	1.53	1.52	1.47	1.45
1.50	−4.		−1.52	−1.56	−1.65	−1.78	−1.87	−1.95	−2.16	−2.28
	0.		−1.11	−1.16	−1.24	−1.35	−1.44	−1.50	−1.69	−1.79
	4.		−0.69	−0.75	−0.82	−0.92	−0.99	−1.04	−1.20	−1.29
	8.		−0.27	−0.33	−0.39	−0.47	−0.53	−0.58	−0.72	−0.79
	12.		0.15	0.09	0.05	−0.02	−0.07	−0.11	−0.22	−0.29
	16.		0.58	0.53	0.49	0.44	0.40	0.37	0.28	0.23
	20.		1.01	0.97	0.94	0.91	0.88	0.85	0.79	0.75
	24.		1.47	1.44	1.43	1.40	1.38	1.36	1.32	1.29

Activity level 120 kcal/m²/hour

Clothing clo	Ambient temp. °C	Relative velocity (m/s) <0.10	0.10	0.15	0.20	0.30	0.40	0.50	1.00	1.50
0	16.			−1.88	−2.22					
	18.			−1.34	−1.63					
	20.			−0.79	−1.05					
	22.			−0.23	−0.44					
	24.			0.34	0.17					
	26.			0.91	0.78					
	28.			1.49	1.40					
	30.			2.07	2.03					
0.25	14.			−1.31	−1.52	−1.85	−2.12	−2.34		
	16.			−0.89	−1.08	−1.37	−1.61	−1.81	−2.49	
	18.			−0.47	−0.63	−0.89	−1.10	−1.27	−1.87	−2.26
	20.			−0.05	−0.19	−0.41	−0.58	−0.73	−1.24	−1.58
	22.			0.39	0.28	0.09	−0.05	−0.17	−0.60	−0.88
	24.			0.84	0.74	0.60	0.48	0.39	0.05	−0.17
	26.			1.28	1.22	1.11	1.02	0.95	0.70	0.53
	28.			1.73	1.69	1.62	1.56	1.51	1.35	1.24
0.50	12.			−0.97	−1.11	−1.34	−1.51	−1.65	−2.12	−2.40
	14.			−0.62	−0.76	−0.96	−1.11	−1.24	−1.65	−1.91
	16.			−0.28	−0.40	−0.58	−0.71	−0.82	−1.19	−1.42
	18.			0.07	−0.03	−0.19	−0.31	−0.41	−0.73	−0.92
	20.			0.42	0.33	0.20	0.10	0.01	−0.26	−0.43
	22.			0.78	0.71	0.60	0.52	0.45	0.22	0.08
	24.			1.15	1.09	1.00	0.94	0.88	0.70	0.59
	26.			1.52	1.47	1.41	1.36	1.32	1.19	1.11
0.75	10.			−0.71	−0.82	−0.99	−1.11	−1.21	−1.53	−1.71
	12.			−0.42	−0.52	−0.67	−0.79	−0.88	−1.16	−1.33
	14.			−0.13	−0.22	−0.36	−0.46	−0.54	−0.79	−0.94
	16.			0.16	0.08	−0.04	−0.13	−0.20	−0.42	−0.56
	18.			0.45	0.38	0.28	0.20	0.14	−0.05	−0.17
	20.			0.75	0.69	0.60	0.54	0.49	0.32	0.22
	22.			1.06	1.01	0.94	0.88	0.84	0.70	0.62
	24.			1.37	1.33	1.27	1.23	1.20	1.09	1.02
1.00	6.			−0.78	−0.87	−1.01	−1.12	−1.20	−1.45	−1.60
	8.			−0.54	−0.62	−0.75	−0.85	−0.92	−1.15	−1.29
	10.			−0.29	−0.37	−0.49	−0.57	−0.64	−0.86	−0.98
	12.			−0.04	−0.11	−0.22	−0.29	−0.36	−0.55	−0.66
	14.			0.21	0.15	0.06	−0.01	−0.07	−0.24	−0.34
	16.			0.47	0.41	0.33	0.27	0.22	0.07	−0.02
	18.			0.73	0.68	0.60	0.55	0.51	0.38	0.30
	20.			0.98	0.94	0.88	0.84	0.80	0.69	0.62
1.25	−4.			−1.46	−1.56	−1.72	−1.83	−1.91	−2.17	−2.32
	0.			−1.05	−1.14	−1.27	−1.37	−1.44	−1.67	−1.80
	4.			−0.62	−0.70	−0.81	−0.90	−0.96	−1.16	−1.27
	8.			−0.19	−0.26	−0.35	−0.42	−0.48	−0.64	−0.74
	12.			0.25	0.20	0.12	0.06	0.02	−0.12	−0.20
	16.			0.70	0.66	0.60	0.55	0.52	0.41	0.35
	20.			1.16	1.13	1.08	1.05	1.02	0.94	0.90
	24.			1.65	1.63	1.60	1.57	1.56	1.51	1.48
1.50	−8.			−1.44	−1.53	−1.67	−1.76	−1.83	−2.05	−2.17
	−4.			−1.07	−1.15	−1.27	−1.35	−1.42	−1.61	−1.72
	0.			−0.70	−0.77	−0.87	−0.94	−1.00	−1.17	−1.27
	4.			−0.31	−0.37	−0.46	−0.53	−0.57	−0.72	−0.80
	8.			0.07	0.02	−0.05	−0.10	−0.14	−0.27	−0.34
	12.			0.47	0.43	0.37	0.33	0.29	0.19	0.14
	16.			0.88	0.85	0.80	0.77	0.74	0.66	0.62
	20.			1.29	1.27	1.24	1.21	1.19	1.13	1.10

Activity level 150 kcal/m²/hour

Clothing clo	Ambient temp. °C	Relative velocity (m/s) <0.10	0.10	0.15	0.20	0.30	0.40	0.50	1.00	1.50
0	14.				−1.92	−2.49				
	16.				−1.36	−1.87				
	18.				−0.80	−1.24				
	20.				−0.24	−0.61				
	22.				0.34	0.04				
	24.				0.93	0.70				
	26.				1.52	1.36				
	28.				2.12	2.02				
0.25	12.				−1.19	−1.53	−1.80	−2.02		
	14.				−0.77	−1.07	−1.31	−1.51	−2.21	
	16.				−0.35	−0.61	−0.82	−1.00	−1.61	−2.02
	18.				0.08	−0.15	−0.33	−0.48	−1.01	−1.36
	20.				0.51	0.32	0.17	0.04	−0.41	−0.71
	22.				0.96	0.80	0.68	0.57	0.21	−0.03
	24.				1.41	1.29	1.19	1.11	0.83	0.64
	26.				1.87	1.78	1.71	1.65	1.45	1.32
0.50	10.				−0.78	−1.00	−1.18	−1.32	−1.79	−2.07
	12.				−0.43	−0.64	−0.79	−0.92	−1.34	−1.60
	14.				−0.09	−0.27	−0.41	−0.52	−0.90	−1.13
	16.				0.26	0.10	−0.02	−0.12	−0.45	−0.65
	18.				0.61	0.47	0.37	0.28	0.00	−0.18
	20.				0.96	0.85	0.76	0.68	0.45	0.30
	22.				1.33	1.24	1.16	1.10	0.91	0.79
	24.				1.70	1.63	1.57	1.53	1.38	1.28
0.75	6.				−0.75	−0.93	−1.07	−1.18	−1.52	−1.72
	8.				−0.47	−0.64	−0.76	−0.86	−1.18	−1.36
	10.				−0.19	−0.34	−0.45	−0.54	−0.83	−1.00
	12.				0.10	−0.03	−0.14	−0.22	−0.48	−0.63
	14.				0.39	0.27	0.18	0.11	−0.12	−0.26
	16.				0.69	0.58	0.50	0.44	0.24	0.12
	18.				0.98	0.89	0.82	0.77	0.59	0.49
	20.				1.28	1.20	1.14	1.10	0.95	0.87
1.00	−6.				−1.68	−1.88	−2.03	−2.14	−2.50	−2.70
	−2.				−1.22	−1.39	−1.52	−1.62	−1.94	−2.12
	2.				−0.74	−0.90	−1.01	−1.10	−1.37	−1.53
	6.				−0.26	−0.39	−0.49	−0.56	−0.80	−0.93
	10.				0.22	0.12	0.04	−0.02	−0.22	−0.33
	14.				0.73	0.64	0.58	0.53	0.38	0.29
	18.				1.24	1.18	1.13	1.09	0.97	0.91
	22.				1.77	1.73	1.69	1.67	1.59	1.54
1.25	−8.				−1.36	−1.52	−1.64	−1.73	−2.00	−2.15
	−4.				−0.95	−1.10	−1.20	−1.28	−1.52	−1.65
	0.				−0.54	−0.66	−0.75	−0.82	−1.03	−1.15
	4.				−0.12	−0.22	−0.30	−0.36	−0.54	−0.64
	8.				0.31	0.22	0.16	0.11	−0.04	−0.13
	12.				0.75	0.68	0.63	0.59	0.47	0.40
	16.				1.20	1.15	1.11	1.08	0.98	0.93
	20.				1.66	1.62	1.59	1.57	1.50	1.46
1.50	−10.				−1.13	−1.26	−1.35	−1.42	−1.64	−1.76
	−6.				−0.76	−0.87	−0.96	−1.02	−1.21	−1.32
	−2.				−0.39	−0.49	−0.56	−0.62	−0.79	−0.88
	2.				−0.01	−0.10	−0.16	−0.21	−0.36	−0.44
	6.				0.38	0.30	0.25	0.21	0.08	0.01
	10.				0.76	0.70	0.66	0.62	0.52	0.46
	14.				1.17	1.12	1.09	1.06	0.98	0.93
	18.				1.58	1.54	1.52	1.50	1.44	1.40

AUTHOR INDEX

SUBJECT INDEX